NANOPLASMONICS

ADVANCED DEVICE APPLICATIONS

Devices, Circuits, and Systems

Series Editor

Krzysztof Iniewski
CMOS Emerging Technologies Research Inc.,
Vancouver, British Columbia, Canada

FORTHCOMING TITLES:

MIMO Power Line Communications: Narrow and Broadband Standards, EMC, and Advanced Processing
Lars Torsten Berger, Andreas Schwager, Pascal Pagani, and Daniel Schneider

Mobile Point-of-Care Monitors and Diagnostic Device Design
Walter Karlen and Krzysztof Iniewski

Nanoelectronics: Devices, Circuits, and Systems
Nikos Konofaos

Nanomaterials: A Guide to Fabrication and Applications
Gordon Harling and Krzysztof Iniewski

Nanopatterning and Nanoscale Devices for Biological Applications
Krzysztof Iniewski and Seila Selimovic

Nanoscale Semiconductor Memories: Technology and Applications
Santosh K. Kurinec and Krzysztof Iniewski

Radio Frequency Integrated Circuit Design
Sebastian Magierowski

Semiconductor Device Technology: Silicon and Materials
Tomasz Brozek and Krzysztof Iniewski

Smart Grids: Design, Strategies, and Processes
David Bakken and Krzysztof Iniewski

Soft Errors: From Particles to Circuits
Jean-Luc Autran and Daniela Munteanu

Technologies for Smart Sensors and Sensor Fusion
Kevin Yallup and Krzysztof Iniewski

VLSI: Circuits for Emerging Applications
Tomasz Wojcicki and Krzysztof Iniewski

NANOPLASMONICS

ADVANCED DEVICE APPLICATIONS

EDITED BY

James W. M. Chon
Krzysztof Iniewski

CRC Press
Taylor & Francis Group
Boca Raton London New York

CRC Press is an imprint of the
Taylor & Francis Group, an **informa** business

CRC Press
Taylor & Francis Group
6000 Broken Sound Parkway NW, Suite 300
Boca Raton, FL 33487-2742

First issued in paperback 2017

© 2014 by Taylor & Francis Group, LLC
CRC Press is an imprint of Taylor & Francis Group, an Informa business

No claim to original U.S. Government works

ISBN-13: 978-1-4665-1426-3 (hbk)
ISBN-13: 978-1-138-07263-3 (pbk)

Library of Congress Cataloging-in-Publication Data

Nanoplasmonics : advanced device applications / editors, James W.M. Chon, Krzysztof Iniewski.
pages cm -- (Devices, circuits, and systems)
Includes bibliographical references and index.
ISBN 978-1-4665-1426-3 (hardcover : alk. paper)
1. Nanoelectronics--Materials. 2. Plasmons (Physics) 3. Surface plasmon resonance. I. Chon, James W. M., editor of compilation. II. Iniewski, Krzysztof, 1960- editor of compilation.

TK7874.84.N383 2014
621.381--dc23

2013033039

Visit the Taylor & Francis Web site at
http://www.taylorandfrancis.com

and the CRC Press Web site at
http://www.crcpress.com

Contents

Preface

Nanoplasmonics is a field of science that focuses on control and manipulation of plasmons (collective oscillation of free electron cloud at optical frequencies) at nanometer dimensions. It combines the strength of electronics (easy chip integration and reduced operational dimensions) and photonics (speed at optical frequencies), thereby replacing the existing integrated circuits and photonic devices in the future. No doubt, it is one of the fastest growing fields of science that has a potential to impart a significant impact on our daily lives. The areas of application cover telecommunication, consumer electronics, data storage, medical diagnostics, energy, and environment.

However, all of these applications are at varying degrees of their developmental stages, and there are still many scientific and technological challenges that need to be overcome. Despite its large attention, there are only limited number of dedicated reference books that summarize the up-to-date progress and perspectives in this field. *Nanoplasmonics: Advanced Device Applications* is one such reference book that provides a scientific and technological background of a particular nanoplasmonic application and outlines the progress and challenges of the application. The areas of application are chosen for the practical aspects rather than the lopsided fundamental or impractical futuristic aspects. The areas include optical data storage (Chapter 1), optical elements and interconnects (telecommunications; Chapters 4 and 6), photovoltaics and photocatalysts (energy and environment; Chapter 3), nanofabrication and metamaterials (telecommunications; Chapters 2 and 5), and biomedical sensors (medical diagnostics; Chapters 7 and 8). Each chapter will provide a balanced scientific review and technological progress of how these areas of application are shaping up into the future.

Contributions to this book have been made by world-class experts in this field, from the United States, Australia, Japan, Korea, and the Netherlands. Readership is expected from both industry and academia, and from graduate students to experts and professors.

MATLAB® is a registered trademark of The MathWorks, Inc. For product information, please contact:

The MathWorks, Inc.
3 Apple Hill Drive
Natick, MA 01760-2098 USA
Tel: 508 647 7000
Fax: 508-647-7001
E-mail: info@mathworks.com
Web: www.mathworks.com

Editors

James W. M. Chon is an associate professor at Swinburne University of Technology in Melbourne, Australia, and is an Australian Research Council Future Fellow. He is an expert in gold nanorod microscopy and spectroscopy, optical data storage, and nanoparticle-based photonic applications. Dr. Chon has published 50 research papers in multidisciplinary international journals and conferences and is a member of OSA and ACS. His recent research interests are in nanoplasmonic applications of gold nanorods.

Krzysztof (Kris) Iniewski is managing R&D at Redlen Technologies Inc., a start-up company in Vancouver, Canada. Redlen's revolutionary production process for advanced semiconductor materials enables a new generation of more accurate, all-digital, radiation-based imaging solutions. Kris is also an Executive Director of CMOS Emerging Technologies (www.cmoset.com), a series of high-tech events covering communications, microsystems, optoelectronics, and sensors. In his career, Dr. Iniewski has held numerous faculty and management positions at the University of Toronto, University of Alberta, SFU, and PMC-Sierra Inc. He has published more than 100 research papers in international journals and conferences. He holds 18 international patents granted in the USA, Canada, France, Germany, and Japan. He is a frequent invited speaker and has been consulted for multiple organizations at the international level. He has written and edited several books for Wiley, CRC Press, McGraw-Hill, Artech House, and Springer. His personal goal is to contribute to sustainability through innovative engineering solutions. In his leisure time, Kris can be found pursuing his hobbies such as hiking, sailing, or biking in beautiful British Columbia. He can be reached at kris.iniewski@gmail.com.

Contributors

Phillip Blake
Deparment of Chemical Engineering
University of Arkansas
Fayetteville, Arkansas

Brent D. Cameron
Department of Bioengineering
University of Toledo
Toledo, Ohio

James W. M. Chon
Centre for Micro-Photonics
Swinburne University of
    Technology
Hawthorn, Victoria, Australia

Drew DeJarnette
Microelectronics and Photonics
    Graduate Program
University of Arkansas
Fayetteville, Arkansas

Xiao Ming Goh
School of Physics
The University of Melbourne
Melbourne, Victoria, Australia

Braden Harbin
Microelectronics and Photonics
    Graduate Program
University of Arkansas
Fayetteville, Arkansas

Haroldo T. Hattori
School of Engineering and Information
    Technology
The University of New South
    Wales
Canberra, Australia

Saulius Juodkazis
Centre for Micro-Photonics
Swinburne University of Technology
Hawthorn, Victoria, Australia
and
Melbourne Centre for Nanofabrication
Melbourne, Victoria, Australia

Byoungho Lee
Inter-University Semiconductor
    Research Center and School of
    Electrical Engineering
Seoul National University
Seoul, Korea

Seung-Yeol Lee
Inter-University Semiconductor
    Research Center and School of
    Electrical Engineering
Seoul National University
Seoul, Korea

Ziyuan Li
School of Engineering and Information
    Technology
The University of New South Wales
Canberra, Australia

Yoshiaki Nishijima
Department of Electrical and
    Computer Engineering
Yokohama National University
Yokohama, Japan

Malin Premaratne
Department of Electrical and Computer
    Systems Engineering
Monash University
Melbourne, Victoria, Australia

Ann Roberts
School of Physics
The University of Melbourne
Melbourne, Victoria, Australia

D. Keith Roper
Deparment of Chemical Engineering
and
Microelectronics and Photonics
    Graduate Program
University of Arkansas
Fayetteville, Arkansas

Lorenzo Rosa
Centre for Micro-Photonics
Swinburne University of
    Technology
Hawthorn, Victoria, Australia

Ivan D. Rukhlenko
Department of Electrical and
    Computer Systems Engineering
Monash University
Melbourne, Victoria, Australia

Takuo Tanaka
RIKEN Advanced Science Institute
Metamaterials Laboratory
Wako, Saitama, Japan

and

Research Institute of Electronic Science
Hokkaido University
Kita-Ward, Sapporo, Japan

Adam B. Taylor
Centre for Micro-Photonics
Swinburne University of Technology
Hawthorn, Victoria, Australia

Rui Zheng
Department of Bioengineering
University of Toledo
Toledo, Ohio

Peter Zijlstra
Department of Applied Physics
Eindhoven University of Technology
Eindhoven, The Netherlands

1 Plasmonic Nanorod-Based Optical Recording and Data Storage

James W. M. Chon, Adam B. Taylor, and Peter Zijlstra

CONTENTS

1.1 INTRODUCTION

Metallic nanoparticles that can support plasmon resonances are called plasmonic nanoparticles and these are the important building blocks in nanoplasmonic applications. Their surface plasmon resonances (SPR)—collective free electron charge oscillations—provide tunable extinction in the visible to near-infrared range that are dependent on geometric size, shape, and environment. Owing to their highly sensitive resonances, many applications have been proposed, including biosensors [1–3], nanoantennas [4–7], and nonlinear biomarkers [8–11]. Further, their strong and weak SPR coupling property is proposed to be used as a "plasmonic ruler" for detecting DNA folding and unfolding [12] and nanometer-scale plasmonic circuit elements [13–15].

In terms of device application potential, the most promising and reliable property of plasmonic nanoparticles perhaps is their photothermal properties [16–21]. At SPR, the photon absorption is maximized, and large proportion of the absorbed energy is used in heating electrons and lattices of the nanoparticle. If plasmonic nanoparticles are embedded into nonabsorbing dielectric media such as silica or polymer, nanoparticles effectively become point heat sources, releasing heat to the surrounding by phonon–phonon coupling [16–18]. When the absorbed energy is large enough, the lattice of the nanoparticle can be melted to form a more thermodynamically stable shape [19]. Photothermal properties of plasmonic nanoparticles have been demonstrated to be very important in cancer treatment, where *in vivo* injection of nanoparticles in live animals had resulted in cancerous cell death or apoptosis [20–23].

Another application that can fully benefit from photothermal properties is optical data storage (ODS) [24]. It is well known that the current optical storage media such as phthalocyanine dyes [25], azo-type dyes, and GeSbTe (GST) phase-change alloy [26–28] utilize photothermal energy conversion to induce a phase change or decomposition on the recording marks. As in the cancer therapy, the nanoparticles can facilitate the photothermal energy conversion in the recording medium and improve the recording efficiency. The nanoparticles are utilized as a passive, photothermal-sensitizing element in this case. However, they can also assume an active role and become a recording material themselves by either a shape relaxation [29–32], explosive boiling [33], or fragmentation [34,35]. Subsequent changes in extinction can be the mechanism for readout. Figure 1.1 shows the various potential ODS applications that can arise from plasmonic nanoparticles (in this case gold nanorods).

Consumer ODS platforms such as CD/DVD/Blu-Ray DVD have been used in content distribution for music/movie industry. This is a big market with a room for improvement. Many different next-generation ODS technologies were proposed, for example, from volumetric methods such as holographic/micro holographic storage [36–39], 3D/multilayered bit-by-bit techniques [40–45], to 2D methods such as near-field recording [46,47]. Optical storage is also expanding its horizon to archival/

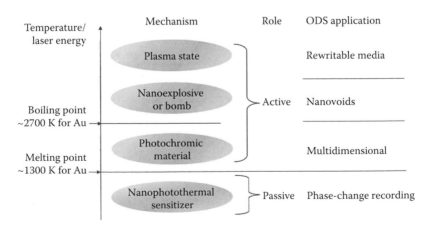

FIGURE 1.1 Photothermal properties of gold nanoparticles/nanorods for various ODS applications.

permanent storages [48,49], due to its data stability and energy efficiency in data retention. Magnetic disks consume large amount of energy for retaining data for longer periods [50], whereas tape-based storages have slow retrieval speed with no random access capability [51]. However, the proposed optical storage techniques also have many challenges to overcome in the data density [48,49]. Plasmonic nanorod-based optical storage could hopefully fill this gap, with its stability and capability to record in high density. Given that plasmonic nanorods are generally made with noble metals, they are expected to have much longer lifetime than the existing dye or polymer-based materials, which are suitable for long-term archival storage.

In this chapter, we introduce the core physical concepts of SPR of plasmonic nanoparticles and their photothermal properties and summarize the potential recording/reading proposals for future archival/consumer data storage applications. We explore various temperature states that the nanorods can reach up to and also consider the practical requirements for the proposals for a viable storage technique.

1.1.1 OPTICAL PROPERTIES OF PLASMONIC NANOPARTICLES AND NANORODS

Absorption and scattering of small metallic nanoparticles are well described by many in the literature. Here, we capture the core equations and concepts without venturing into the details of derivation. The first full exact analytical solutions to electromagnetic problem of light scattering by small-to-large spherical particles were provided by Mie [52]. The extinction cross-section σ_{ext} is given by

$$\sigma_{\text{ext}}(\lambda) = \frac{2\pi}{k^2} \sum_{n=1}^{n=\infty} (2n + 1)\operatorname{Re}[a_n + b_n], \tag{1.1}$$

$$\sigma_{\text{sca}}(\lambda) = \frac{2\pi}{k^2} \sum_{n=1}^{n=\infty} (2n + 1)[|a_n|^2 + |b_n|^2], \tag{1.2}$$

where σ_{ext} and σ_{sca} are the extinction and scattering cross-sections, respectively, $k = 2\pi\sqrt{\varepsilon_{\text{m}}}/\lambda$, and a_n and b_n are the scattering coefficients that are expressed as

$$a_n = \frac{m\psi_n(mx)\psi_n'(x) - \psi_n(x)\psi_n'(mx)}{m\psi_n(mx)\xi_n'(x) - \xi_n(x)\psi_n'(mx)}, \tag{1.3}$$

$$b_n = \frac{\psi_n(mx)\psi_n'(x) - m\psi_n(x)\psi_n'(mx)}{\psi_n(mx)\xi_n'(x) - m\xi_n(x)\psi_n'(mx)}, \tag{1.4}$$

where ψ and ξ are Riccati–Bessel functions of order n, $x = kR$ is a size parameter (R is the radius of the particle), and $m = \sqrt{\varepsilon_{\text{p}}/\varepsilon_{\text{m}}}$ is the square root of the ratio of the dielectric functions of the particle and of the medium. The dielectric function for the particles of varying sizes can be calculated from the bulk values using

$$\varepsilon(\omega) = 1 - \frac{\omega_{\text{p}}^2}{\omega^2 + i\Gamma\omega} + \varepsilon_{\text{ib}}(\omega), \tag{1.5}$$

where ω_p is the plasma frequency, Γ is the damping constant, and ε_{ib} is the interband contribution to the dielectric function. The damping constant Γ accounts for the size effects

$$\Gamma = \Gamma_{bulk} + \frac{Av_F}{L_{eff}} + \frac{\hbar \kappa V}{\pi}. \tag{1.6}$$

Details of the correction factors can be found in Refs. [53,54].
The absorption cross-section σ_{abs} can be calculated by

$$\sigma_{abs} = \sigma_{ext} - \sigma_{sca}. \tag{1.7}$$

For very small particles, that is, $R \ll \lambda$, the scattering cross-section is negligible, that is $\sigma_{abs} \sim \sigma_{ext}$. Also there is negligible retardation effect within the particle, and only the first few terms are important from a_n and b_n, and the extinction cross-section takes the simple form

$$\sigma_{abs}(\lambda) \approx \sigma_{ext}(\lambda) = \frac{24\pi^2 R^3 \varepsilon_m^{3/2}}{\lambda} \frac{\varepsilon''}{(\varepsilon' + 2\varepsilon_m')^2 + \varepsilon''^2}, \tag{1.8}$$

where ε' and ε'' are the real and imaginary parts of ε_p, and ε_m is the dielectric function of the surrounding medium. One can see that the absorption will be maximum when

$$\varepsilon' = -2\varepsilon_m. \tag{1.9}$$

Gans extended Mie's theory for calculating extinction characteristics of ellipsoidal nanoparticles with size much smaller than the wavelength of light [55]. The cross-sections are related to the polarizability α by

$$\alpha_{x,y,z} = \frac{4\pi abc(\varepsilon_p - \varepsilon_m)}{3\varepsilon_m + 3L_{x,y,z}(\varepsilon_p - \varepsilon_m)} = \frac{V(\varepsilon_p - \varepsilon_m)}{\varepsilon_m + L_{x,y,z}(\varepsilon_p - \varepsilon_m)}, \tag{1.10}$$

where V is the volume of the ellipsoid (where a, b, and c are their dimensions in x, y, z-axis, respectively. We assume that a is the length, and $b = c$ are the width) and $L_{x,y,z}$ is the depolarization factor in the respective axis. L is given by

$$L_x = \frac{1 - e^2}{e^2}\left(-1 + \frac{1}{2e}\ln\frac{1 + e}{1 - e}\right), \tag{1.11}$$

$$L_{y,z} = \frac{1 - L_x}{2}, \tag{1.12}$$

$$e = \sqrt{1 - \frac{b^2}{a^2}}. \tag{1.13}$$

The maximum absorption, scattering, and extinction cross-sections are then

$$\sigma_{abs}(\lambda) = k \, \mathrm{Im}[\alpha_x] \approx \sigma_{ext}(\lambda),$$ (1.14)

$$\sigma_{sca}(\lambda) = \frac{k^4}{6\pi} |\alpha_x|^2 .$$ (1.15)

The cross-sections have their resonances when the denominator of the polarizability α_x goes to zero and this happens when

$$\varepsilon' = \left(\frac{1 - L_x}{L_x} \right) \varepsilon_m.$$ (1.16)

When the shape of the nanorods deviates from the prolate spheroids, the correction comes into play in the L factor, depending on whether the end cap geometry is a hemisphere, or hemioblate spheroids. Either way, the correction factors can be found in reference [56]. The calculated absorption cross-section spectrum of varying aspect ratio of nanorods is shown in Figure 1.2. One can see the pronounced peaks over visible and near-infrared region.

Neglecting the heat coupling to the environment, the temperature increase in individual nanoparticles during a laser pulse absorption is proportional to the absorption cross-section divided by the volume of the nanoparticle, or

$$\Delta T \sim \frac{\sigma_{abs}(\lambda) I}{c\rho V},$$ (1.17)

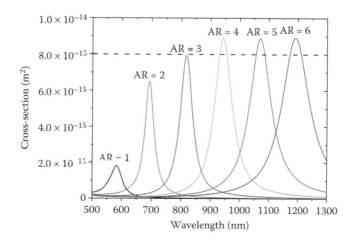

FIGURE 1.2 Absorption cross-sections of gold nanorods with fixed volume (~6500 nm³) and varying aspect ratio. Horizontal dashed line represents the melting point of gold in the case of laser irradiance of 0.22 mJ/cm².

where I is the laser pulse irradiance (J/cm²), c is the specific heat (~129 kJ/kg K for gold), ρ is the density (~19,300 kg/m³ for gold), and V is the volume of the individual nanoparticle. For gold nanorod of aspect ratio 3, with a fixed volume of ~6500 nm³, and for $I = 0.22$ mJ/cm², the temperature can reach the melting point of gold (~1300 K), as shown in Figure 1.2 (dashed line). It shows how high the temperature of individual nanoparticles of varying aspect ratio, from 1 to 6, but fixed volume, can reach up to. As can be seen, five times increase in temperature can be reached with an aspect ratio of 6, compared to a spherical particle, that is, aspect ratio 1.

1.1.2   Photothermal Excitation and Melting of Plasmonic Nanorods

In the previous section, we described the enhanced light absorption of nanorods by surface plasmon resonance. Here, we briefly summarize the process of photon energy conversion into thermal energy.

Typically, the excitation dynamics of plasmonic nanorods are observed by ultra-fast pump–probe spectroscopy [16–18, 57–60], where the pump pulse excites the SPR and then the probe pulse with 2–3 orders magnitude lower power probes the transient absorption. The probe pulse is delayed with respect to the pump pulse to record the time trace of the transient SPR spectrum. Earlier work by El-Sayed and coworkers [16–18] measured excitation dynamics of nanoparticle ensemble in a solution but recently, Orrit and coworkers [57, 58] demonstrated single-particle microscopy and spectroscopy and a clear dynamics could be observed. An observed transient pump–probe trace of a single gold nanorod (low excitation power) is shown in Figure 1.3. Within the pump pulse duration of ~500 fs, the surface plasmon dephases and thermalizes the electrons by scattering. This causes the transmittance at the SPR wavelength to increase. The heated electrons then couple to phonons and transfer their energy to the lattice within 1–2 ps. This process restores the transmittance back to ~80% of the original value, but the heated lattice does not dissipate its heat to the

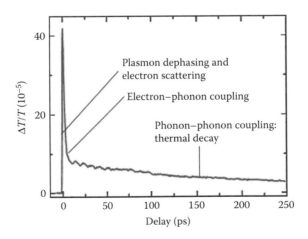

FIGURE 1.3   Ultrafast plasmon decay dynamics of gold nanoparticle showing the plasmon dephasing, electron–phonon coupling, and phonon–phonon coupling processes.

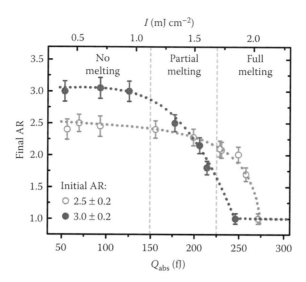

FIGURE 1.4 Aspect ratio of nanorods against the absorbed laser energy, after femtosecond-pulsed laser irradiations. The two vertical dashed lines correspond to the energies required to reach the melting point (~150 fJ) and the added latent heat of fusion (~230 fJ).

surrounding material as quickly, thereby exhibiting the longer tail of the transient transmittance up to nanosecond regime.

At high enough power, the shape transformation due to melting of the lattice can occur. This shape transformation causes strong SPR shifts that can be detected and used for optical recording. Zijlstra et al. [61] showed the decrease of gradual aspect ratio in single gold nanorods due to laser-induced shape transformation (Figure 1.4) and demonstrated that it is dependent on absorbed energy. When the absorbed energy was enough to raise the temperature of the lattice close to bulk melting point, the decrease was observed to be larger, and when the energy was enough to include latent heat of fusion, complete collapse to the spherical shape was observed. Below the melting point, the SPR shift is minimal, indicating that the structural integrity is being maintained. This work confirms that the melting point of the nanorods is similar to that of the bulk values. Earlier, Petrova et al. [62] observed reshaping of nanorods at constant temperatures down to ~500 K inside an oven. This difference is believed to be due to surface melting or diffusion, which can be the predominant reshaping mechanism for constant temperature heating. This effect is minimal for nanorods subjected to a heat spike, such as in ultrafast laser excitations.

1.2 PLASMONIC NANORODS AS NANOPHOTOTHERMAL SENSITIZERS IN OPTICAL STORAGE MEDIA

Since plasmonic nanorods have a low heat capacity, high absorption cross-section, and fast electron–phonon thermalization, they can be effectively used as nano-photothermal energy converters or sensitizers for optical storage media such as phthalocyanine dyes [25] and GeSbTe (GST) phase-change alloys [26,27,63–68]. These utilize photothermal energy conversion to induce a phase change or decomposition on the recording marks. In this section, we present how these nanorods can be used to enhance the recording efficiency of these recording media.

1.2.1 GOLD NANORODS DOPED IN OPTICAL RECORDING DYES

Optical recording dyes, such as phthalocyanine, used in typical CD-R or DVD-R disks use photothermal decomposition to record marks [25]. Azo-type dyes are another popular dyes in write-once optical disks, and they utilize *trans–cis* isomerization to change the absorbance and refractive index to cause recording [69]. When gold nanorods that are tuned to match the recording wavelength are mixed with these dyes, two different recording results could be observed. Heat-based recording materials such as phthalocyanine dyes show almost three times more sensitive recording when mixed with nanorods than by itself (Figure 1.5a and b). The temperature analysis showed that at the end of 1-µs pulse with 16-mW power, the nanorod temperature would increase more than five times the temperature of the surrounding dyes, inducing significant changes in material and optical properties of the recording dyes. It is well known that the photothermal decomposition of the dye is the recording mechanism for cyanine-based materials [25]. The role of nanorods in this case would be to sensitize and enhance the photothermal coupling to the recording media, hence increasing the efficiency and contrast of the recording. In the case of nanorod-doped azo-type recording media, Figure 1.5c and d show that recording is impossible when nanorods are present. This could be understood by the fact that the recording mechanism for azo-type media is not based on photothermal decomposition, but rather by *trans–cis* photoisomerization that causes the birefringence of the recorded region (i.e., change in absorbance and complex index) [69]. However, it is interesting to note that the recorded azo-dye molecules in *cis* form can relax back to *trans* state thermally. This means that the nanorods in the azo media enhance photothermal *erasing* of recorded bits.

1.2.2 GOLD NANORODS ON PHASE-CHANGE MEDIA

The phase-change media such as GST have been under heavy investigation [63–68] due to their importance in the next-generation optical and electronic memory applications such as near-field optical disks [68] and phase-change random access memory [70–72]. The intriguing aspect of photothermal GST recording lies in its complex change in reflectance upon transitions between various crystalline phases (as-deposited, primed, melt-quenched amorphous, and crystalline). As an as-deposited GST sample is in its amorphous phase, it requires a prelaser treatment before recording (known as initialization), which heats the sample just above the glass transition temperature to switch it to crystalline phase (high reflectance). The recording laser pulse of intense power would then melt the crystalline GST, which subsequently quenches to an amorphous phase (melt-quenched amorphous, low reflectance). This switching from a crystalline state to an amorphous state would constitute a written bit.

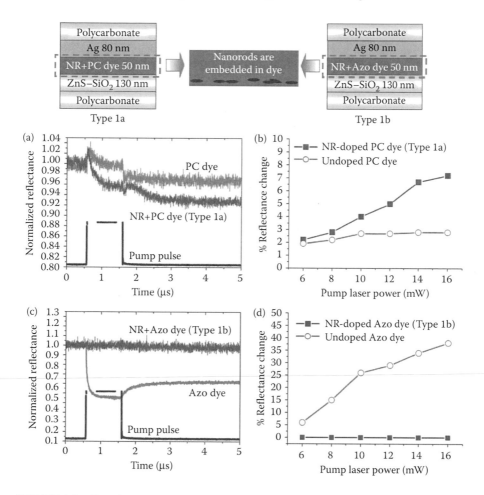

FIGURE 1.5 Transient reflectance pump–probe results of nanorod (NR) mixed with phthalocyanine (PC) dyes (a,b) and azo-type dyes (c,d). The NR increases the sensitivity of the recording in type 1a, demonstrating the metallic nanorods as efficient thermal sensitizer for recording. For azo-type dyes, NRs act as photothermal erasers of written bits, because of thermal activation of isomerizations.

However, there is an intermediate amorphous state called primed amorphous state, which is essentially a pulse-treated, as-deposited amorphous state. This peculiar state has an identical reflectance profile as the as-deposited state, but shows very different characteristics during initialization. In short, the crystallization onset time and the rate are shorter and faster than those of the as-deposited state, hence provide faster initialization and increased recording sensitivity [63–65]. It is thought that the priming pulse creates crystalline nucleation sites or cells that lower the energy barrier for the crystal growth [64,65]. Without a priming pulse, the nucleation rate would be slower due to the high-energy barrier.

Despite its clear advantages, the priming has not been widely used in initialization process due to a cumbersome and inefficient prepulse treatment. In order to retain the advantages of priming without sacrificing the efficiency in the initialization process, a photothermal-sensitizing element in the sample accessible to extra thermal energy to climb the energy barrier is highly desirable. In this respect, incorporating plasmonic nanorods (NRs) into GST would provide a sensitive, efficient recording process that removes the need for priming.

In Figure 1.6, a typical transient reflectance change during a pulsed recording is shown for the NR-doped GST sample and for the reference sample which does not contain NRs (GST). Stark difference in reflectance change profiles can be observed. The reference GST sample exhibits the typical profile of an as-deposited amorphous GST sample [63] characterized by initial reflectance drop at the arrival of a pulse, followed by a reflectance increase at the temperature exceeding the critical crystallization point ($T_c \sim 430°C$) [65]. At shorter recording pulse widths (<100 ns) and at smaller powers (<1 mW), the decrease in initial reflectance recovers to the original value, confirming the reversibility. This phenomenon was previously explained by the variation in the optical constants of the sample with the rising temperature [63,65].

On the other hand, the NR-doped GST sample (NR+GST) shows a dramatic reduction in crystallization onset time (t_{onset}), no initial drop in reflectance, and a two-fold increase in final reflectance to the reference GST sample at the same recording power. These observations are represented in Figure 1.7, where the onset

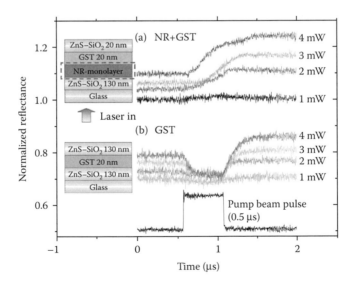

FIGURE 1.6 Transient reflectance pump–probe results of the nanorod-coated GST layer. The sample structures used in this study are shown on the left. (a) The NR monolayer coated on the phase-change material, GeSbTe, and (b) GST sample without NRs. The difference in the reflectance profile of the two types of sample is clearly manifested in the graphs; the GST shows a dip in reflectance with longer onset time for the reflectance increase, whereas the NR-coated GST does not show such trend, instantly increasing the reflectance.

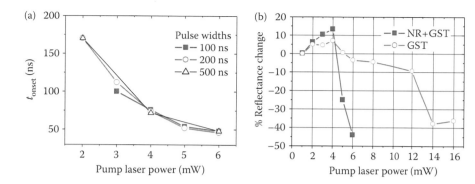

FIGURE 1.7 (a) The reflectance increase onset time of the NR+GST sample for various pump pulse powers and pulse widths, showing the decay with power. The onset time for GST layer without NRs is found to be the same as the pulse width of the pump beam (i.e., 100, 200, and 500 ns not shown)—much slower than what is shown above. (b) Percent reflectance change with respect to the pump power. NR-coated GST shows the efficiency of the system, only requiring 6 mW to completely show the full profile, compared to 16 mW for GST by itself.

time for various pulse widths and powers (Figure 1.7a) as well as the reflectance change with various powers at a fixed pulse width (Figure 1.7b) are shown for the two samples in the current study. Figure 1.7a confirms the dramatic decrease in crystallization onset time for NR+GST sample, while Figure 1.7b shows a two-fold increase in crystallization reflectance and less than a half of the power required to reach the melt-quenched amorphous state for NR+GST sample. Both figures confirm that the NR+GST sample is significantly more efficient than the reference GST sample in both recording speed and power.

The observed profile of NR+GST sample bears a resemblance to the reflectance profile of the already primed GST sample, typically characterized by a dramatic reduction in crystallization onset time (t_{onset}), no initial drop in reflectance, increase in final reflectance, and increased crystallization rate [65]. The only difference between the NR+GST and primed GST samples is the increased crystallization rate. For the reflectance increase rate, which corresponds to the crystallization rate, the NR+GST sample shows an overall value of ~10 ($\Delta R\%/\Delta t$ ns) comparable to that of the reference GST sample, whereas the primed GST sample in the past had shown more than a two-fold increase in crystallization rate to that of the as-deposited GST sample [65]. The implication of the observation is that we could neglect the priming for NR+GST sample altogether, since it already behaves as a primed sample.

A qualitative comparison between the pump–probe profiles of NR+GST sample and the already primed GST sample provides more insights into the photothermal-sensitizing role of NRs. Previously, both Wright et al. [64] and Khulbe et al. [65] utilized the nucleation theory of liquid droplets from supersaturated vapor by Frenkel [73] to explain the underlying mechanism of the GST priming. The model proposed that as a priming pulse heats up an as-deposited sample, it produces localized basic crystalline units of lower free energy called embryos and that the cluster of g embryos dynamically increases or decreases the free energy according to the following equation:

$$\Delta F_g = Ag^{2/3} - kT \ln(S)g \qquad (1.18)$$

where A is a constant, k is the Boltzmann constant, and S is the supersaturation ratio—the density of crystalline embryos to the density of embryos at saturation. The plot of free energy ΔF_g with respect to g results in an energy barrier at $g = g^* = [2A/3kT \ln(S)]^3$, the critical embryo cluster size. The cluster of size $g > g^*$ would grow larger, but $g < g^*$ would shrink in an effort to reduce the free energy of the system. Since g^* is inversely proportional to the saturation S, which itself is proportional to the density of embryos, the role of priming pulse is to create as many embryos as possible by reaching the critical temperature T_c (i.e., S is increased). The created embryos survive even after the end of the priming pulse and they reduce the critical cluster size g^* and lowers the energy barrier height of ΔF_g for the next recording pulse. Hence, for the next pulse it is easier to increase g beyond g^* and start crystal growth at shorter onset time t_{onset}. It is also notable that the crystallization rate is inversely proportional to the g^* size, and therefore the crystallization rate is increased as g^* becomes smaller. This successfully explains the dramatic reduction in t_{onset} and increased crystallization rate of the primed GST samples.

However, the current *pseudo-primed* observations in Figure 1.6 of NR+GST sample cannot be explained by the above model, due to the absence of priming pulse. In this case, the barrier height would stay intact without reduction, and there is no existence of precreated embryos that reduce the size g^*. The NRs, however, could instead provide extra heat energy to the surrounding GST as we have seen in Figure 1.7. It is this extra activation energy that forces the nominal clusters of g embryos in the sample to effectively climb up the initial free energy barrier height and go beyond the critical size g^* to start the crystal growth. The difference in crystallization rate of the NR+GST sample to the primed GST sample can then be explained by the nonexistence of the precreated embryos in NR+GST sample. As stated earlier, the crystallization rate is inversely proportional to the size of g^*. The lack of precreated embryos in NR+GST sample does not increase the crystallization rate.

1.2.3 NANORODS DOPED IN POLYMER MATRIX FOR MICRO-VOID FABRICATION

Laser fabrication by microexplosion and void generation in a solid dielectric material such as glass or polymer has widely been employed for the fabrication of photonic devices such as waveguide structures, photonic crystals, optical storage, and microfluidic channels [74–79]. In particular, polymer materials possess a relatively low thermal conductivity resulting in significant build-up of heat upon subsequent laser pulse absorption. Due to low decomposition and melting temperature (400–600 K), the photothermal bond breakage and consequent void generation are achievable even at low power densities ($\sim$MW/cm^2). The major advantage of polymers over dielectric glasses is that they can be doped with functional materials which can facilitate void generation.

When a common polymer such as polyvinyl alcohol (PVA) is doped with gold NRs (AuNRs) for void generation, the threshold laser power for generation of the voids decreases by one order of magnitude lower than an undoped matrix or a matrix doped with nanoparticles in which the SPR absorption does not match the laser wavelength. The transmission optical and scanning electron microscopy images of the void patterns produced in AuNR-doped and -undoped PVA samples (identical recording condition) are shown in Figure 1.8. The incomplete void patterning in the undoped PVA matrix (Figure 1.8d) can be seen compared to the matrix doped with NPs (Figure 1.8a–c). When the laser irradiation wavelength was matched the SPR band of the NPs (NR 4 case, Figure 1.8a), the efficiency of the void generation increases and a complete patterning can be produced. For the non-SPR matching conditions, the efficiency decreases to 46% for NR 2 (Figure 1.6b) and 40% for NS 40 (Figure 1.6c).

The dependence of efficiency for void generation with respect to exposure time for SPR-matched and -unmatched cases shows a clear difference between the two—only a fraction of the exposure time is required to reach the same efficiency of void fabrication in unmatched cases (i.e., 40-ms exposure is required to produce 50% efficiency in the NR 4 case, whereas the other doped samples required >100 ms). This difference increases to one order of magnitude when the SPR-matched NR 4 sample is compared with the undoped PVA sample.

The fact that the void generation has a definite threshold in exposure time indicates that it is a multiple pulse process. In the undoped polymer matrix, possible mechanisms for void generation are self-focusing, photochemical decomposition [80, 81], multiphoton ionization [82,83], and thermal decomposition [80]. Of these,

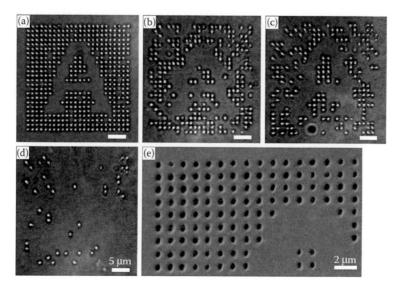

FIGURE 1.8 Void patterns fabricated in PVA doped with metallic NPs imaged using a transmission optical microscope: (a) NR 4 sample, (b) NR 2, (c) NS 40, and (d) PVA only (100-ms exposure per bit at 25 mW of average power for all experiments). The presence of voids was verified using an (e) SEM image. (Adapted from Choi, K. et al., *Advanced Functional Materials,* 18(15), 2237–2245, 2008.)

self-focusing, photochemical decomposition, and multiphoton ionization are single-pulse processes, in which the pulse energy determines the possibility of void generation. Typical pulse energies required for these processes at NIR wavelengths are >5 nJ, assuming a pulse width of 100 fs [82]. With a typical pulse energy of <0.3 nJ used, these effects are negligible.

Thermal decomposition, on the other hand, is produced by applying heat directly above the decomposition temperature of PVA for a prolonged period of time [84,85]. PVA is known to be partially or fully hydrolyzed at 180–190°C and decomposes rapidly above 200°C as it can undergo pyrolysis. Currently, there are two mechanisms known to cause decomposition in PVA—one is elimination of water molecules only, which leaves a C=C backbone with dark coloring [84], and the other is through a random scission of polymer which produces more volatile products such as acetaldehyde, aldehydes, and ketones [85]. This process is capable of reducing its weight by more than 80% without leaving any residue. The void created using femtosecond pulses accompanies a large volume loss without any dark residue indicating the decomposition of PVA into volatile products. The source of heat in this case is heat accumulated by sequential pulse absorption.

In the undoped samples, void generation is still observed even though there is no direct absorption band present at the laser wavelength. Since the absorption spectrum of the undoped PVA shows absorption onset at around 350 nm, the absorption is most probably due to three-photon absorption. The energy deposition through this nonlinear absorption is extremely small for single-pulse irradiation, but becomes significant as the number of irradiated pulses increases. For undoped PVA, the onset exposure time (50% void generation point; Figure 1.9) reaches 300 ms, which is equivalent to ~25 million irradiated pulses and absorbed heat.

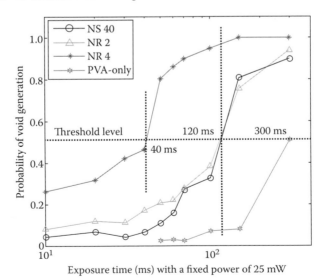

FIGURE 1.9 The probability of void generation for different exposure times at a fixed laser power (25 mW). The dotted lines indicate the threshold at which 50% efficiency is achieved. (Adapted from Choi, K. et al., *Advanced Functional Materials,* 18(15), 2237–2245, 2008.)

The temperature evolution of the focal volume in the undoped PVA can be estimated using a simple analytical solution to the heat conduction equation [86]:

$$\Theta(r,t) = T(r,t) - T_0 = \sum_{i}^{\# \text{pulses}} \frac{Q}{8\rho c_p (\pi \alpha(t - t_0 - \tau_i)^{3/2})} \exp\left[\frac{-(r - r_0)^2}{4\alpha(t - t_0 - \tau_i)}\right],$$

(1.19)

where T_0 is the initial temperature (293 K), Q is the amount of thermal energy absorbed by the multiphoton process, ρ is the density of PVA (1200 kg m^{-3}), c_p is the specific heat capacity of PVA [8] (1650 J kg^{-1} K^{-1}), α equal to k/c_p where k is the thermal conductivity of PVA [74] (0.21 W m^{-1} K^{-1}), τ_i is the time at which the ith pulse arrives (i.e., i = pulse separation $\times \tau_i$), t_0 the term specific to the Gaussian heat profile of the irradiated spot and determines the width of profile, and $r - r_0$ the radial distance away from the center of Gaussian beam. For a large number of pulses (>1000), the summation notation in Equation 1.19 can be replaced with an integral, and $\Theta(t)$ can now be expressed as

$$\Theta(r,t) = \frac{Q}{4\pi k(r - r_0)} \text{erfc}\left[\frac{r - r_0}{2(\alpha(t - t_0))^{1/2}}\right].$$

(1.20)

This could be further simplified by noting that erfc(0) = 1,

$$\Theta(r,\infty) = \frac{Q}{4\pi k(r - r_0)},$$

(1.21)

for large values of t. From the simulations, the steady state is reached for $t < 1$ ms. This means that during a laser exposure for void generation (>40 ms), the majority of the time is spent on the steady-state heat profile, and void generation is controlled by how long this steady-state heat profile is sustained above the decomposition temperature rather than an instantaneous temperature profile [87].

When NPs are doped in the matrix, the decomposition of NPs themselves by evaporation [27] or ablation [88,89] could be an additional mechanism for void generation, along with the decomposition of PVA mediated by photothermal heat transfer from the rods (i.e., the NR 4 particles in a passive photothermal role). Given the high absorption cross-section of the NR 4 sample, the pulse energy of 0.3 nJ (~0.035 J/cm^2) is enough to induce the evaporation of NRs.

The transmission image of the voids generated using single-pulse irradiation (pulse energy of 5.5 nJ) is shown in Figure 1.10, where incomplete patterning at 5.5 nJ of laser pulse energy (<20% void generation) is observed. Below 5 nJ, no voids were generated. Given that the void generation of PVA via single pulse can only be achieved above 10-nJ pulse energy [83], the likely mechanism for the void generation in this case is explosive boiling of NRs. Also, the produced voids are unevenly spaced and smaller in size compared to the multiple pulse irradiation case, which

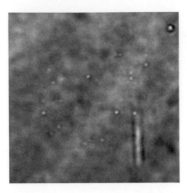

FIGURE 1.10   Voids created by single pulse irradiation (pulse energy 5.5 nJ, l = 780 nm, NA 1.4) on gold nanorods-doped PVA.

reflects that the pressure induced by decomposition of nanosized mass of gold is less than the pressure induced by the photothermal decomposition of PVA.

1.3   NANORODS AS AN ACTIVE OPTICAL RECORDING MEDIA AND 5D OPTICAL STORAGE

So far, the review focused on the use of plasmonic nanorods as passive recording media, where nanorods facilitate the photothermal recording process. In this section, we focus on the cases where nanorods themselves are active recording medium. In Section 1.1, the mechanism for laser-induced shape transformation of plasmonic nanorods and subsequent SPR spectrum change was presented. This section will focus on various optical recording and encoding techniques using photothermal melting.

1.3.1   OPTICAL RECORDING USING PHOTOTHERMAL MELTING OF PLASMONIC NANORODS

When plasmonic nanorods are irradiated with laser with enough energy, the nanorods will melt and shape transform into shorter rods or spheres. At the same time, their longitudinal SPR peak will shift to blue or bleach. If the laser with the same wavelength but with much lower power is irradiated for the readout, the transmission would increase due to the bleaching of SPR band. Such mechanism can be the basis for optical recording and data storage. Previously, Wilson et al. [30] showed the feasibility of optical recording using photothermal melting on silver nanorod-doped polymer film, and at the similar time, Niidome et al. [31] also performed optical recording on gold nanorods. Figure 1.11 shows the concept of photothermal optical recording on nanorods that are randomly dispersed inside a dielectric medium [24]. Figure 1.12 shows the experimental demonstration of the concept, where gold nanorod-doped thin silica film was irradiated with femtosecond laser resonant to SPR peak to induce shortening of the rods, and therefore when read out using transmission mode with laser at SPR wavelength (~760 nm), the recorded area

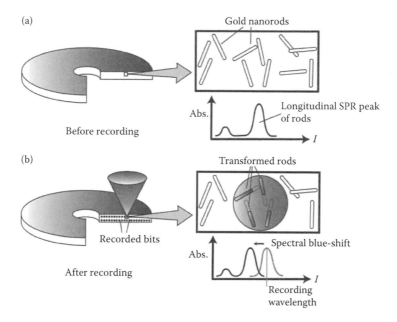

FIGURE 1.11 The concept of photothermal melting as optical recording on plasmonic nanorods. (a) Before recording, the nanorods are randomly distributed in a dielectric matrix with absorbance spectrum showing a longitudinal SPR peak at longer wavelength. (b) After recording by pulsed laser irradiation tuned at the longitudinal SPR bands, the transformation of the rod shape is induced, which results in spectral shift of the SPR peak. If read out at the same wavelength at lower power density, the absorbance change could be detected. (Adapted from Chon, J. W. M. et al. *Advanced Function Materials*, 17(6), 875–880, 2007.)

appears brighter due to less absorption, and at shorter wavelength (~690 nm), the area appears darker due to increased absorption.

1.3.2 Five-Dimensional Multiplexing Using Spectral, Polarization, and Spatial Encoding

The advantage of photothermal optical recording on plasmonic nanorods is that multiplexed optical recording is possible. Previously, spectral multiplexing on electron-beam-fabricated plasmonic nanostructures was demonstrated by Ditlbacher et al. [90] and polarization and spectral recordings were conducted independently by several groups on plasmonic nanorods [24,30–32]. Later, integration of these encoding methods into a single technique called five-dimensional optical storage was achieved by Zijlstra et al. [91]. The five-dimensional optical recording on plasmonic nanorods utilizes the fact that nanorods of different orientation and aspect ratio respond to particular laser wavelength and polarization, making optically selective recording possible. When the laser at a particular wavelength and a polarization irradiates a distributed sample, only a subpopulation of the rods that are in resonant condition with the laser beam will interact and the unaffected rods in the same volume can be used later for multiplexed recording. Furthermore, the multilayered

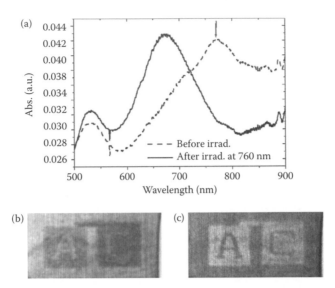

FIGURE 1.12 (a) Absorbance spectrum of (aspect ratio ~3) gold nanorods doped in a silica sol–gel matrix, before and after laser irradiation at 760 nm. Transmission images of letters (scale: 64 μm × 64 μm) recorded at 760 nm, reading at (b) 760 nm and (c) 690 nm. (Adapted from Chon, J. W. M. et al. *Advanced Function Materials*, 17(6), 875–880, 2007.)

recording in a thick or layered sample is possible due to the threshold melting of nanorods. If the laser power density is not high enough, the nanorods will not be affected and stay intact, thus allowing nanorods in other layers to be recorded later.

Since multiplexed encoding can be achieved by either wavelength dimension (variation in aspect ratio) or polarization dimension (orientation) as well as three spatial dimensions (multilayering), five-dimensional multiplexing could be achieved. The concept of five-dimensional recording using plasmonic nanorods is shown in Figure 1.13.

In the proof-of-concept experimental demonstration, the multilayer readout was achieved by two-photon luminescence (TPL) of gold nanorods. TPL of plasmonic nanorods is an interesting phenomenon that is yet to be fully understood. Generally, extremely low-efficient interband luminescence of bulk metal ($\delta_{gold} \sim 10^{-10}$ [92]) is greatly enhanced by plasmon field of the nanorods. The action cross-section of the TPL ($\delta_{2P}\sigma_{2P}$) for gold nanorods was measured to be ~10^6 GM, which is one of the highest that was observed for TPL agents [91]. The quantum yield in gold nanorods is reported to be ~10^{-6} [93] and exciting at SPR shows a remarkable improvement in efficiency for TPL. Since the TPL intensity is increased to the square of laser power density, it can be excited only at the focal region of the laser beam but not at other layers, reducing cross-talk between layers. Moreover, the nonlinear property of TPL makes the spectral and polarization response much sharper than the linear readout methods (i.e., scattering), thereby further suppressing cross-talk that might arise between recording channels. Figure 1.14a and b shows the comparison of the spectral and polarization responses of TPL and scattering readout.

When photothermal melting of nanorods causes the SPR to bleach, the associated TPL also loses its efficiency, and the detection of the TPL on the recorded spot is

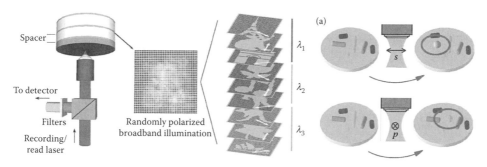

FIGURE 1.13 Concept of five-dimensional optical storage. When a multilayered plasmonic nanorod-doped recording layers are illuminated with randomly polarized broadband illumination, the multiplexed patterns are not recognizable. However, when laser light with particular wavelength and polarization is used, the multiplexed images are discernible. (a) The plasmonic nanorods inside the recording medium are distributed in aspect ratio and orientation. However, laser beam with particular wavelength and polarization conditions can target only the resonant nanorods and melt them without affecting the rest of the rods. This will allow the multiplexing in five dimensions. (Adapted from Zijlstra, P., J. W. M. Chon, and M. Gu, *Nature*, 459(7245), 410–413, 2009.)

suppressed, appearing as a dark spot. This is what is being observed in the scanned TPL image of the recorded region (Figure 1.14c). When the scanning wavelength or polarization is changed, the TPL contrast is lost, allowing multiplexed recording to be performed in the same volume (Figure 1.14d–f). In Figure 1.15, a full five-dimensional multiplexing is demonstrated with three-state spectral and two-state polarization encodings per layer. The data bit density was measured to be 1.1 TBits/cm^3, equivalent to 1.6 TB per DVD-sized disk. The data density can further be improved by multiplexing three polarization states and by increasing layer numbers (by decreasing the spacer layer thickness). Such a technique of increasing data density could be very useful for future archival and cloud-level data storages.

However, for practical applications of the technique, two issues need to be resolved. Firstly, the bulky ultrafast pulsed lasers that are used in recording and readout should be replaced with cheap, compact, continuous-wave diode lasers for more efficient device integration [48]. Secondly, the nanorods need to be homogeneously distributed without getting closer between themselves within the dielectric medium to reduce noise originating from plasmon coupling [94]. The use of continuous wave lasers for recording and readout could be a solution to the first issue, and the use of electron beam lithography technique for prefabricated array structure of nanorods could resolve the second issue.

1.4 CONTINUOUS WAVE READOUT USING DETUNED SPR SCATTERING METHOD

The first hurdle for the realization of multidimensional optical storage using plasmonic nanorods is demonstrating a continuous-wave (cw) laser as an effective alternative to femtosecond pulsed lasers in recording and readout. Previously, cw

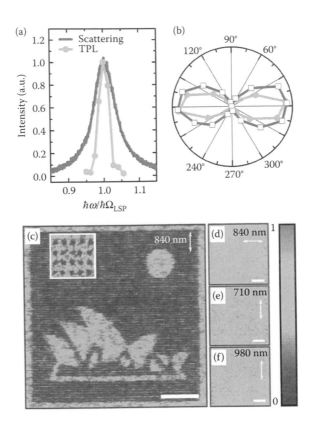

FIGURE 1.14 Readout using TPL. (a) Normalized white light scattering spectrum (solid line) and TPL excitation profile (filled circles) as a function of photon energy $\hbar\omega$ for individual gold nanorods with approximate dimensions of 90×30 nm. The profiles were centered around the longitudinal SPR energy $\hbar\Omega_{LSP}$. (b) Polarized scattering (open squares) and TPL intensity (filled circles) versus the polarization of the excitation light. (c) Normalized TPL raster scan of an image patterned using single-laser pulses per pixel with a wavelength of 840 nm and vertical polarization. (Panels c–f are labeled with laser wavelength and a double-headed arrow indicating polarization direction.) The TPL was excited with the same wavelength and polarization as was used for the patterning. The pattern is 75×75 pixels, with a pixel spacing of 1.33 μm. Inset, high-magnification image of the recording (size 7×7 μm). (d–f) Images obtained when the TPL is excited at 840 nm with horizontal polarization (d), at 710 nm with vertical polarization (e), and at 980 nm with vertical polarization (f). Scale bars, 20 μm. (Adapted from Zijlstra, P., J. W. M. Chon, and M. Gu, *Nature*, 459(7245), 410–413, 2009.)

laser recording and readout in multilayer setting was demonstrated [44,45]. For nanorods, cw laser cannot produce nonlinear signals due to the low peak power and therefore the TPL readout has to be replaced with linear scattering from nanorods. However, the cw laser causes heavy beam extinction in multilayered structures and it is more problematic for readout than recording, due to the two-pass requirement for the scattering signal readout (i.e., readout laser beam extinction plus the signal beam extinction). One way of reducing the heavy beam extinction is to detune the readout beam from the SPR peak as the readout layer becomes deeper in the medium. This

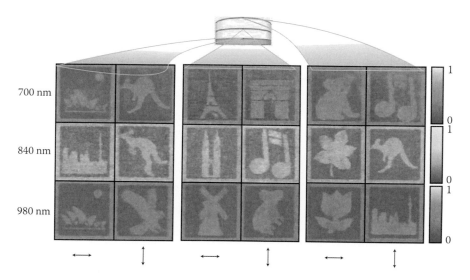

FIGURE 1.15 Five-dimensional recording and readout. Normalized TPL raster scans of 18 patterns encoded in the same area using two laser light polarizations and three different laser wavelengths. Patterns were written in three layers spaced by 10 μm. The recording laser pulse properties are indicated (wavelength at left, polarization at bottom). The recordings were retrieved by detecting the TPL excited with the same wavelength and polarization as employed for the recording. The size of all images is 100 × 100 μm, and the patterns are 75 × 75 pixels. (Adapted from Zijlstra, P., J. W. M. Chon, and M. Gu, *Nature*, 459(7245), 410–413, 2009.)

is called detuned SPR scattering method, and Taylor et al. [95] successfully demonstrated a cw recording and readout on a 16-layer gold nanorod sample. By detuning from the peak, the extinction becomes smaller and accessing deeper layers become easier. Such detuned readout could improve the signal strength by orders of magnitude and readout with linear scattering accessible.

In order to understand the detuned SPR method, the extinction spectrum of nanorod recording medium has to be acquired, and the concentration distribution function of nanorods is extracted from the spectrum by fitting the function

$$\text{Ext}(\lambda) = \frac{l}{2} \int_{R=1}^{R_{max}} c(R)\sigma_e(\lambda,R)\,dR \tag{1.22}$$

to an experimental extinction spectra. Then for a recording layer consisting of distributed nanorods embedded in a dielectric medium, the signal power to input power ratio ξ can be described by

$$\xi(\lambda,n) = \frac{P_s}{P_i} = \frac{l}{2} \int_{R=1}^{R_{max}} c(R)\sigma_s(\lambda,R)\,dR \left[\exp\left(-\frac{l}{2} \int_{R=1}^{R_{max}} c(R)\sigma_e(\lambda,R)\,dR \right) \right]^{2(n-1)},$$

$$\tag{1.23}$$

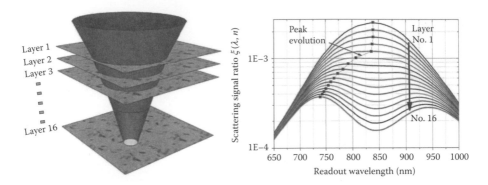

FIGURE 1.16 A 16-layer sample with distributed, randomly oriented AuNRs. Right: Log plot of scattering signal ratio $\xi(\lambda,n)$ with gold nanorod concentration in recording layer polymer of ~40 nM, mean aspect ratio of 3.135, and standard deviation of 0.48. Note the peak evolution at deeper layers, indicating the best readout can be achieved at detuned, off-SPR resonant wavelengths. The required detuning is more for a distributed sample than a nondistributed sample. (Adapted from Taylor, A. B., J. Kim, and J. W. M. Chon, *Optics Express*, 20(5), 5069–5081, 2012.)

where σ_s and σ_e are the scattering and extinction cross-sections of the nanorods, respectively, l the recording layer thickness, $c(R)$ a concentration distribution function of nanorods with respect to the aspect ratio R, and 1/2 factor at the front of concentration is required to take into account the contribution from 2D random orientation of nanorods. The term inside the square bracket represents the transmittance through each layer, which is assumed to be constant. The $2(n-1)$ power index represents a number of layers that the readout beam traverses. The term in front of the square bracket represents scattering signal generated from the readout layer in concern. In Figure 1.16, ratio ξ is plotted for normally distributed gold nanorods with a mean aspect ratio of 3.14, standard deviation of 0.5, fixed width at 14 nm with $c \sim 40$ nM, and $l \sim 1$ µm. The calculation was based on assumptions that nanorods are randomly oriented, and the impinging light has linear polarization. The cross-sections are calculated using Mie–Gans theory [55] (i.e., Equations 1.13 and 1.14), with radiation, surface scattering damping corrections [54], and end cap geometry corrections (as cylinders rather than ellipsoids [56]). One could clearly notice that the beam extinction can be reduced at deeper layers (i.e., layer number > 4) if the readout beam was detuned from the SPR.

In Figure 1.17, the demonstration of cw laser recording and readout operation on a 16-layer gold nanorod recording sample is shown. The cw readout of patterned images of a total of 16 layers using detuned wavelengths is shown, where consistent image qualities are observed till the 16th layer. In Figure 1.17b, cw laser readout power at detuned and on-resonant wavelengths used for constant readout signal level is plotted against the layer number. Overlaid is the theoretical readout power, calculated using Equation 1.23. One can see a good agreement between the theory and experiment, with more than one order of magnitude reduction in detuned readout power at layers above 11 compared to on-resonant readout powers. The recordings were checked and confirmed to be due to photothermal reshaping of resonant

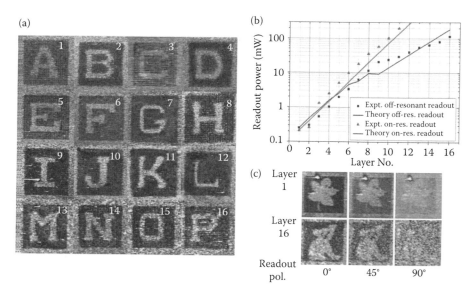

FIGURE 1.17 (a) cw readout images (50×50 μm^2) on a 16-layer sample, with readout power and contrast optimized experimentally. (b) cw laser power used for constant readout signal level. The readout of the pattern was conducted using two wavelengths, one at detuned (varied between 833 and 765 nm) and one at extinction peak (833 nm). (c) The patterns recorded were all confirmed to be due to photothermal melting of gold nanorods, by checking the scattering images with readout beam polarization orthogonal to that of the recording beam. Cases for layers 1 and 16 are shown in the figure, which show little traces of the original recordings. (Adapted from Taylor, A. B., J. Kim, and J. W. M. Chon, *Optics Express*, 20(5), 5069–5081, 2012.)

nanorods—by rotating the readout beam polarization 90° to the recording polarization, we confirmed the image disappearance (Figure 1.17c).

The detuned readout technique can also be achieved in *polarization* domain. For this to be possible, the plasmonic nanorods in the recording medium need to be all aligned, and the angle between the readout beam polarization and the nanorod alignment direction is the detuning parameter. However, in a randomly oriented nanorod sample, this would not be possible because the scattering at SPR peak is omni-directional, that is, no peak exists in polarization domain. With nanofabricating, perfectly aligned nanorods in large scale and in multilayers using electron beam and focused ion beam lithography techniques, detuning in polarization may be possible. In fact, detuned readout using polarization control would be a preferred method, given difficulties associated with wavelength control of lasers.

1.5 SUMMARY AND OUTLOOK

In this chapter, a number of optical recording techniques using the photothermal properties of plasmonic nanorods are discussed. As a nanophotothermal-sensitizing element, nanorods could facilitate and enhance the necessary heat intake from the laser. Optical recording materials, such as phthalocyanine and azo dyes, as well as phase-change alloy materials can benefit from these schemes. As nanorods play an active photothermal role, their reshaping due to melting could provide a photochromic optical contrast. Multiplexed optical recording using this method was possible due to polarization and spectral selectivity of nanorods, that is, recording laser polarization and wavelength can target specific nanorods with narrow aspect ratio and orientation binning. Using this method, the five-dimensional optical recording and readout was demonstrated. Continuous wave laser operation of recording and readout was also possible on plasmonic nanorods, demonstrating recording and readout on a 16-layer gold nanorod sample.

While all these optical recording techniques using plasmonic nanorods are promising, there are many challenges lie ahead for a commercial-level device. The biggest challenge is developing a reliable, reproducible, scaled-up production method of plasmonic nanorods. Currently, wet-chemically synthesized nanorods are produced only in small amounts, have huge distribution in size and shape, and could be aggregated inside the medium, thus causing huge optical noise. Perhaps modern nanofabrication methods such as electron beam lithography could resolve this issue by producing perfectly aligned, identical nanorod arrays. However, the e-beam techniques also have limitations in producing large-area arrays. For a DVD-sized disk with potentially hundreds of layers, the process may take days to produce a disk. Nevertheless, the use of noble metals as plasmonic nanorods is a great advantage owing to its stability and data longevity. With the current climate of data being lost or having to reproduce/relocate every 3–5 years, the plasmonic nanorod-based optical storage could be an ideal, permanent, archival storage with high data density.

REFERENCES

1. Raschke, G., S. Kowarik, T. Franzl, C. Sönnichsen, T. A. Klar, J. Feldmann, A. Nichtl, and K. Kürzinger, Biomolecular recognition based on single gold nanoparticle light scattering, *Nano Letters*, **3**(7), 935–938, 2003.
2. Anker, J. N., W. P. Hall, O. Lyandres, N. C. Shah, J. Zhao, and R. P. Van Duyne, Biosensing with plasmonic nanosensors, *Nature Materials*, **7**(6), 442–453, 2008.
3. Mayer, K. M. and J. H. Hafner, Localized surface plasmon resonance sensors, *Chemical Reviews*, **111**(6), 3828, 2011.
4. Crozier, K. B., A. Sundaramurthy, G. S. Kino, and C. F. Quate, Optical antennas: Resonators for local field enhancement, *Journal of Applied Physics*, **94**(7), 4632–4642, 2003.
5. Schuck, P. J., D. P. Fromm, A. Sundaramurthy, G. S. Kino, and W. E. Moerner, Improving the mismatch between light and nanoscale objects with gold bowtie nanoantennas, *Physical Review Letters*, **94**(1), 017402, 2005.
6. Curto, A. G., G. Volpe, T. H. Taminiau, M. P. Kreuzer, R. Quidant, and N. F. van Hulst, Unidirectional emission of a quantum dot coupled to a nanoantenna, *Science*, **329**(5994), 930–933, 2010.
7. Novotny, L. and N. van Hulst, Antennas for light, *Nature Photonics*, **5**(2), 83–90, 2011.
8. Wang, H., T. B. Huff, D. A. Zweifel, W. He, P. S. Low, A. Wei, and J.-X. Cheng, In vitro and in vivo two-photon luminescence imaging of single gold nanorods, *Proceedings of the National Academy of Sciences of the United States of America*, **102**(44), 15752–15756, 2005.

9. Ramakrishna, G., O. Varnavski, J. Kim, D. Lee, and T. Goodson, Quantum-sized gold clusters as efficient two-photon absorbers, *Journal of the American Chemical Society*, **130**(15), 5032–5033, 2008.

10. Imura, K., T. Nagahara, and H. Okamoto, Near-field two-photon-induced photoluminescence from single gold nanorods and imaging of plasmon modes, *The Journal of Physical Chemistry B*, **109**(27), 13214–13220, 2005.

11. Durr, N. J., T. Larson, D. K. Smith, B. A. Korgel, K. Sokolov, and A. Ben-Yakar, Two-photon luminescence imaging of cancer cells using molecularly targeted gold nanorods, *Nano Letters*, **7**(4), 941–945, 2007.

12. Sönnichsen, C., B. M. Reinhard, J. Liphardt, and A. P. Alivisatos, A molecular ruler based on plasmon coupling of single gold and silver nanoparticles, *Nature Biotechnology*, **23**(6), 741–745, 2005.

13. Ditlbacher, H., A. Hohenau, D. Wagner, U. Kreibig, M. Rogers, F. Hofer, F. R. Aussenegg, and J. R. Krenn, Silver nanowires as surface plasmon resonators, *Physical Review Letters*, **95**(25), 257403, 2005.

14. Barnes, W. L., A. Dereux, and T. W. Ebbesen, Surface plasmon subwavelength optics, *Nature*, **424**(6950), 824–830, 2003.

15. Ozbay, E., Plasmonics: merging photonics and electronics at nanoscale dimensions, *Science*, **311**(5758), 189–193, 2006.

16. Link, S., C. Burda, M. B. Mohamed, B. Nikoobakht, and M. A. El-Sayed, Laser photothermal melting and fragmentation of gold nanorods: energy and laser pulse-width dependence, *The Journal of Physical Chemistry A*, **103**(9), 1165–1170, 1999.

17. Link, S., Z. L. Wang, and M. A. El-Sayed, How Does a Gold Nanorod Melt? *The Journal of Physical Chemistry B*, **104**(33), 7867–7870, 2000.

18. Link, S., C. Burda, B. Nikoobakht, and M. A. El-Sayed, Laser-induced shape changes of colloidal gold nanorods using femtosecond and nanosecond laser pulses, *The Journal of Physical Chemistry B*, **104**(26), 6152–6163, 2000.

19. Chang, S.-S., C.-W. Shih, C.-D. Chen, W.-C. Lai, and C. R. C. Wang, The shape transition of gold nanorods, *Langmuir*, **15**(3), 701–709, 1999.

20. Letfullin, R. R., C. Joenathan, T. F. George, and V. P. Zharov, Laser-induced explosion of gold nanoparticles: potential role for nanophotothermolysis of cancer, *Nanomedicine*, **1**(4), 473–480, 2006.

21. Hirsch, L. R., R. J. Stafford, J. A. Bankson, S. R. Sershen, B. Rivera, R. E. Price, J. D. Hazle, N. J. Halas, and J. L. West, Nanoshell-mediated near-infrared thermal therapy of tumors under magnetic resonance guidance, *Proceedings of the National Academy of Sciences*, **100**(23), 13549–13554, 2003.

22. Huang, X., I. H. El-Sayed, W. Qian, and M. A. El-Sayed, Cancer cell imaging and photothermal therapy in the near-infrared region by using gold nanorods, *Journal of the American Chemical Society*, **128**(6), 2115–2120, 2006.

23. Huang, X., P. K. Jain, I. H. El-Sayed, and M. A. El-Sayed, Gold nanoparticles: interesting optical properties and recent applications in cancer diagnostics and therapy, *Nanomedicine*, **2**(5), 681–693, 2007.

24. Chon, J. W. M., C. Bullen, P. Zijlstra, and M. Gu, Spectral encoding on Gold Nanorods Doped in a Silica Sol–Gel Matrix and Its Application to High-Density Optical Data Storage, *Advanced Functional Materials*, **17**(6), 875–880, 2007.

25. Hamada, E., Y. Arai, Y. Shin, and T. Ishiguro, Optical information recording medium, *Google Patents*, 1992.

26. Ovshinsky, S. R., Reversible electrical switching phenomena in disordered structures, *Physical Review Letters*, **21**(20), 1450–1453, 1968.

27. Yamada, N., E. Ohno, K. Nishiuchi, N. Akahira, and M. Takao, Rapid-phase transitions of GeTe-Sb 2 Te 3 pseudobinary amorphous thin films for an optical disk memory, *Journal of Applied Physics*, **69**(5), 2849–2856, 1991.

28. Ohta, T., Phase-change optical memory promotes the DVD optical disk, *Journal of Optoelectronics and Advanced Materials*, **3**(3), 609–626, 2001.

29. Sugiyama, M., S. Inasawa, S. Koda, T. Hirose, T. Yonekawa, T. Omatsu, and A. Takami, Optical recording media using laser-induced size reduction of Au nanoparticles, *Applied Physics Letters*, **79**(10), 1528–1530, 2001.

30. Wilson, O., G. J. Wilson, and P. Mulvaney, Laser writing in polarized silver nanorod films, *Advanced Materials*, **14**(13–14), 1000–1004, 2002.

31. Niidome, Y., S. Urakawa, M. Kawahara, and S. Yamada, Dichroism of poly (vinylalcohol) films containing gold nanorods induced by polarized pulsed-laser irradiation, *Japanese Journal of Applied Physics*, **42**, 1749, 2003.

32. Pérez-Juste, J., B. Rodríguez-González, P. Mulvaney, and L.-M. Liz-Marzán, Optical Control and Patterning of Gold-Nanorod–Poly (vinyl alcohol) Nanocomposite Films, *Advanced Functional Materials*, **15**(7), 1065–1071, 2005.

33. Choi, K., P. Zijlstra, J. W. M. Chon, and M. Gu, Fabrication of Low-Threshold 3D Void Structures inside a Polymer Matrix Doped with Gold Nanorods, *Advanced Functional Materials*, **18**(15), 2237–2245, 2008.

34. Podlipensky, A., A. Abdolvand, G. Seifert, and H. Graener, Femtosecond laser assisted production of dichroitic 3D structures in composite glass containing Ag nanoparticles, *Applied Physics A*, **80**(8), 1647–1652, 2005.

35. Stalmashonak, A., G. Seifert, and H. Graener, Spectral range extension of laser-induced dichroism in composite glass with silver nanoparticles, *Journal of Optics A: Pure and Applied Optics*, **11**(6), 065001, 2009.

36. Ashley, J., M. P. Bernal, G. W. Burr, H. Coufal, H. Guenther, J. A. Hoffnagle, C. M. Jefferson, B. Marcus, R. M. Macfarlane, and R. M. Shelby, Holographic data storage technology, *IBM Journal of Research and Development*, **44**(3), 341–368, 2000.

37. Eichler, H. J., P. Kuemmel, S. Orlic, and A. Wappelt, High-density disk storage by multiplexed microholograms, *Selected Topics in Quantum Electronics, IEEE Journal of*, **4**(5), 840–848, 1998.

38. Orlic, S., S. Ulm, and H. J. Eichler, 3D bit-oriented optical storage in photopolymers, *Journal of Optics A: Pure and Applied Optics*, **3**(1), 72, 2001.

39. McLeod, R. R., A. J. Daiber, M. E. McDonald, T. L. Robertson, T. Slagle, S. L. Sochava, and L. Hesselink, Microholographic multilayer optical disk data storage, *Applied Optics*, **44**(16), 3197–3207, 2005.

40. Parthenopoulos, D. A. and P. M. Rentzepis, Three-dimensional optical storage memory, *Science*, **245**(4920), 843–845, 1989.

41. Strickler, J. H. and W. W. Webb, Three-dimensional optical data storage in refractive media by two-photon point excitation, *Optics Letters*, **16**(22), 1780–1782, 1991.

42. Cumpston, B. H., S. P. Ananthavel, S. Barlow, D. L. Dyer, J. E. Ehrlich, L. L. Erskine, A. A. Heikal, S. M. Kuebler, I. Y. S. Lee, and D. McCord-Maughon, Two-photon polymerization initiators for three-dimensional optical data storage and microfabrication, *Nature*, **398**(6722), 51–54, 1999.

43. Kawata, Y., H. Ishitobi, and S. Kawata, Use of two-photon absorption in a photorefractive crystal for three-dimensional optical memory, *Optics Letters*, **23**(10), 756–758, 1998.

44. Ichimura, I., K. Saito, T. Yamasaki, and K. Osato, Proposal for a multilayer read-only-memory optical disk structure, *Applied Optics*, **45**(8), 1794–1803, 2006.

45. Mitsumori, A., T. Higuchi, T. Yanagisawa, M. Ogasawara, S. Tanaka, and T. Iida, Multilayer 500 Gbyte optical disk, *Jpn J Appl Phys*, **48**(3), 03A055–03A055, 2009.

46. Tominaga, J., T. Nakano, and N. Atoda, An approach for recording and readout beyond the diffraction limit with an Sb thin film, *Applied Physics Letters*, **73**(15), 2078–2080, 1998.

47. Tsai, D. P. and W. C. Lin, Probing the near fields of the super-resolution near-field optical structure, *Applied Physics Letters*, **77**(10), 1413–1415, 2000.
48. Optical memory roadmap report, International Symposium on Optical Memory, 2006, Takamatsu, Japan.
49. Possibilities of Optical Disc Archives, *Fujiwara-Rothschild Ltd Market Analysis Report*, 2012.
50. The Diverse and Exploding Digital Universe, *IDC White Paper*, 2008.
51. *International Magnetic Tape Storage Roadmap*. Information Storage Industry Consortium, Section 2.3, 90–106, 2012.
52. Mie, G., Beiträge zur Optik trüber Medien, speziell kolloidaler Metallösungen, *Annalen der Physik*, **330**(3), 377–445, 1908.
53. Sönnichsen, C., T. Franzl, T. Wilk, G. von Plessen, J. Feldmann, O. Wilson, and P. Mulvaney, Drastic reduction of plasmon damping in gold nanorods, *Physical Review Letters*, **88**(7), 077402, 2002.
54. Novo, C., D. Gomez, J. Perez-Juste, Z. Zhang, H. Petrova, M. Reismann, P. Mulvaney, and G. V. Hartland, Contributions from radiation damping and surface scattering to the linewidth of the longitudinal plasmon band of gold nanorods: a single particle study, *Phys. Chem. Chem. Phys.*, **8**(30), 3540–3546, 2006.
55. Gans, R., Über die form ultramikroskopischer goldteilchen, *Annalen der Physik*, **342**(5), 881–900, 1912.
56. Prescott, S. W. and P. Mulvaney, Gold nanorod extinction spectra, *Journal of Applied Physics*, **99**(12), 123504–123504, 2006.
57. Zijlstra, P., A. L. Tchebotareva, J. W. M. Chon, M. Gu, and M. Orrit, Acoustic oscillations and elastic moduli of single gold nanorods, *Nano Letters*, **8**(10), 3493–3497, 2008.
58. van Dijk, M. A., M. Lippitz, and M. Orrit, Detection of acoustic oscillations of single gold nanospheres by time-resolved interferometry, *Physical Review Letters*, **95**(26), 267406, 2005.
59. Pelton, M., M. Liu, S. Park, N. F. Scherer, and P. Guyot-Sionnest, Ultrafast resonant optical scattering from single gold nanorods: Large nonlinearities and plasmon saturation, *Physical Review B*, **73**(15), 155419, 2006.
60. Lamprecht, B., J. R. Krenn, A. Leitner, and F. R. Aussenegg, Resonant and off-resonant light-driven plasmons in metal nanoparticles studied by femtosecond-resolution third-harmonic generation, *Physical Review Letters*, **83**(21), 4421–4424, 1999.
61. Zijlstra, P., J. W. M. Chon, and M. Gu, White light scattering spectroscopy and electron microscopy of laser induced melting in single gold nanorods, *Physical Chemistry Chemical Physics*, **11**(28), 5915–5921, 2009.
62. Petrova, H., J. P. Juste, I. Pastoriza-Santos, G. V. Hartland, L. M. Liz-Marzán, and P. Mulvaney, On the temperature stability of gold nanorods: comparison between thermal and ultrafast laser-induced heating, *Physical Chemistry Chemical Physics*, **8**(7), 814–821, 2006.
63. Coombs, J. H., A. Jongenelis, W. van Es-Spiekman, and B. A. J. Jacobs, Laser-induced crystallization phenomena in GeTe-based alloys. I. Characterization of nucleation and growth, *Journal of Applied Physics*, **78**(8), 4906–4917, 1995.
64. Wright, E. M., P. K. Khulbe, and M. Mansuripur, Dynamic Theory of Crystallization in GeSbTe Phase-Change Optical Recording Media, *Applied Optics*, **39**(35), 6695–6701, 2000.
65. Khulbe, P. K., E. M. Wright, and M. Mansuripur, Crystallization behavior of as-deposited, melt quenched, and primed amorphous states of GeSbTe films, *Journal of Applied Physics*, **88**(7), 3926–3933, 2000.
66. Kolobov, A. V., P. Fons, A. I. Frenkel, A. L. Ankudinov, J. Tominaga, and T. Uruga, Understanding the phase-change mechanism of rewritable optical media, *Nature Materials*, **3**(10), 703–708, 2004.

67. Wełnic, W., A. Pamungkas, R. Detemple, C. Steimer, S. Blügel, and M. Wuttig, Unravelling the interplay of local structure and physical properties in phase-change materials, *Nature Materials*, **5**(1), 56–62, 2006.
68. Inasawa, S., M. Sugiyama, and Y. Yamaguchi, Laser-induced shape transformation of gold nanoparticles below the melting point: the effect of surface melting, *The Journal of Physical Chemistry B*, **109**(8), 3104–3111, 2005.
69. Kippelen, B., N. Peyghambarian, S. R. Lyon, A. B. Padias, and H. K. Hall Jr, Dual-grating formation through photorefractivity and photoisomerization in azo-dye-doped polymers, *Optics Letters*, **19**(1), 68–70, 1994.
70. Tyson, S., G. Wicker, T. Lowrey, S. Hudgens, and K. Hunt. Nonvolatile, high density, high performance phase-change memory, *Proc SPIE*, **3891**, 2–9, 1999.
71. Lai, S. and T. Lowrey. OUM-A 180 nm nonvolatile memory cell element technology for stand alone and embedded applications, *IEDM'01 Tech Digest*, 803–806, 2001.
72. Lankhorst, M. H. R., B. W. Ketelaars, and R. A. M. Wolters, Low-cost and nanoscale non-volatile memory concept for future silicon chips, *Nature Materials*, **4**(4), 347–352, 2005.
73. Frenkel, J., *Kinetic Theory of Liquids*. Vol. 488. 1955: Dover New York.
74. Glezer, E. N. and E. Mazur, Ultrafast-laser driven micro-explosions in transparent materials, *Applied Physics Letters*, **71**(7), 882–884, 1997.
75. Yamasaki, K., S. Juodkazis, M. Watanabe, H. B. Sun, S. Matsuo, and H. Misawa, Recording by microexplosion and two-photon reading of three-dimensional optical memory in polymethylmethacrylate films, *Applied Physics Letters*, **76**(8), 1000–1002, 2000.
76. Day, D. and M. Gu, Formation of voids in a doped polymethylmethacrylate polymer, *Applied Physics Letters*, **80**(13), 2404–2406, 2002.
77. Ventura, M. J., M. Straub, and M. Gu, Void channel microstructures in resin solids as an efficient way to infrared photonic crystals, *Applied Physics Letters*, **82**(11), 1649–1651, 2003.
78. Ventura, M. J., M. Straub, and M. Gu, Planar cavity modes in void channel polymer photonic crystals, *Optics Express*, **13**(7), 2767–2773, 2005.
79. Zhou, G., M. J. Ventura, M. R. Vanner, and M. Gu, Use of ultrafast-laser-driven micro-explosion for fabricating three-dimensional void-based diamond-lattice photonic crystals in a solid polymer material, *Optics Letters*, **29**(19), 2240–2242, 2004.
80. Lippert, T. and J. T. Dickinson, Chemical and spectroscopic aspects of polymer ablation: special features and novel directions, *Chemical Reviews-Columbus*, **103**(2), 453–486, 2003.
81. Vogel, A. and V. Venugopalan, Mechanisms of pulsed laser ablation of biological tissues, *Chemical Reviews*, **103**(2), 577–644, 2003.
82. Juodkazis, S., K. Nishimura, S. Tanaka, H. Misawa, E. G. Gamaly, B. Luther-Davies, L. Hallo, P. Nicolaï, and V. T. Tikhonchuk, Laser-induced microexplosion confined in the bulk of a sapphire crystal: evidence of multimegabar pressures, *Physical Review Letters*, **96**(16), 166101, 2006.
83. Gamaly, E. G., S. Juodkazis, K. Nishimura, H. Misawa, B. Luther-Davies, L. Hallo, P. Nicolai, and V. T. Tikhonchuk, Laser-matter interaction in the bulk of a transparent solid: Confined microexplosion and void formation, *Physical Review B — Condensed Matter and Materials Physics*, 2006. **73**(21).
84. Holland, B. J. and J. N. Hay, The thermal degradation of poly (vinyl alcohol), *Polymer*, **42**(16), 6775–6783, 2001.
85. Gilman, J. W., S. Lomakin, T. Kashiwagi, D. L. VanderHart, and V. Nagy, Characterization of Flame-retarded Polymer Combustion Chars by Solid-state 13C and 29Si NMR and EPR, *Fire and Materials*, **22**(2), 61–67, 1998.
86. Bejan, A., Heat Transfer. 1993: Wiley, New York.

87. Chen, H., X. Liu, H. Muthuraman, J. Zou, J. Wang, Q. Dai, and Q. Huo, Direct laser writing of microtunnels and reservoirs on nanocomposite materials, *Advanced Materials*, **18**(21), 2876–2879, 2006.

88. Plech, A., V. Kotaidis, M. Lorenc, and J. Boneberg, Femtosecond laser near-field ablation from gold nanoparticles, *Nature Physics*, **2**(1), 44–47, 2005.

89. Plech, A., V. Kotaidis, S. Gresillon, C. Dahmen, and G. Von Plessen, Laser-induced heating and melting of gold nanoparticles studied by time-resolved x-ray scattering, *Physical Review B*, **70**(19), 195423, 2004.

90. Ditlbacher, H., B. Lamprecht, A. Leitner, and F. R. Aussenegg, Spectrally coded optical data storage by metal nanoparticles, *Optics Letters*, **25**(8), 563–565, 2000.

91. Zijlstra, P., J. W. Chon, and M. Gu, Five-dimensional optical recording mediated by surface plasmons in gold nanorods, *Nature*, **459**(7245), 410–413, 2009.

92. Mooradian, A., Photoluminescence of metals, *Physical Review Letters*, **22**(5), 185–187, 1969.

93. Yorulmaz, M., S. Khatua, P. Zijlstra, A. Gaiduk, and M. Orrit, Luminescence Quantum Yield of Single Gold Nanorods, *Nano Letters*, **12**(8), 4385–4391, 2012.

94. Funston, A. M., C. Novo, T. J. Davis, and P. Mulvaney, Plasmon coupling of gold nanorods at short distances and in different geometries, *Nano Letters*, **9**(4), 1651–1658, 2009.

95. Taylor, A. B., J. Kim, and J. W. M. Chon, Detuned surface plasmon resonance scattering of gold nanorods for continuous wave multilayered optical recording and readout, *Optics Express*, **20**(5), 5069–5081, 2012.

2 Integrated Plasmonic Nanodevices

Haroldo T. Hattori, Ivan D. Rukhlenko,
Ziyuan Li, and Malin Premaratne

CONTENTS

2.1 GENERAL BACKGROUND THEORY

Surface plasmon polaritons (SPPs) are waves that propagate along an interface between a dielectric and a metal if the permittivity of the interface materials satisfies certain conditions. At low frequencies, metals are considered as good conductors; however, they behave very much like dielectrics at optical frequencies [1]. Owing to the presence of intrinsic losses associated with material dispersion, Kramers–Kronig relations suggest that it is necessary to consider both the real and imaginary parts of metal permittivity at optical frequencies.

When an oscillating optical field couples to the electron gas at the vicinity of the metal surface, the resulting collective oscillations of photons and electrons are widely referred to as surface plasmons [2]. The propagating counterparts of these surface plasmons are known as the SPPs. The existence of SPPs was originally predicted by Ritchie [3] and demonstrated experimentally by several research groups worldwide. SPPs are currently being used in the design of very compact devices discussed in this chapter.

2.1.1 Extraordinary Transmission of Light in Simple and Chirped Structures

Electronic devices are generally smaller than optical devices. For example, nowadays, transistors can be made with dimensions close to 20 nm, whereas optical devices struggle to achieve dimensions of <50 nm. The main limiting factor is the diffraction limit of light, which states that light (with wavelength around 1 μm) cannot be stored or guided in dimensions much smaller than half of its wavelength. Even though fundamentally this aspect is always true for wave phenomena, we could bypass the diffraction limit notionally by varying the wavelength. The reality of such an approach was first noted when the extraordinary transmission of light was discovered by Ebbesen et al. in 1998 [4]. They showed that the excitation of plasmonic waves with wavelengths in the nanometer range can make transmission of light through holes with dimensions considerably smaller than the wavelength of light. This opened up the possibility to create very compact devices with dimensions much smaller than the wavelength of light. In this chapter, we would like to elucidate a variety of such ideas by discussing the operating principles of some devices that take advantage of these features.

The transmission of light through small apertures in metallic films had been studied a long time ago. Initial studies [5] assumed that the metallic film was a perfect conductor, and diffraction theory was employed to study the transmission of light through small holes. As an example, the transmittance of both TE- and TM-polarized light through a small cylindrical aperture of radius r in an infinite conductor plane, under normal incidence, is characterized by the transmission coefficient [5]

$$T = \frac{1024\pi^2}{27}\left(\frac{r}{\lambda}\right)^4.$$

(2.1)

This expression clearly shows that the amount of light that can get through a hole with a diameter much smaller than the wavelength of light λ is significantly small. After leaving the hole, light get dispersed due to diffraction at the exit aperture.

Bethe's approach [5] to the study of transmission of light through small holes has several shortcomings at optical frequencies. Specifically, the thickness of the metal films is not infinitely small, and metals are not perfect conductors at optical frequencies. In 1998, Ebbesen et al. [4] showed that a considerable enhancement in the transmission of light through small holes could be achieved by structuring the metallic screen with a periodic lattice. In this way, the periodic lattice could excite SPPs and funnel light through the aperture. The transmission coefficient could even be greater than unity, since the incident power density through certain areas could be compressed through the tiny hole by the excitation of plasmonic waves. This does not suggest that the principle of conservation of energy is violated in this experiment. Rather, it emphasizes the fact that energy can be collected from areas outside the aperture and get through the hole, implying some pseudo-amplification phenomena.

Figure 2.1 shows the transmission spectrum through a single aperture. The diameter of the aperture, which is etched all the way through the metal film, is 250 nm.

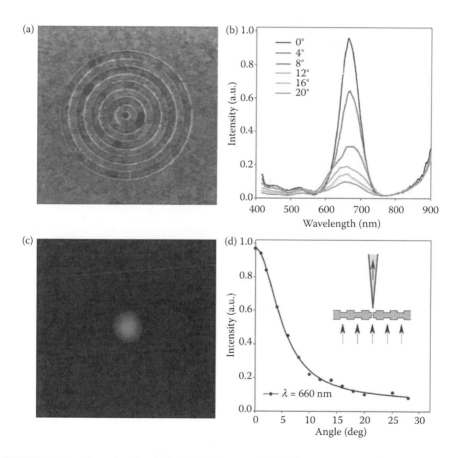

FIGURE 2.1 (See color insert.) (a) FIB image of a bull's eye structure in a suspended silverfilm and (b) its transmission spectra for different collection angles. The tail beyond 800 nm is an artifact of the measurement. The structure is illuminated under normal incidence with unpolarized light; the period of the grating is 500 nm, the groove depth is 60 nm, the hole diameter is 250 nm, and the film thickness is 300 nm. (c) Optical image and (d) angular intensity distribution at the wavelength of maximum transmission. (After H. J. Lezec et al., Beaming light from a subwavelength aperture, *Science 297*, 820–822, 2002. © AAAS 2002, with permission.)

The circular grating has a period of 500 nm and a filling factor of 50%. The trenches in the grating are not etched through the metal film; their depth is 60 nm, while the film thickness is 300 nm. This circular grating helps to excite plasmonic waves and consequently aids transmission of light through the hole. The wavelength of maximum transmission occurs when there is a phase matching between the wavevectors of the incident wave, grating, and plasmonic wave. It is clear that the transmission maximum occurs at a wavelength larger than the diameter of the air hole, in sharp contrast with the predictions from Equation 2.1. Moreover, Lezec et al. [6] have shown that the transmission of light through the hole is directional (it is more intense in the normal direction), in contrast with the prediction that the emerging light would be diffracted in different directions. It is interesting to note that this directionality is aided by the phase matching of the incident, plasmonic, and grating wavevectors.

Minovich et al. [7] have studied the transmission of light through different arrangements of holes, both experimentally and theoretically. It is instructive to consider the behavior of light transmitted through these types of complex structures.

The first device we analyze is the Thue Morse structure, which consists of a sequence of holes corresponding to the bits sequence 01101001, where bit "1" corresponds to a hole of size 300×300 nm^2, and bit "0" corresponds to an unpatterned area of similar size. Arrays of Thue Morse holes are shown in the inset in Figure 2.2a. The separations of different structures are 12 and 2 μm in the horizontal and vertical directions, respectively. It is interesting to note that the transmission spectrum of the Thue Morse structure is nearly flat in the wavelength range between 700 and 1700 nm, with little variation between 700 and 1300 nm.

The second device of interest is a chirped structure shown in Figure 2.2b, which has a lattice period of 2 μm and a hole size varying in the horizontal direction from 300 to 800 nm (there is an increase in dimension of 50 nm when moving from left to right). The transmission spectrum is also modified by the chirped structure, and high transmission can be observed from 800 to 1300 nm.

These two structures show the viability of tailoring the transmission spectra by modifying the size of the holes in the metallic screen. It is also clear that light is

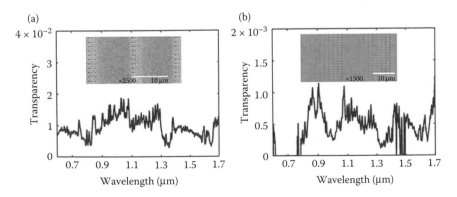

FIGURE 2.2 Transmission spectra of (a) Thue Morse sequence and (b) chirped holes. (After A. Minovich et al., *Opt. Commun.* **282**, 2023–2027, 2009. © ESD 2009, with permission.)

transmitted at wavelengths considerably larger than the size of the holes. Light can also be transmitted through very small apertures, which means that it could also be confined at subwavelength regions. This idea boosted the field of plasmonics and opened the doors to designing nanometric devices with dimensions much smaller than the optical wavelength.

2.1.2 FUNDAMENTALS OF SURFACE-ENHANCED RAMAN SCATTERING

Interest in devices of sizes comparable to the wavelength of light stems from their potential usability in applications, such as optical communications systems and all-optical computing. In contrast, if light is confined into tiny spaces, strong electric fields get created. These strong localized electric fields, or "hot spots," can be used to enhance nonlinear effects, such as the surface-enhanced Raman scattering (SERS). SERS is widely used in designing micro/nanosensors, as discussed below.

In order to understand SERS, it is necessary to recall the fundamentals of the Raman effect. The Raman effect describes the inelastic scattering of incident photons at molecules or atoms. When incident photons reach a set of molecules, part of their energy can be either transferred or absorbed by the molecules (leading to the creation of phonons), meaning that the scattered photons will emerge with either lower or higher energy. If the energy of the incident photon is $\hbar\omega_{in}$ and the energy of the phonon is $\hbar\omega_{ph}$, then the energy of the scattered photon is

$$\hbar\omega_{sc} = \hbar\omega_{in} \pm \hbar\omega_{ph}.$$

In general, the energy of the scattered phonon is high and the frequency of the output photons is translated by several THz. However, the phonon energy is spread over a certain bandwidth depending on the properties of the material: different materials will have different phonon energies leading to different translated spectra. This means that Raman spectra can be used to identify the presence of a particular substance, since a given Raman spectrum is a fingerprint of that material.

SERS is similar to the Raman effect, but occurs only at the surface of a suitable material. In contrast to the Raman effect, SERS is surface-specific and highly selective. This means that SERS can give information about the conditions of the surface of a material, by passing the capabilities of the conventional Raman sensing which utilizes information coming from different regions of the bulk material. Since plasmonics can excite large electric fields at the surface of a metal and lead to enhancement of SERS, it acts as a perfect masking medium to suppress Raman signals generated inside the bulk of the material. This enhancement is achieved by utilizing two fundamental agents: (i) electrical and (ii) chemical. The enhancement of SERS due to chemical effects can be explained by the fact that molecules can be adsorbed to the surface of metal leading to a change in the electronic states of the material, whereas the electric field enhancement occurs through the excitation of plasmonic waves at the boundary between the metal and the dielectric. It is generally believed that the electromagnetic field enhancement plays the main role to get stronger SERS due to the generation of strong localized surface plasmon resonances on both the incident and Raman scattered waves. The SERS enhancement factor can be estimated using the expression [8,9]

$$EF_{SERS} = \frac{|E_{loc}(\omega_{in})|^2}{|E_0(\omega_{in})|^2} \frac{|E_{loc}(\omega_{sc})|^2}{|E_0(\omega_{sc})|^2},$$

where $E_{loc}(\omega_{in})$ is the amplitude of the enhanced local electric field at the incident frequency, $E_0(\omega_{in})$ is the amplitude of the incident electric field at the incident frequency, $E_{loc}(\omega_{sc})$ is the amplitude of the enhanced local electric field at the Raman scattered frequency, and $E_0(\omega_{sc})$ is the amplitude of the incident electric field at the Raman scattered frequency.

2.1.3 IMPEDANCE THEORY

Since many plasmonic devices are considerably shorter than the corresponding excitation optical wavelength, circuit models can be effective in describing their behavior [10,11]. In fact, plasmonic devices have been coined as a way to make integrated circuits for optoelectronic systems. Examples of such plasmonic devices include the plasmonic analogs of electronic counterparts: resistors, capacitors, and inductors.

It is true that numerical methods, such as the finite-difference time-domain (FDTD) method, can provide highly accurate descriptions of optical devices. However, FDTD simulations are computationally expensive and time-consuming. They do not provide valuable physical insights essential for understanding the operational characteristics and design optimizations. Thus, being analogous to familiar electronic components, circuit equivalents of optical devices can provide a much better understanding of optical devices and an intuitive base to perform complex designs. However, metals at optical frequencies are very dispersive and equivalent circuits cannot be directly derived from scattering measurements [10]. Therefore, detailed numerical analysis underpins the synthesis of optical equivalents of electronic components.

Central to the analysis of optical circuits is the concept of impedance, which is defined as [12]

$$Z_{load} = \frac{V_{eq}}{I_{eq}},$$

where V_{eq} is the average equivalent potential difference between the terminals of the device due to the residual electric field, which is obtained by subtracting the incident field from the total internal field, and I_{eq} is the equivalent current flowing into the element.

In the case of waveguides, nanoantennas, and apertures, impedance can be calculated by integrating the main components of electric and magnetic fields along certain paths. In the case of TE modes with main field components along the x- and y-directions, the impedance can be calculated as a ratio [13]

$$Z_{\text{load}} = \frac{\int_{-\infty}^{\infty} E_x \, dx}{\int_{-\infty}^{\infty} H_y \, dy},$$

where the integrals are taken along the central cuts of the structure.

Given the load impedance to a certain waveguide with characteristic impedance Z_{W}, the reflectivity can be calculated as

$$R = \left| \frac{Z_{\text{load}} - Z_{\text{W}}}{Z_{\text{load}} + Z_{\text{W}}} \right|^2.$$

Impedance theory will be used when analyzing nanoantennas and plasmonic-waveguide-based devices, as it provides a simple way to interpret the numerical results generated by FDTD simulations.

2.2 EXCITATION OF PLASMONIC NANOANTENNAS BY MICROLASERS

In the previous section, we briefly touched on the concepts required to understand and use plasmonics. We learned that the extraordinary transmission of light opened up the possibility to transmit light through very small apertures, enabling a richer set of technologies to steer and confine light into subwavelength regions. We also learned that it is possible to generate high-intensity electric fields at these local sites since the electric field intensity is proportional to $\sqrt{P_{\text{in}}/A_{\text{eff}}}$, where P_{in} is the incident power through the aperture and A_{eff} is the effective area of the aperture. However, what is really required is that the concentrated light be delivered to preselected sites/molecules at will, to initiate reactions or excite energy levels. Another requirement is to collect the dispersed light to a subwavelength area for multitudes of uses. Both these tasks can be fulfilled by using specially designed structures called nanoantennas, which have the ability to concentrate light into small regions and create high-intensity localized electric fields (hot spots). These high-intensity localized electric fields could be used to excite nonlinear effects [14], to visualize nanoparticles [10], and even to detect specific chemicals and biological substances [12]. Recently, nanoantennas have also been used to collimate the far-field emission from semiconductor lasers [13]. A general review of the main properties of nanoantennas can be found in Ref. [14].

The simplest nanoantenna resembles a dipole antenna, consisting of two metallic regions separated by a dielectric gap (typically air). The easiest way to excite a nanoantenna is to use a large-area source and excite the nanoantenna from the top. In the following section, we present more efficient ways to couple light into a nanoantenna.

2.2.1 COUPLING OF LIGHT INTO NANOANTENNAS BY USING NANOTAPERS

Nanoantennas are small devices that are capable of generating high-intensity electric fields. Since these nanoantennas have small footprints, they cannot handle much power. This means that they are good candidates for being excited by microlasers, such as microdisks and triangular, square, and photonic crystal lasers. These lasers emit power over an exceedingly large range, which spans from several nanowatts to a few milliwatts.

The epitaxially layered structure that is used to create the integrated microlaser and nanoantenna is shown in Figure 2.3a. If the quantum heterostructure consists of quantum wells, three layers of 7.3-nm $In_{0.2}Ga_{0.8}As$ separated by 6-nm GaAs confinement barriers can emit light at 980 nm with a 40-nm linewidth. On the other hand, three layers of $In_{0.5}Ga_{0.5}As$ quantum dots can emit light at 1160 nm with a linewidth of nearly 100 nm. The choice of the active medium depends upon the application: quantum dots generally provide a larger gain bandwidth and can lead to a device that is less temperature-dependent than quantum wells, but it is generally easier to achieve lasing with quantum wells.

Initially, light from a quantum dot microdisk laser is coupled to a dipole nanoantenna, as shown in Figure 2.3b. Microdisk lasers generally emit power in the range

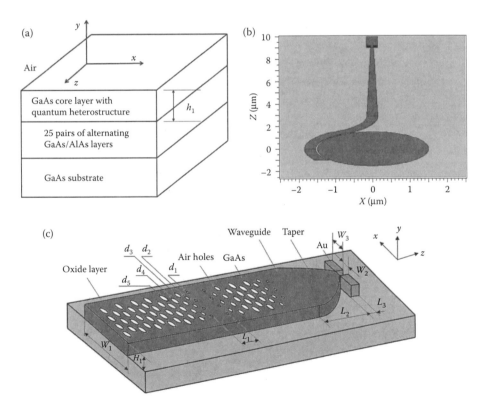

FIGURE 2.3 **(See color insert.)** (a) Schematic of the epitaxially layered structure and (b) tapered coupling of light from a microdisk laser to a nanoantenna. (After H. T. Hattori et al., Coupling of light from microdisk lasers into plasmonic nano-antennas, *Opt. Express* **17**, 20878–20884, 2009. © OSA 2009, with permission.) (c) Schematic of light coupling from a photonic crystal laser into a nanoantenna by using a parabolic taper. (After Z. Li et al., Merging photonic wire lasers and nanoantennas, *J. Lightw. Technol.* **29**, 2690–2697, 2011. © IEEE 2011, with permission.)

of several microwatts to few milliwatts. In the configuration shown in Figure 2.3b, the microdisk resonator has a diameter of 3 μm and the thickness of the GaAs core layer with quantum dots is $h_1 = 170$ nm [15]. The main resonant mode of the microdisk laser appears at $\lambda = 1166$ nm, corresponding to the main resonant whispering gallery mode $TE_{17,1}$.

This microdisk laser couples about 38% of the emitted power to a single-mode waveguide. The power coming from the single-mode laser is then funneled to a nanoantenna by using a linear nanotaper. The dipole nanoantenna consists of two golden regions of width 125 nm and length 100 nm. The gap between the metallic regions is 100 nm. By using a nanotaper, about 12% of the incident power reaches the nanoantenna, leading to an electric field of 1.4 MV/m. If lower power is needed to excite a nanoantenna, photonic crystal lasers are a good option. Photonic crystal lasers emit light in the range of nanowatts to microwatts, depending on the size of the photonic crystal cavity. An interesting laser structure is the photonic crystal wire laser in which light is confined laterally and vertically by total internal reflection, while light is confined in the longitudinal direction by a photonic crystal operating in its bandgap region. This is illustrated by the photonic crystal wire laser shown in Figure 2.3c. In this structure, the parameters of the photonic wire resonator are $W_1 = 1.2$ μm, $H_1 = 140$ nm, $d_1 = 50$ nm, $d_2 = 80$ nm, $d_3 = 110$ nm, $d_4 = 170$ nm, $d_5 = 180$ nm, and $L_1 = 810$ nm [16]. The main photonic crystal works in its bandgap region, preventing light to move backward (in the $-z$-direction) while concentrating light in the defect region (unpatterned region between the large holes). The size of holes is tapered to allow for a high transmission to the nanotaper. Interestingly, in this configuration, about 70% of light is coupled to the nanotaper.

The main resonant mode appears at 987.6 nm, close to the peak gain wavelength of the quantum wells at 980 nm. This laser will excite a dipole nanoantenna consisting of two golden regions with parameters $W_3 = 200$ nm, $L_3 = 100$ nm, and $W_2 = 50$ nm. In order to analyze the performance of this structure, the electric field enhancement η is defined as $\eta = E_{nano}/E_{wav}$, where E_{nano} is the electric field in the gap of the nanoantenna and E_{wav} is the electric field at the laser output close to the taper.

After optimizing the parabolic taper of the nanoantenna, which is more efficient than a linear taper [16], field enhancements as high as 7 are achieved. It is estimated that about 45% of the emitted power is available to the nanoantenna. By using FDTD simulations, the impedance of the nanoantenna was estimated as $Z_{nano} = (157 - 124i)$ Ω, while the impedance of the waveguide was estimated as $Z_{wav} = 676$ Ω. The parabolic taper reduces the impedance discontinuity between the nanoantenna and the waveguide and, at the same time, attempts to focus light into the gap of the dipole antenna. In this particular laser, electric field as high as 1.3 MV/m can be achieved. If lower intensities of electric fields are needed, the size of the defect cavity can be reduced in the photonic crystal wire cavity, making photonic crystal wire quite versatile.

2.2.2 Driving Plasmonic Nanoantennas Using Triangular Lasers and Slot Dielectric Waveguides

One problem with the previous coupling scheme is that a significant amount of power may be reflected back to the laser. Reflection of light back into the laser source can lead to multimode lasing and mode competition. Moreover, reflected light can be coupled with the lasing modes causing their phases to vary, leading to an increased laser noise. One way to reduce this noise is by using slot dielectric waveguide, as will be discussed in this section.

The basic schematic is shown in Figure 2.4a. The core layer consists of GaAs with 7.3-nm $In_{0.2}Ga_{0.8}As$ quantum wells. An underneath oxidized $Al_{0.98}Ga_{0.02}As$ (450-nm thick before oxidation takes place) and air layers concentrate light in the core GaAs layer by total internal reflection. In the GaAs layer, an equilateral resonator is fabricated with a side length of 3.7 μm. The main peak appears at 968.1 nm corresponding to the mode $TE_{36,0}$ and is the only high quality factor peak in the gain region of the quantum wells (gain linewidth of 40 nm). About 42% of the generated light from the triangular resonator is then coupled into a single-mode waveguide.

In order to couple light from a single-mode waveguide into the dielectric slot waveguide, a 9-μm-long nanotaper is used [17]. The resulting combination of single-mode waveguide, nanotaper, and slot waveguide can lead to about 36% of the generated power being coupled into the slot waveguide. The width of the gap in the slot waveguide is assumed to be $G_1 = 50$ nm (see Figure 2.4b for details). The nanoantenna consists of two gold regions of width $W_1 = 200$ nm and length $L_1 = 100$ nm.

The end of the slot waveguide is tapered to further reduce reflections. After optimizing the slot waveguide with $L_2 = 100$ nm, a field enhancement of 3.8 (with respect to the electric field in the single-mode waveguide) is achieved with reflectivity of 8%. This low reflectivity is achieved because the slot waveguide brings light to the nanoantenna efficiently but the scattered light from the nanoantenna is not directional and is poorly confined in the slot waveguide: most of the scattered nondirectional light leaks away from the slot waveguide. The impedance of the nanoantenna is about $(160 - 115i)$ Ω, while the impedance of the slot waveguide is about 210 Ω. These values would lead

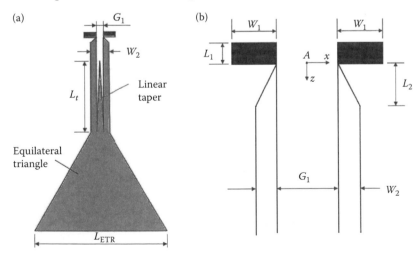

FIGURE 2.4 (a) Top view of the triangular laser combined with the nanotaper, slot waveguide, and nanoantenna and (b) closer view of the slot waveguide close to the nanoantenna. (After H. T. Hattori, Z. Li, and D. Liu, Driving plasmonic nanoantennas with triangular lasers and slot waveguides, *Appl. Opt.* **50**, 2391–2400, 2011. © OSA 2011, with permission.)

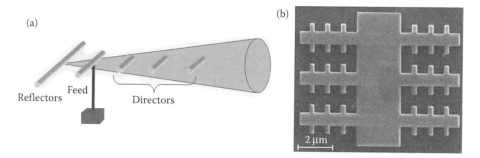

FIGURE 2.5 (**See color insert.**) (a) Typical geometry of a five-element RF Yagi–Uda antenna and (b) SEM image of Charnia-like structure.

to a reflectivity of about 10%, which is close to the actual 8% reflectivity. Higher field enhancements could be produced by reducing the width of the antenna and air gaps, but the fabrication of this device would become more challenging.

2.2.3 DIRECTIONAL AND BROADBAND NANOANTENNAS

Dipole antennas cannot emit light in a directive way. In certain applications, it is useful to concentrate light in a certain direction. This can be achieved, at optical frequencies, by mimicking radio-frequency Yagi–Uda antennas and adapting their main properties to the optical frequencies [18–22]. In this sense, light is coupled to an element called feed and the directionality of light is achieved by parasitic elements. A reflector prevents light from moving backwards, while the directors focus light in the forward direction as illustrated in Figure 2.5a. These antennas can make the emission of light from molecules, quantum dots, or other localized emitters more directional.

Other broadband applications, such as multiplexed sensor or plasmonic Charnia-like structures (see Figure 2.5b), may require devices that operate uniformly over a wide range of frequencies. The structure shown in Figure 2.5b was fabricated by using a focused ion beam milling that patterned 300-nm-thick gold on top of the quartz substrate.

When this structure was simulated using the FDTD method, in the wavelength region spanning from 700 to 800 nm, a transmission coefficient of 45% was observed with a standard deviation of 14%. In the same wavelength region, the electric field enhancement was observed to live between 8.8 and 11 at the hot spot regions. In the wavelength range between 800 and 900 nm, the electric field enhancement was observed to be about 8.8. Beyond 900 nm, the electric field enhancement decreases with the wavelength but still stays above 6 up to 1100 nm. The total SERS enhancement factor measured for this structure was 2.56×10^5.

2.3 PLASMONIC-WAVEGUIDE-BASED DEVICES IN THE LIGHT OF TRANSMISSION-LINES ANALOGY

Over the last five years, a vast variety of nanodevices based on plasmonic gap waveguides (PGWs) have been proposed, theoretically analyzed, and experimentally fabricated [23–27]. Owing to the analogy between PGWs and microwave transmission lines, such devices can be conveniently described using the standard tools of the network analysis [28–33]. This approach enjoys wide application, as it allows device characterization to be performed about 10,000 times faster than with the full-blown FDTD calculations [34]. This section presents the essentials of the transmission-line method and discusses its application to a number of widespread geometries.

2.3.1 Modes of PGWs

It is common to model PGWs by a plane dielectric layer sandwiched between two equal metal slabs, whose thicknesses substantially exceed the skin depth of the metal (~20 nm) [8]. If the gap width is comparable to or smaller than the evanescent decay length of the fields penetrating the dielectric (~0.2–2 μm), then a pair of surface plasmon waves guided by the two metal–dielectric interfaces constitute a single SPP mode of the waveguide. It is well known that PGWs support only TM modes, which may be described by the complex propagation constant $\beta = \beta' + i\beta''$. These modes are either symmetric or antisymmetric with respect to the reflection of their transverse electric (or magnetic) field from the medium plane of the waveguide.

In the deep subwavelength regime, realized where the gap width $2h$ is much smaller than the wavelength λ of light in a vacuum ($2h \ll \lambda$), the antisymmetric SPP mode ceases to exist and a PGW operates as a single-mode plasmonic waveguide [35]. The propagation constant of the symmetric mode obeys the dispersion relation [8,36,37]

$$\tanh(k_d h) = -\left(\frac{\varepsilon_m}{\varepsilon_d}\right)\left(\frac{k_d}{k_m}\right), \tag{2.2}$$

in which $k_j = (\beta^2 - \varepsilon_j k^2)^{1/2}$, $k = 2\pi/\lambda$, and $\varepsilon_j = \varepsilon_j' + i\varepsilon_j''$ is the permittivity of the dielectric ($j = d$) or metal ($j = m$). A simple analysis of this relation in the limit of small h reveals the two important features of the symmetric SPP mode: (i) its propagation length $L_{SPP} = 1/2\beta''$ decreases with h like $L_{SPP} \sim h \, |\varepsilon_m|^2/2\varepsilon_d'\varepsilon_m''$, while (ii) its effective index $n_{eff} = \beta/k$ diverges like $n_{eff} \sim \varepsilon_d'/|\varepsilon_m'| \, kh$. Since the transverse localization of SPPs in PGWs improves with increasing β', these relations imply that one should always find a compromise between the attenuation rate of the guided mode and its confinement to the gap [38].

2.3.2 Analogy with Transmission Lines

As it was mentioned earlier, the description of PGW-based devices is substantially facilitated by the analogy between plasmonic waveguides and microwave transmission lines. This analogy is based on similar evolution of the transverse electric (magnetic) field in a single-mode PGW and polarity of the voltage (current) in a transmission line, which occurs as long as $h \ll \lambda$. The fact of the similarity is embodied in the same time–space dependency

$$U(z,t) = [U_f \exp(i\beta z) + U_b \exp(-i\beta z)]\exp(-i\omega t)$$

of the field ($U = E_x$, H_y) and circuit ($U = V$, I) variables. In writing this expression in the form shown, we have assumed that the phase fronts propagate in the $+z$-direction ($\beta' > 0$) and marked the amplitudes of the waves traveling forward and backward by the subscript f and b. It is easy to show that an infinite PGW may be represented by an infinite transmission line characterized by the complex impedance [33]

$$Z_{\mathrm{PGW}}(h) = \frac{2\beta(h)h}{\varepsilon_d k} Z_0, \tag{2.3}$$

where Z_0 is the impedance of free space and β is a function of h, as it follows from Equation 2.2. The total impedance of a finite waveguide section (stub) depends on the section length d and the reflection properties of its end face. The reflection-induced phase shift of the plasmonic wave may be allowed for by taking the amplitude reflectance of SPPs at the end face to be given by the Fresnel's formula [39]

$$\Gamma = \frac{\sqrt{\varepsilon_d} - \sqrt{\varepsilon_e}}{\sqrt{\varepsilon_d} + \sqrt{\varepsilon_e}},$$

in which ε_e is the permittivity of the medium bounding the stub. Notice that the sign of Γ is dictated by the conservation of the voltage polarity upon the mode reflection from the short-circuited line (which corresponds to the limit $|\varepsilon_e| \to \infty$), rather than by the actual behavior of SPPs upon the reflection from the perfect electric conductor. A simple algebra then yields the stub impedance of the form [23]

$$Z_s(d,w) = \frac{\sqrt{\varepsilon_d} - i\sqrt{\varepsilon_e}\ \tan(\beta_s d)}{\sqrt{\varepsilon_e} - i\sqrt{\varepsilon_d}\ \tan(\beta_s d)} Z_{\mathrm{PGW}}(w), \tag{2.4}$$

where $\beta_s \equiv \beta(w)$ is the SPP propagation constant in an infinite PGW of the width $2w$, equal to the stub width (see Figure 2.6b).

Using Equations 2.2 through 2.4, together with the transfer matrix method, allows calculation of the transmission spectra for different assembles of the plasmonic waveguides and stubs. To find expressions for these spectra in a closed form, one needs to establish an equivalent circuit for a given waveguide structure. This can be done via the replacement of stubs by their impedances, which then should be connected either in series or in parallel to the transmission lines representing the waveguides. The type of connection depends on the structure topology and is to be selected in accordance with Figure 2.6a. Since the stub in this figure is bounded by metal, one should set in Equation 2.4 $\varepsilon_e = \varepsilon_m$.

2.3.3 LINEAR PLASMONIC FILTERS

The transmittance of a straight, linear PGW is determined by the SPP propagation length, which grows monotonously with the wavelength. Coupling one or more stubs perpendicular to the PGW substantially alters its transmission dispersion and makes it heavily dependent on the stub parameters (such as length, width, and dielectric

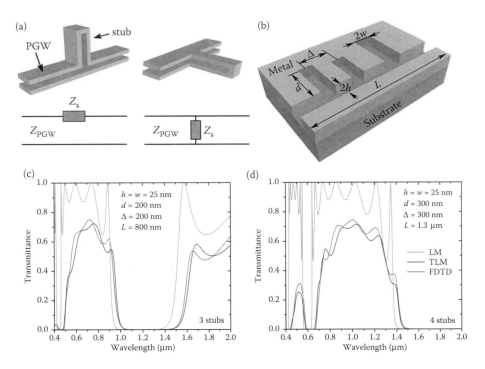

FIGURE 2.6 (a) Two ways of stub connection to a PGW and transmission-line models of the resulting configurations; (b) schematic of a PGW-based plasmonic filter and the notation employed. Lower panels show transmittance of a plasmonic filter made of silver for (c) three and (d) four equal stubs coupled perpendicular to the bus waveguide ($\varepsilon_d = 1$); simulation parameters are shown beside the plots. Different curves correspond to the results obtained by ignoring absorption losses (LM) and using transmission-line model (TLM) and FDTD simulations. (After A. Pannipitiya et al., Improved transmission model for metal–dielectric–metal plasmonic waveguides with stub structure, *Opt. Express* **18**, 6191–6204, 2010. © OSA 2010.)

filling), thus enabling a challenging application of PGWs as the wavelength-selective plasmonic filters [26,29,34].

If the plasmonic filter is realized by coupling a single stub to the bus waveguide of length L, then its transmittance is given by the formula [23,34]

$$T(\lambda) = \left| 1 + \frac{1}{2} \left(\frac{Z_s}{Z_{PGW}} \right)^{\pm 1} \right|^{-2} \exp\left(-\frac{L}{L_{SPP}} \right), \qquad (2.5)$$

in which ± signs correspond to the left and right stub configurations in Figure 2.6a. Here, the expression under the modulus allows for the interference between the incident plasmonic wave and the wave reflected from the stub, while the exponential describes attenuation of SPPs on the length of the bus waveguide. The damping of SPPs along the stub is included into the complex impedance Z_s given in Equation 2.4.

By varying the number of stubs and their relative positions Δ in the design of a typical plasmonic filter shown in Figure 2.6b, one may engineer its transmission properties. For multiple stubs, the function $T(\lambda)$ has a structure similar to that of Equation 2.5, with additional terms proportional to $\exp(2i\beta_s\Delta)$ under the modulus describing interference between the waves reflected from different stubs. This function gives reasonably accurate transmission spectra of the plasmonic filters, provided that SPPs from different stubs do not interact with each other along the stubs' boundaries. In the geometry of Figure 2.6b, this condition will be met if the thickness of metal separating the stubs is much larger than the skin depth of metal.

Figure 2.6c and d shows the examples of the transmission spectra calculated for silver PGWs coupled to three and four identical stubs. The widths of the stubs are set equal to the width of the bus waveguide ($h = w$). It is seen that the spectra obtained using the transmission-line model agree well with the numerical FDTD data for the entire wavelength range. For reference, green curves show the transmittance neglecting metal losses (assuming $\varepsilon''_m = 0$) and the phase shift due to the reflection from the end face of the stub (assuming $\varepsilon'_m \to \infty$). The comparison of these curves with the red spectra indicates the significance of the proper account for the complex-valued permittivity of metal at optical frequencies.

2.3.4 Nonlinear Plasmonic Switches

Among the disadvantages of linear plasmonic filters is the inability to alter their transmission properties after the filters are fabricated. One may enable the dynamic variation of the transmittance by introduction of an optically active medium into the design of a plasmonic filter. For instance, in a single-stub configuration, a nonlinear dielectric may be embedded in the bus waveguide or the stub. In either case, the resulting plasmonic filter may operate in the linear regime, when the propagation of a relatively weak signal is controlled by the strong pump altering the permittivity of the nonlinear medium, or in the nonlinear regime, in which the filter transmittance is governed by the strong signal itself.

The nonlinear regime is of particular interest to us, as it gives rise to the phenomenon of optical bistability, which can be employed to develop plasmonic memories and switches [40]. In order to illustrate the possibility of optical switching with nonlinear PGWs, we consider a single-stub plasmonic filter shown in Figure 2.6a and suppose that its stub is filled with a Kerr medium of permittivity $\varepsilon_{NL} = \varepsilon_d + \chi^{(3)}|E|^2$, where $\chi^{(3)}$ is the nonlinear optical coefficient and E is the electric field. The length of the stub and the coefficient $\chi^{(3)}$ should be such that the nonlinear change in the optical path length is sufficient for the noticeable alteration of the transmittance. Since the size of the switch is limited to the submicrometer range by the miniaturization demands, the third-order nonlinear coefficient must be relatively high— in the order of 10^{-16} (m/V)^2. The nonlinear coefficients of such magnitudes are exhibited by metallic nanoparticles embedded in glass [41].

By employing the transmission-line analogy and solving the problem of SPP scattering at the junction of the bus waveguide and the stub, we found the input–output characteristic of the above plasmonic switch in the parametric form, with the SPP intensity averaged over the stub length as the parameter [28]. The typical bistability curve, obtained analytically, is shown in Figure 2.7b. One can see that the switching of the output signal occurs at the practically achievable electric field strengths of 650

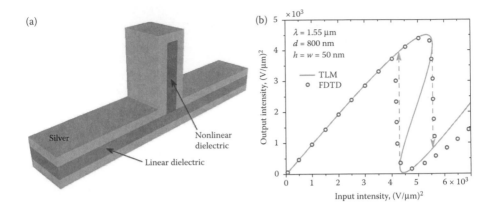

FIGURE 2.7 (a) Nonlinear plasmonic switch and (b) its input–output characteristic featuring optical bistability. Solid curve shows data obtained using transmission-line model (TLM), while open circles represent the result of FDTD simulations. The stub is assumed to be coupled right to the middle of a 500-nm-long PGW; $\varepsilon_d = 2.25$ for both linear and nonlinear dielectrics and $\chi^{(3)} = 2 \times 10 \times 4$ $(\mu m/V)^2$; the rest of the simulation parameters are given in the figure. (After A. Pannipitiya, I. D. Rukhlenko, and M. Premaratne, Analytical theory of optical bistability in plasmonic nanoresonators, *J. Opt. Soc. Am. B*, accepted for publication. © OSA 2011.)

and 750 kV/cm. Also noteworthy is that the transmittance calculated analytically adequately describes the real bistable behavior of the plasmonic switch, as can be concluded by its comparison with the numerical data shown by open circles.

2.3.5 JUNCTIONS OF PLASMONIC WAVEGUIDES

The transmission-line method finds another application in analytical modeling of the PGW junctions, which is needed for the rapid optimization of their transmittance [25,30,32,42–45]. The transmittance of a given wavelength of light may be enhanced by either coupling a stub to the input PGW (see Figure 2.8a) or introducing an intermediate gap section between the different waveguides (see Figure 2.8c). As such, the enhancement is due to the destructive interference between the SPPs reflected from different parts of the structure. When the parameters of the stub and gap section are optimized, the reflectance at the junction almost vanishes and the transmittance peaks. The approximate values of the optimal parameters in Figure 2.8a and c can be defined from Figure 2.8b and d, where we plot the transmittance of a 100-fs Gaussian pulse with its spectrum centered at the wavelength of 1.55 μm. They correspond to the points M_1 (165, 435), M_2 (450, 225), and M (60, 0.285), which indicate the principal maxima of the density plots.

It is not hard to derive approximate expressions for the optimal parameters, if the absorption losses are ignored. In the case of the stub coupler, the result is [23]

$$\Delta_n^{\pm} = \frac{\lambda_1}{2}\left(n \pm \frac{1}{4} \mp \frac{1}{2\pi}\tan^{-1}\left(\frac{\Omega}{2}\right)\right), \qquad (2.6a)$$

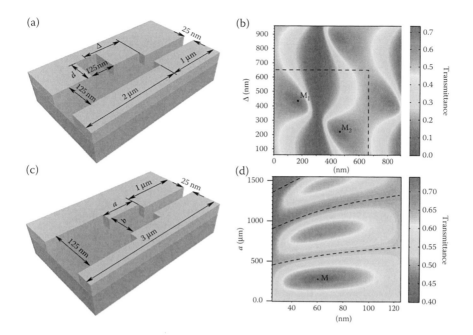

FIGURE 2.8 Reflection-compensated PGW junctions coupled to (a) a perpendicular stub and (c) an intermediate gap section, and [(b) and (d)] their analytically calculated transmittance in the domains (Δ, d) and (a, b) of the stub and gap parameters. The principal maxima of the density plots are marked by the points M_1, M_2, and M; dashed are the boundaries of the quasi-unit cells. The dimensions of the two junctions, made of silver and filled with air (ε_d = 1), are shown in the figure. For other parameters, refer to text. (After A. Pannipitiya, I. D. Rukhlenko, and M. Premaratne, Analytical modeling of resonant cavities for plasmonic-slot-waveguide junctions, *IEEE Photonics J.* **3**, 220–223, 2011. © IEEE 2011.)

$$d_m^\pm = \frac{\lambda_s}{2}\left[m \pm \frac{1}{\pi}\tan^{-1}\left(\frac{Z_1\Omega\sqrt{|\varepsilon_m|} \mp Z_s\sqrt{\varepsilon_d}}{Z_1\Omega\sqrt{|\varepsilon_d|} \mp Z_s\sqrt{\varepsilon_m}}\right)\right].$$ (2.6b)

The parameters entering these expressions are all real and defined as follows: $n, m \in \mathbb{N}_0$, $\lambda_v = 2\pi/\beta_v$, $\beta_v = \beta(h_v)$, $\Omega = \sqrt{Z_2/Z_1} - \sqrt{Z_1/Z_2}$, $Z_{1,2} = Z_{PGW}(h_{1,2})$, and Z_s is given in Equation 2.4.

The subscripts 1, 2, and s denote the quantities related to the input PGW, output PGW, and stub, respectively. According to Equation 2.6, the optimal values of Δ and d are spaced at $\lambda_1/2$ and $\lambda_s/2$, that is, at the halves of the SPP wavelengths guided by the input waveguide and stub. The optimal values (172, 432) and (444, 225), corresponding to the principal maxima M_1 and M_2 and calculated using Equation 2.6, are close to the exact ones.

The optimal dimensions of the intermediate gap section are of the form [23]

$$a_n = \left(\frac{\lambda_g}{4}\right)(2n + 1), \quad b = \frac{\chi}{k},$$ (2.7)

where $n \in \mathbb{N}_0$, $\lambda_g = 2\pi/\beta(b)$, and the parameter χ is to be found from the transcendental equation

$$|\varepsilon_m| \tan(\varepsilon_d \chi^2 - \rho)^{1/2} = \varepsilon_d \left[\frac{\rho - |\varepsilon_m| \chi^2}{\varepsilon_d \chi^2 - \rho} \right]^{1/2},$$

in which $\rho = \beta_1 \beta_2 h_1 h_2 / 4$. As it might be expected from the microwave theory, the optimal length of the gap section is a multiple integer of the quarter of the guided wavelength [44], which is about 58 nm. For the parameters of Figure 2.8c, the optimal width given in Equation 2.7 is about 286 nm.

To summarize this section, we wish to reiterate that borrowing concepts from transmission-line theory seems to be extremely helpful in the analysis of PGW-based nanodevices.

2.4 DESIGN OPTIMIZATION OF COMPOSITE PLASMONIC NANOWIRES

Planar plasmonic waveguides are suitable for guiding linearly polarized light, but become much less efficient, even useless, for applications where polarization is random or changing. If light is arbitrarily polarized, then its transfer in the form of SPPs requires an axially symmetric analog of the planar plasmonic waveguide, which is a composite plasmonic nanowire (CPN) [46–50]. The performance of CPNs, much like the performance of PGWs, heavily depends on geometrics and is limited by metal losses. This section investigates the problem of the optimal design of passive and active CPNs from the viewpoints of the longest propagation and strongest localization of their guided modes.

2.4.1 Modes of CPNs

Consider an infinitely long CPN, which consists of a homogeneous core of radius R_1 and permittivity ε_1, covered by a homogeneous shell of radius R_2 and permittivity ε_2. Just as planar plasmonic waveguides, CPNs support only TM modes with their magnetic field vector in the azimuth direction. The radial dependency of the electromagnetic field of the fundamental, azimuthally symmetric SPP modes may be described using the modified Bessel functions $I_\mu(z)$ inside the core, using modified Bessel functions $K_\mu(z)$ outside the nanowire, and using linear superposition of these two types of functions inside the shell. Assuming that the CPN is surrounded by a dielectric of permittivity ε_3, we find the following dispersion relation for its fundamental modes [51]:

$$\frac{\varepsilon_1}{\varepsilon_2} \frac{k_2}{k_1} \frac{I_1(k_1 R_1)}{I_0(k_1 R_1)} = -\frac{\varepsilon_3 k_2 K_1(k_3 R_2)\psi_{00} + \varepsilon_2 k_3 K_0(k_3 R_2)\psi_{10}}{\varepsilon_3 k_2 K_1(k_3 R_2)\psi_{01} + \varepsilon_2 k_3 K_0(k_3 R_2)\psi_{11}}, \tag{2.8}$$

where, as before, $k_j = (\beta^2 - \varepsilon_j k^2)^{1/2}$ and we have defined the four parameters $\psi_{\mu\nu}$ as

$$\psi_{\mu\nu} = I_\mu(k_2 R_2) K_\nu(k_2 R_1) - (-1)^{\mu+\nu} I_\nu(k_2 R_1) K_\mu(k_2 R_2).$$

It is instructive to examine this relation in the absence of retardation (when $c \to \infty$ and $k_j = \beta$), as this will allow us to ascertain the limiting behavior of the plasmon modes. If the shell thickness is much smaller than the nanowire radius, that is, $h = R_2 - R_1 \ll R_1 \approx R_2 = \zeta/\beta$, then Equation 2.8 in the nonretarded limit reduces to

$$\frac{\varepsilon_1}{\varepsilon_2} \frac{I_1(\zeta)}{I_0(\zeta)} = \frac{(\varepsilon_3 - \varepsilon_2)K_0(\zeta)}{\varepsilon_3 I_0(\zeta)K_1(\zeta) + \varepsilon_2 I_1(\zeta)K_0(\zeta)} - \frac{K_0(\zeta)}{K_1(\zeta)} + O(R_2 - R_1).$$

The fact that the solution of this equation is a complex constant $\zeta = \zeta' + i\zeta''$ means that the propagation length of SPPs grows in inverse proportion to their confinement. Indeed, $L_{SPP} \sim R_2/2\zeta''$ and $\beta' \sim \zeta'/R_2$, so that $L_{SPP} \propto 1/k_3'$. Hence, similar to the case of a PGW, the diameter of a CPN determines the balance between its guiding and confinement efficiencies.

This conclusion will also hold for the shell of an arbitrary thickness, provided that h is kept independent of R_2.

If the diameter of the CPN be preset, the performance of the nanowire can still be altered by varying the dimensions of its core and shell. In doing so, it may be practical to choose the shell thickness that corresponds to the weakest electric field inside the metal. As we shall see later, in the case of a high index contrast between the surrounding dielectric and the dielectric of the nanowire, the nontrivial optimal thickness ($h \neq 0$, R_2) exists for a relatively broad range of the radii R_1 and R_2. If the optimized CPN is made optically active (e.g., by doping its constituent dielectric with rare-earth ions) and pumped to provide gain [52], then it may serve as the spaser of the least threshold power.

2.4.2 SINGLE-CORE PLASMONIC NANOWIRES

Let us first discuss the CPNs with a single dielectric core shown in Figure 2.9a. The nanowires of this type support at least two fundamental SPP modes below the plasma frequency of metal. The lowest frequency mode does not exhibit a cutoff in the reciprocal space and is therefore the most challenging for applications. A performance study of the CPN exploiting this mode was first conducted in Ref. [53]. It was revealed that both the relative mode power residing in the metallic shell and its dissipation rate simultaneously reach their minimum values in the active, 200-nm-thick GaAs–Ag CPN with $h \approx 17$ nm. Owing to this feature, the propagation length of the plasmon mode in this nanowire peaks.

The decay of SPPs along the CPN may be partially or fully suppressed by pumping the nanowire core. Without specifying the model of the active medium, this possibility can be allowed for by introducing the gain parameter γ into the permittivity of the core as $\varepsilon_1 = \varepsilon_d - i\gamma\sqrt{\varepsilon_d}/k$, where ε_d is the core permittivity in the absence of pumping. When γ takes a certain "critical" value γ_c, the gain fully overcomes ohmic losses and SPPs are guided by the CPN without attenuation. In Figure 2.9b, we plot γ_c as a function of the relative shell thickness $q = 1 - R_1/R_2$ for three CPNs of different radii. One can see that each CPN has some minimal critical gain γ_0, which is larger, the thinner the nanowires. This behavior is due to the additive effect of

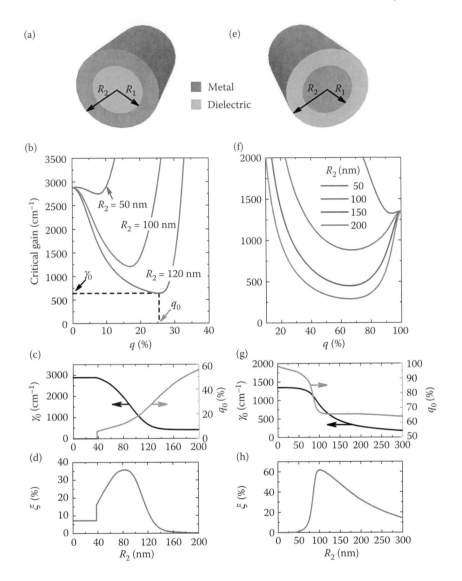

FIGURE 2.9 (a) Single-core and (e) single-shell plasmonic nanowires, their [(b) and (f)] critical gains, [(c) and (g)] minimal critical gains and optimal shell thicknesses, and [(d) and (h)] penetration factors. The first nanowire consists of a GaAs core and silver shell, while the second nanowire has a silver core and ZnO shell; both nanowires are located in the air ($\varepsilon_3 = 1$). It was assumed that $\lambda = 2.58$ μm for the panels (b)–(d) and that $\lambda = 1.55$ μm for the panels (f)–(h). See text for details and notation. (After D. Handapangoda et al., Optimal design of composite nanowires for extended reach of surface plasmon-polaritons, *Opt. Express* **19**, 16058–16074, 2011; D. Handapangoda et al., Optimization of gain-assisted waveguiding in metal–dielectric nanowires, *Opt. Lett.* **35**, 4190–4192, 2010. © OSA 2010, 2011.)

the decrease in size of the active region, and the increase in absorption losses with decreasing shell thickness [54]. Interestingly, the minimal value of γ_c is exhibited by the same CPN that provides the longest propagation of the guided modes in the case of a fixed pump power. This result can be understood by noticing that typically $\gamma_c \ll k$ and, hence, pumping the core dielectric has little effect on the dispersion of β' and the electromagnetic field pattern of the SPP mode. The dependencies of γ_0 and the optimal shell thickness q_0 on the CPN radius are presented in Figure 2.9c. As may be inferred from the figure, the optimized CPNs that are thinner than 200 nm require critical gains below 10^3 cm^{-1}. The gains on this order were demonstrated experimentally and are not difficult to achieve in practice [55]. One may also see that γ_0 becomes a monotonous function of the CPN radius for $R_2 < 40$ nm, with its minimum at $q_0 = 0$. For shell thicknesses of a few nanometers, this prediction becomes inaccurate, as it ignores the process of SPP dephasing due to the elastic scattering of electrons on the nanowire interfaces [8,56].

For the stable operation of a photonic circuit, it is often necessary to minimize the interaction between its nearby components. This weak-interaction requirement, as applied to CPNs, means that the guided mode may not extend far beyond the nanowires. We characterize the SPP confinement to the nanowire using the penetration factor ξ, which is the ratio of the power outside the CPN to the total mode power. Figure 2.9d shows how ξ varies with the radius of the nanowire that has an optimal shell thickness. It is seen that the penetration factor exhibits a 35% maximum for $R_2 \approx 80$ nm and becomes negligibly small for R_2 exceeding 150 nm.

Hence, CPNs thicker than 200 nm should be used if one needs to suppress SPP damping or amplify SPPs with minimal pump power. CPNs thinner than 100 nm are also suitable for the lossless guiding of optical energy, but at the expense of high pump powers.

2.4.3 SINGLE-SHELL PLASMONIC NANOWIRES

Now suppose that the CPN has a single dielectric shell, as shown in Figure 2.9e. As before, only the lowest frequency TM mode is of interest to us, because it exists regardless of the nanowire parameters. Using the example of an optically active Ag–ZnO CPN in air, we recently showed that the damping of this mode may theoretically be minimized by varying the shell thickness [51]. The critical gain is minimized simultaneously with the damping and decreases with the nanowire diameter. We also demonstrated that the strongest coupling of the guided mode to the environment is attained for a finite radius of the CPN with an optimum shell thickness.

These points are illustrated by Figure 2.9f–h, where we plot the dependencies analogous to those shown in Figure 2.9b–d, but for a slightly different set of material parameters. In contrast to the previous example, we now assume that the CPN dielectric is ZnO and the SPP wavelength is 1.55 μm. By comparing the two sets of figures, one may arrive at the following conclusions. First, the single-shell CPN exhibits lower critical gains than the single-core CPN of the same diameter. This is a consequence of both the difference in topology of the two nanowires and the lower refractive index of ZnO ($n_d \approx 2$) when compared to GaAs ($n_d \approx 3.4$). For instance, if $R_2 = 50$ nm, then the SPP damping can be entirely compensated by gain

$\gamma_c \approx 2750$ cm^{-1} in a single-core nanowire, whereas the single-shell nanowire would require the gain that is less than half of that value, $\gamma_c \approx 1320$ cm^{-1}. Second, the dependency of the optimal ratio q_0 on the nanowire radius has opposite trends in the two types of CPNs; however, in both cases, it corresponds to the decrease in the filling factor of metal with R_2. More importantly, for a single-shell CPN, q_0 is nearly independent of the nanowire size if R_2 is above 100 nm but does not exceed 300 nm. This feature facilitates fabrication of the high-performance CPNs and constitutes an explicit design rule. Third, the longest propagation of SPPs can be combined with the relatively strong mode confinement only inside a CPN whose radius differs by several tens of nanometers from the certain "least preferred" value. This value depends on the nanowire type and is in the order of 100 nm.

We would like to emphasize that the optimization of the single-core and single-shell nanowires was possible because they involved amplifying media. In the CPNs that consist of only passive dielectrics, the propagation length of SPPs monotonously grows with the increase in the content of metal. Consequently, the slowest attenuation of SPPs along the nanowire of a given radius is exhibited in the event that the nanowire is pure metallic.

2.4.4 METAL–DIELECTRIC–METAL NANOWIRES

One of the main practical problems in using CPNs that have been discussed so far is associated with the penetration of their mode fields into regions adjacent to the nanowires. As we saw from Figure 2.9d and h, up to 35% and 60% of the mode power may be carried outside the single-core and single-shell nanowires, respectively. This may cause undesirable interactions between different parts of the photonic circuit and degrade its performance.

It is not very difficult, however, to almost fully localize the electromagnetic fields of the SPP mode within a given cylindrical region by plating the single-shell nanowire with a sufficiently thick layer of metal. The resulting metal–dielectric–metal (MDM) CPN, shown in Figure 2.10a, supports two fundamental SPP modes without a cutoff [51]. The lower-frequency mode is predominantly localized within the dielectric gap of the CPN (see Figure 2.10b), so that more than 98.5% of the mode's energy is transmitted within the nanowire cross-section regardless of its diameter. A similar confinement of the SPP fields can be achieved in CPNs with a metallic core and two dielectric shells (e.g., made of SiO_2 and Si), but only at the expense of larger nanowire radii [46].

To minimize the propagation loss for an MDM nanowire of a fixed diameter, one needs to optimally choose the radius R_1 of its metallic core and the thickness $R_2 - R_1$ of the inner dielectric shell. Figure 2.10c shows the critical gain as a function of these two parameters for a 200-nm-thick MDM CPN. The minimal gain value can be directly compared with the grey curve in Figure 2.9f, because we assumed that both CPNs are made of the same materials. It is seen that the MDM nanowire may feature critical gains below 550 cm^{-1} if $R_1 \approx R_2 \approx 40$ nm, while the single-shell CPN of equal radius does not support undamped modes for gains below $\gamma_0 \approx 880$ cm^{-1}. Thus, the MDM geometry not only enables extremely tight light confinement, but also alleviates the gain requirement when compared to simpler structures. One can

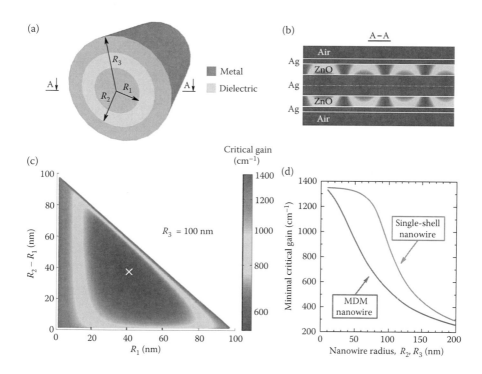

FIGURE 2.10 (a) Schematic of an MDM plasmonic nanowire and (b) electric field distribution for an Ag–ZnO–Ag nanowire located in the air ($\varepsilon_3 = 1$); the nanowire radii are $R_1 = 40$ nm, $R_2 = 80$ nm, and $R_3 = 100$ nm. Panel (c) shows density plot of the critical gain for a 200-nm-thick nanowire (white cross marks the minimum point), and panel (d) illustrates the variation of the minimal critical gain with the size of the MDM and single-shell nanowires. All the simulations were performed for $\lambda = 1.55$ μm. (After D. Handapangoda et al., Optimal design of composite nanowires for extended reach of surface plasmon-polaritons, *Opt. Express* **19**, 16058–16074, 2011; © OSA 2011.)

also see from this figure that γ_c remains relatively small as long as $R_1 + R_2 \leq 90$ nm and each of the inner radii exceed 25 nm. The comparison between the minimal critical gains in a single-shell nanowire and an MDM nanowire is illustrated in Figure 2.10d.

2.5 CONCLUSION

In this chapter, we have discussed different nanodevices whose operation principle is based on the excitation of surface plasmon waves. By the way of a first example, we showed how positioning of air holes in a metallic screen can tailor its transmission spectrum. Next, we showed how to achieve efficient light coupling from a microlaser into a nanoantenna and analyzed different ways to minimize the reflection and improve the coupling. By using the analogy with microwave waveguides, we designed different plasmonic filters, couplers, and switches. In the last section, we analyzed plasmonic nanowires that are, in principle, polarization-insensitive. As a highlight, we discussed some compact devices that exploit surface plasmons to control light interaction at the nanoscale. It is clear from our analysis that the vibrant research field of nanoplasmonics promises novel types of ultracompact devices for optoelectric applications.

ACKNOWLEDGMENTS

The work of H. T. Hattori, I. D. Rukhlenko, and M. Premaratne was sponsored by the Australian Research Council (ARC) through its Discovery Grant scheme under grant DP110100713.

REFERENCES

1. D. Halliday, R. Resnick, and J. Walker, *Fundamentals of Physics*, 5th ed. John Wiley and Sons, New York, 1997.
2. D. Pines and D. Bohm, A collective description of electron interactions: II. Collective vs individual particle aspects of the interactions, *Phys. Rev.* **85**, 338–353, 1952.
3. R. H. Ritchie, Plasma losses by fast electrons in thin films, *Phys. Rev.* **106**, 874–881, 1957.
4. T. W. Ebbesen, H. J. Lezec, H. F. Ghaemi, T. Thio, and P. A. Wolff, Extraordinary optical transmission through sub-wavelength hole arrays, *Nature* **391**, 667–669, 1998.
5. H. A. Bethe, Theory of diffraction by small holes, *Phys. Rev.* **66**, 163–182, 1944.
6. H. J. Lezec, A. Degiron, E. Devaux, R. A. Linke, L. Martin-Moreno, F. J. Garcia-Vidal, and T. W. Ebbesen, Beaming light from a subwavelength aperture, *Science* **297**, 820–822, 2002.
7. A. Minovich, H. T. Hattori, I. McKerracher, H. H. Tan, D. N. Neshev, C. Jagadish, and Y. S. Kivshar, Enhanced transmission of light through periodic and chirped lattices of nanoholes, *Opt. Commun.* **282**, 2023–2027, 2009.
8. S. A. Maier, *Plasmonics: Fundamentals and Applications*, Springer, New York, 2007.
9. K. Kneipp, H. Kneipp, I. Itzkan, R. R. Dasari, and M. S. Feld, Surface-enhanced Raman scattering and bio-physics, *J. Phys.: Cond. Matt.* **14**, R597–R624, 2002.
10. A. Alù and N. Engheta, Input impedance, nanocircuit loading, and radiation tuning of optical nanoantennas, *Phys. Rev. Lett.* **101**, 043901, 2008.
11. N. Engheta, A. Salandrino, and A. Alù, Circuit elements at optical frequencies: Nanoinductors, nanocapacitors, and nanoresistors, *Phys. Rev. Lett.* **95**, 095504, 2005.
12. R. E. Collin, *Foundations for Microwave Engineering*, 2nd ed. Wiley-IEEE Press, New York, 2000.
13. W. Cai, W. Shin, S. Fan, and M. L. Brongersma, Elements for plasmonic nanocircuits with three-dimensional slot waveguides, *Adv. Mater.* **22**, 5120–5124, 2010.
14. N. Yu, E. Cubukcu, L. Diehl, D. Bour, S. Corzine, J. Zhu, G. Höfler, K. B. Crozier, and F. Capasso, Bowtie plasmonic quantum cascade laser antenna, *Opt. Express* **15**, 13272–13281, 2007.
15. H. T. Hattori, Z. Li, D. Liu, I. D. Rukhlenko, and M. Premaratne, Coupling of light from microdisk lasers into plasmonic nano-antennas, *Opt. Express* **17**, 20878–20884, 2009.
16. Z. Li, H. T. Hattori, L. Fu, H. H. Tan, and C. Jagadish, Merging photonic wire lasers and nanoantennas, *J. Lightw. Technol.* **29**, 2690–2697, 2011.
17. H. T. Hattori, Z. Li, and D. Liu, Driving plasmonic nanoantennas with triangular lasers and slot waveguides, *Appl. Opt.* **50**, 2391–2400, 2011.
18. L. Novotny and N. van Hulst, Antennas for light, *Nature Photon.* **5**, 83–90, 2011.

19. T. Kosako, Y. Kadoya, and H. F. Hofmann, Directional control of light by a nano-optical Yagi–Uda antenna, *Nature Photon.* **4**, 312–315, 2010.
20. G. Lerosey, Yagi–Uda antenna shines bright, *Nature Photon.* **4**, 267–268, 2010.
21. T. H. Taminiau, F. D. Stefani, and N. F. van Hulst, Enhanced directional excitation and emission of single emitters by a nano-optical Yagi–Uda antenna, *Opt. Express* **16**, 10858–10866, 2008.
22. H. F. Hofmann, T. Kosako, and Y. Kadoya, Design parameters for a nano-optical Yagi–Uda antenna, *New J. Phys.* **9**, 217, 2007.
23. A. Pannipitiya, I. D. Rukhlenko, and M. Premaratne, Analytical modeling of resonant cavities for plasmonic-slot-waveguide junctions, *IEEE Photonics J.* **3**, 220–223, 2011.
24. A. A. Reiserer, J.-S. Huang, B. Hecht, and T. Brixner, Subwavelength broadband splitters and switches for femtosecond plasmonic signals, *Opt. Express* **18**, 11810–11820, 2010.
25. R. A. Wahsheh, Z. Lu, and M. A. G. Abushagur, Nanoplasmonic couplers and splitters, *Opt. Express* **17**, 19033–19040, 2009.
26. J. Tao, X. Huang, X. Lin, Q. Zhang, and X. Jin, A narrow-band subwavelength plasmonic waveguide filter with asymmetrical multiple-teeth-shaped structure, *Opt. Express* **17**, 13989–13994, 2009.
27. Z. Han, L. Liu, and E. Forsberg, Ultra-compact directional couplers and Mach-Zehnder interferometers based on surface plasmon polariton, *Opt. Commun.* **259**, 690–695, 2006.
28. A. Pannipitiya, I. Rukhlenko, and M. Premaratne, Analytical theory of optical bistability in plasmonic nanoresonators, *J. Opt. Soc. Am. B* **28**, 2820–2826, 2011.
29. X. Lin and X. Huang, Numerical modeling of a teeth-shaped nanoplasmonic waveguide filter, *J. Opt. Soc. Am. B* **26**, 1263–1268, 2009.
30. J. Liu, G. Fang, H. Zhao, Y. Zhang, and S. Liu, Surface plasmon reflector based on serial stub structure, *Opt. Express* **17**, 20134–20139, 2009.
31. G. Veronis, S. E. Kocabas, D. A. B. Miller, and S. Fan, Modeling of plasmonic waveguide components and networks, *J. Comput. Theor. Nanosci.* **6**, 1808–1826, 2009.
32. S. E. Kocabas, G. Veronis, D. A. B. Miller, and S. Fan, Transmission line and equivalent circuit models for plasmonic waveguide components, *IEEE J. Sel. Top. Quantum Electron.* **14**, 1462–1472, 2008.
33. G. Veronis and S. Fan, Bends and splitters in metal–dielectric–metal subwavelength plasmonic waveguides, *Appl. Phys. Lett.* **87**, 131102-1–131102-3, 2005.
34. A. Pannipitiya, I. D. Rukhlenko, M. Premaratne, H. T. Hattori, and G. P. Agrawal, Improved transmission model for metal–dielectric–metal plasmonic waveguides with stub structure, *Opt. Express* **18**, 6191–6204, 2010.
35. K. Y. Kim, Y. K. Cho, H.-S. Tae, and J.-H. Lee, Light transmission along dispersive plasmonic gap and its subwavelength guidance characteristics, *Opt. Express* **14**, 320–330, 2006.
36. J. A. Dionne, L. A. Sweatlock, H. A. Atwater, and A. Polman, Plasmon slot waveguides: Towards chipscale propagation with subwavelength-scale localization, *Phys. Rev. B* **73**, 035407-1–035407-9, 2006.
37. B. Prade, J. Y. Vinet, and A. Mysyrowicz, Guided optical waves in planar heterostructures with negative dielectric constant, *Phys. Rev. B* **44**, 13556–13572, 1991.
38. I. D. Rukhlenko, M. Premaratne, and G. P. Agrawal, Nonlinear propagation in silicon-based plasmonic waveguides from the standpoint of applications, *Opt. Express* **19**, 206–217, 2010.
39. M. Born and E. Wolf, *Principles of Optics*, Cambridge University Press, New York, 1999.
40. C. Min, P. Wang, C. Chen, Y. Deng, Y. Lu, H. Ming, T. Ning, Y. Zhou, and G. Yang, All-optical switching in subwavelength metallic grating structure containing nonlinear optical materials, *Opt. Lett.* **33**, 869–871, 2008.

41. R. W. Boyd, *Nonlinear Optics*, Academic Press, Boston, 2003.
42. J. Liu, H. Zhao, Y. Zhang, and S. Liu, Resonant cavity based antireflection structures for surface plasmon waveguides, *Appl. Phys.* B **98**, 797–802, 2009.
43. Y. Matsuzaki, T. Okamoto, M. Haraguchi, M. Fukui, and M. Nakagaki, Characteristics of gap plasmon waveguide with stub structures, *Opt. Express* **16**, 16314–16325, 2008.
44. P. Ginzburg and M. Orenstein, Plasmonic transmission lines: From micro to nano scale with λ/4 impedance matching, *Opt. Express* **15**, 6762–6767, 2007.
45. G. Veronis and S. Fan, Theoretical investigation of compact couplers between dielectric slab waveguides and two-dimensional metal–dielectric–metal plasmonic waveguides, *Opt. Express* **15**, 1211–1221, 2007.
46. D. Chen, Cylindrical hybrid plasmonic waveguide for subwavelength confinement of light, *Appl. Opt.* **49**, 6868–6871, 2010.
47. R. F. Oulton, V. J. Sorger, D. A. Genov, D. F. P. Pile, and X. Zhang, A hybrid plasmonic waveguide for sub-wavelength confinement and long-range propagation, *Nat. Photonics* **2**, 496–500, 2008.
48. V. Krishnamurthy and B. Klein, Theoretical investigation of metal cladding for nanowire and cylindricalmicro-post lasers, *IEEE J. Quantum Electron.* **44**, 67–74, 2008.
49. U. Schröter and A. Dereux, Surface plasmon polaritons on metal cylinders with dielectric core, *Phys. Rev. B* **64**, 125420, 2001.
50. J. Takahara, S. Yamagishi, H. Taki, A. Morimoto, and T. Kobayashi, Guiding of a one-dimensional optical beam with nanometer diameter, *Opt. Lett.* **22**, 475–477, 1997.
51. D. Handapangoda, M. Premaratne, I. D. Rukhlenko, and C. Jagadish, Optimal design of composite nanowires for extended reach of surface plasmon-polaritons, *Opt. Express* **19**, 16058–16074, 2011.
52. M. Premaratne and G. P. Agrawal, *Light Propagation in Gain Media: Optical Amplifiers*, Cambridge University Press, Cambridge, 2011.
53. D. Handapangoda, I. D. Rukhlenko, M. Premaratne, and C. Jagadish, Optimization of gain-assisted waveguiding in metal–dielectric nanowires, *Opt. Lett.* **35**, 4190–4192, 2010.
54. S. Al-Bader and M. Imtaar, TM-polarized surface-plasma modes on metal-coated dielectric cylinders, *J. Lightw. Technol.* **10**, 865–872, 1992.
55. S. Y. Hu, D. B. Young, S. W. Corzine, A. C. Gossard, and L.A. Coldren, High-efficiency and low-threshold InGaAs/AlGaAs quantum-well lasers, *J. Appl. Phys.* **76**, 3932–3934, 1994.
56. P. Zhao, W. Su, R. Wang, X. Xu, and F. Zhang, Properties of thin silver films with different thickness, *Physica E* **41**, 387–390, 2009.

3 Plasmonic Solutions for Light Harvesting in Solar and Sensing Applications

Saulius Juodkazis, Lorenzo Rosa,
and Yoshiaki Nishijima

CONTENTS

3.1   INTRODUCTION

Plasmonics has become an extensive field of research since last decade due to its unique possibilities to engineer material's optical and electrical properties, which can be harnessed for a new generation of high-speed all-optical processing of information and delivering light with nanoscale precision, and intensity enhancement can reach 108 when single-molecular detection becomes possible. It is no surprise that modern science and technology ventured to explore all these fascinating possibilities.

In the following sections, we discuss several topics of light–matter interaction on micro–nanoscale with a particular focus on the possible use of plasmonic effects in solar energy harvesting, photo-catalytic applications, and Raman spectroscopy. Plasmonic solar cells use the properties of plasmonic metal nanoparticles to enhance scattering and absorption of light in thin-film structures, which would absorb less when compared with optically thick cells. Excitation of surface plasmonic resonances produces local field enhancement in the active layer, trapping light and allowing enhanced generation and separation of the electron–hole pairs [1].

3.2   PLASMONIC SOLAR CELLS

Photovoltaics, the conversion of sunlight into electricity, is a promising technology that may allow the generation of electrical power on a very large scale. World-wide photovoltaic production was more than 5 GW in 2008 and is expected to rise above 20 GW by 2015 [2]. Photovoltaic power has the potential to meet the energy needs of our society for the next generations, because the amount of radiation striking the Earth's surface is 1.76×10^5 TW and current world usage is estimated at 14 TW [3]. However, solar electricity is the most expensive to produce, costing 25–50 cents per kW h, at least 5 times higher than the least expensive coal at 1–4 cents per kW h [4]. In order to compete with nonrenewables, it is necessary to reduce the cost of solar electricity, which depends heavily on the technology used. Prospects for the renewable energy development including solar cells are bright and can also be positively affected by international treaties on greenhouse emissions and taxation policies, for example, carbon tax in Australia applied from July 2012.

3.2.1   THICK FILMS

Currently 90% of the photovoltaics market is based on crystalline Si wafer solar cells [1]. Approximately 50% of their cost is borne by Si itself, as it is a relatively weak absorber, requiring thicknesses of 200–300 μm to fully absorb the incident sunlight. It must also be of high-quality and defect-free, so that the generated carriers are not lost before collection, which contribute to drive up the costs [2].

3.2.2   Thin Films

To address the Si wafer cost issues, thin-film solar cells have been developed [1]. The thickness is ~1–2 μm, and low-cost substrates are used, for example, glass, plastic, or stainless steel. Semiconductors used are not limited to Si (crystalline or amorphous) but include direct-bandgap semiconductors such as cadmium telluride (CdTe), copper indium diselenide ($CuInSe_2$), and copper indium gallium selenide (CIGS), in a bid to improve near-bandgap absorption over indirect-bandgap Si. The highest efficiency is obtained by CIGS at 19.6% [5]. Thin-film solar-cell technologies are limited by their low absorbance of near-bandgap light. Moreover, they use scarce elements such as Te and In, whose natural availability is limited [6]. Therefore, structuring the thin-film solar cell in such a way that light is trapped inside to increase the absorbance is fundamental [2].

3.2.3   Light Trapping

Solar cells are structured to increase light trapping to compensate the relatively weak absorbance of silicon, etching 2–10-μm size pyramids in the surface [7], a method not suitable for thin-film cells. A material with refractive index n can have enhanced long-wavelength absorbance, only if the surface is randomly textured in the Lambertian scattering regime, up to $4n^2$, which is known as the Yablonovitch limit [8,9], but the needed feature roughness is in the order of a micron, which easily surpasses the thickness of a thin-film cell. Semiconductor substrates can be textured to increase scattering [10,11], though this increases the surface area and lowers the semiconductor deposition quality, both factors that increase recombination of the generated carrier pairs before collection [1]. Recently, techniques have been devised to achieve the Yablonovitch limit in Si cells. It was numerically shown in amorphous Si cells that it is possible to achieve the limit by depositing on a textured conformal ZnO layer, with a Si:H layer as thin as 250 nm, or down to 100 nm with embedding within a 750-nm-thick layer that matches the refractive index real part [12]. It was also simulated in crystalline Si cells that the limit can be achieved with Si mass equivalent of 1–3-μm films by construction of nanorod arrays designed by a symmetry-breaking technique based on group theory [13].

Texturing the metal counter-reflector to support surface plasmon polaritons (SPPs) has been shown in a 160-nm-thick a-Si:H cell. SPPs are scattered into guided modes of the cell photoactive layer, with photocurrent enhancement controlled by mode coupling. Conversely, the active layer thickness can be reduced at constant current, saving material and increasing the voltage of an open circuit, V_{oc}, with potentially a better long-term stability [14].

It has recently been observed that conventional light-trapping theory, based on ray-optics in thick film cells, is not applicable to nanophotonics. A statistical temporal coupled-mode theory of light trapping has been proposed to overcome the standard absorption enhancement limit, by confining the optical modes in the active layer to deep-subwavelength scale [15]. Another proposed trapping mechanism transforms electromagnetic energy to thermal and back, making use of thermal rectification due to the temperature dependence of electromagnetic resonances. A prototype device has been proposed, made of silicon carbide, SiC, based on the interaction of temperature-dependent surface phonon polaritons in different polymorphs of SiC [16]. Future prospects to harness solar as well as thermal energy which are usually wasted so far now falls under a sharp focus of basic science.

3.2.4 PLASMONIC FIELD ENHANCEMENT

Absorption in thin-film solar cells can also be increased by scattering induced by metal nanostructures, exploiting enhancement by surface plasmons to solve the light trapping problem [2]. Surface plasmons are supported by the oscillations of electrons bound close to a metal–dielectric interface and are intensely studied because of their ability to concentrate light in tiny volumes through a phenomenon known as nanofocusing [17–19].

Standard photovoltaics relies on the thickness of the Si wafer to reach the maximum light absorption. Figure 3.1 shows the standard air mass 1.5 (AM1.5) solar spectrum. To fully exploit Sun's power, the absorber should work from 300 nm to beyond 1500 nm wavelengths. As shown in Figure 3.2, Si is a good absorber for wavelengths up to 600 nm, and then its efficiency tapers off and is negligible beyond 1100 nm [20]. This is partly compensated by enlarging thickness up to 180–300 μm, which causes a different issue: in order for the photogenerated electron–hole pairs to be collected at the electrodes and generate current, the carrier diffusion length must be significantly larger than the substrate thickness, which is difficult to obtain for holes due to their reduced mobility. As the diffusion length $L = [(kT/q)\mu_h \tau]^{1/2}$, where μ_h is the hole mobility (~450 cm²/V s in crystalline Si [21]) and is τ the carrier lifetime (k is the Boltzmann constant, T is the absolute temperature, q is the electron charge). A hole would need ~100 μs lifetime to move the 300 μm thickness of the wafer, which is compatible with the defect recombination probability of crystalline Si. A cheaper material like polycrystalline Si has a mobility at least 10 times lower, which would reduce efficiency for anything thicker than a thin film, requiring a strategy to increase the optical absorption by light trapping.

This can be achieved by proper engineering of metallo-dielectric structures: light can be concentrated and "folded" into a thin semiconductor layer, thereby increasing the absorption [2]. Three mechanisms can be effectively exploited to increase light trapping, with the aim to reduce thickness at constant absorption:

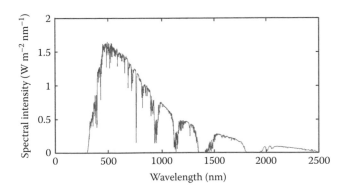

FIGURE 3.1   Standard air mass 1.5 (AM1.5) solar spectrum (ASTMG173-03 Tables).

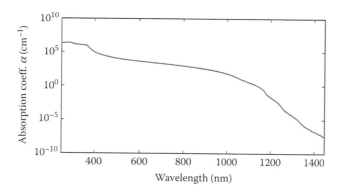

FIGURE 3.2   Absorption coefficient α of intrinsic Si at 300 K.

1. Scattering from the localized surface plasmons (LSPs) excited in metal nanoparticles: Metallic nanoparticles are used as subwavelength scattering elements, by coupling the light in the semiconductor with an angular distribution to increase its path length in the thin absorbing layer. The effect is maximized when the particles provide an additional k-vector contribution to match the incident momentum to the wavevector of waveguide modes inside the thin film, which redirects the light parallel to the absorbing layer and increases the path length up to the cell width [22–25]. LSP resonance also focuses plasmons at specific wavelengths in hot spots along the particle surface, associated with high near-field intensities, leading to an increased absorption. This increases the photogeneration probability and the short-circuit current density J_{sc} [26,27].

2. Direct near-field coupling by tuned plasmonic nanoantennas: Nanoparticles are shaped as subwavelength antennas such that plasmonic focusing generates an enhanced near-field coupled to the thin film, increasing its effective absorption cross-section at the required wavelength (which can be from UV to far-IR and even in terahertz range [28]).

3. SPPs propagating at the metal–semiconductor interface: A metallic film is deposited on the thin-film back surface and patterned to support slow group velocity SPP modes at the metal–semiconductor interface; the k-vector distribution is designed to efficiently couple both to the incident sunlight (plane wave close to normal direction) and to guided modes in the thin film (parallel direction), where the lower group velocity increases the electron–hole pair generation probability. It is shown that SPP mode can efficiently be launched by fine-tuning of the dielectric properties of the interface between the dielectric and uniaxial medium with mixed metal–dielectric properties [29]. Such mode can be used to harness increase of light–matter interaction in a thin photovoltaic layer.

Plasmonic devices for photovoltaic applications have a number of requirements, among which are the following: (1) high scattering cross-section, (2) broadband behavior between 300 and 1600 nm wavelength, and (3) broad angle of incidence sensitivity [3].

A similar propagation effect can be delivered through coupling of LSP resonances in nanoparticle arrays. Broadband Ag arrays on top of a high-refractive index substrate have shown an impedance-matching effect producing broadband coupling of incident light into the substrate. The coupling is determined by competing effects of Fano resonances in the scattering, interparticle coupling, and resonance shifts due to variations in the near-field coupling to the substrate and spacer layers. The particle array showed 50% enhanced in-coupling compared to a bare Si wafer, 8% higher than a standard interference antireflection coating [30].

3.2.4.1 State of the Art

Plasmonic solar cells are most easily obtained by depositing nanoparticles on top of the cell, as the scattering layer deposition can be accomplished without any significant changes to the fabrication procedure. The scatterers can then be deposited with a simpler chemical process, which can make their density uneven; the process must be carefully controlled to avoid clustering and shadowing of the photoactive layer, even though satisfactory scattering can be obtained with as little as 10% particle coverage [18]. The influence of dielectric antireflection coating must be taken into account, as noticed for Si_3N_4 under/overlayers associated to Ag cylinders on Si [18].

After the first experimental work on photodetectors using Ag nanoparticles [22], subsequent experiments obtained enhancements in silicon of 80% on wafer-type cells (Au) [31], 8% on thin films [32], and 33% on silicon-on-insulator (SOI) [26]. While most analyses relate to simple full- or hemispherical shapes easily obtained through chemistry, it largely shows that particle shape optimization is fundamental to increasing light trapping efficiency: path length enhancements in thin films of up to a factor of 30 were found suitable for the optimized shapes [18]. Recent experimental realizations of plasmon-enhanced Si photovoltaics have found photocurrent enhancements of a factor of 18 at 800 nm, and reports in the literature can be found for other inorganic semiconductors such as GaAs, a-Si:H, CdSe/Si, InGaN/GaN, and InP/InGaAsP, for organic semiconductors such as polythiophene (P3HT) and copper phthalocyanine (CuPC), and for hybrid inorganic–organic devices such as dye-sensitized solar cells (DSSC) [3].

As a general rule, particles of cylindrical and hemispherical shapes have a longer contact perimeter with the high-index substrate: as the scattering hot spots are located at the interface, the hot-spot size and the near-field coupling are enhanced, hence they are more effective than spherical particles at trapping light within the substrate [18]. However, recently a thin-film amorphous-Si cell incorporated with 200 nm wet-chemically synthesized nucleated spherical Ag nanoparticles at 10% coverage density demonstrated a performance increase with a 14.3% enhancement in the short-circuit photocurrent density and a 23% enhancement in the energy conversion efficiency [33].

Fabrication can be accomplished by chemical deposition of colloidal Ag and Au nanoparticles as scatterers. The control of size and deposition density can be achieved to satisfactory uniformity, though this method can be used only for essentially random patterns and suffers from aggregation and clustering; it has been

(a) (b)

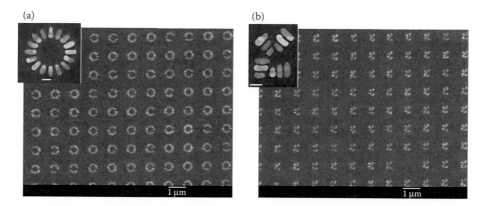

FIGURE 3.3 SEM images of the state-of-the-art EBL samples: gold nanoparticles on glass substrates in "Sun" (a) and "an optical memory bit" (b) patterns; the narrowest nanogaps are 23 nm (a) and 15 nm (b), respectively. Samples prepared by lift-off and 2-nm-thick sublayer of Cr was used. (Courtesy of Dr. Sun Kai and J. W. M. Chon.)

employed on crystalline Si, amorphous Si, and an InP/InGaAsP quantum well p–i–n structures [3]. It is by no means the only way, as other methods have been developed for this purpose, such as techniques for organic solar cells involving island formation on transparent conductive oxides [34] or electrode position, [35] and electron-beam lithography [36]. Also, anodic aluminum oxide (AAO) templates have been used as evaporation masks for the deposition of metallic nanoparticles [27] and have been successfully used with Ag on thin-film (200 nm) GaAs cells with index-matched substrate (not supporting waveguide modes), obtaining an 8% increase in J_{sc} [37]. The AAO templates give several degrees of freedom to reliably control deposition, as anodization voltage controls hole density and etch time controls diameter [38]. Metal is then electrochemically deposited and formed into particles on the cell surface through the regular holes in the template, controlling height via deposition time. A final annealing step permits one to fine-tune the particle shape [39]. Using current electron beam lithography (EBL), nanoparticles can be defined over large areas (meeting the requirements for the solar cell verification size 1×1 cm^2) with sub-10 nm resolution and using plasmonic metal deposition and lift-off patterns of complex shapes with encoded wavelength resonances and polarization sensitivity can be fabricated (Figure 3.3).

3.2.4.2 Fractal Nanoantennas

Nanoantenna-based cells, where the absorption is increased due to the high electromagnetic intensity in the near field of the antenna, have also shown promise. This configuration can strongly increase absorption and open an avenue for unconventional device architectures [3]. In particular, fractal-based patterning of materials is shown to provide devices that can easily be designed, according to the pattern used such as inherently broadband, polarization-independent, and with wide-angle behavior, due to self-similarity. A nanoantenna based on the triangular Sierpinski gasket fractal has shown an optical response ranging from visible wavelengths and expanding toward the THz band, capable of achieving significantly high field enhancement

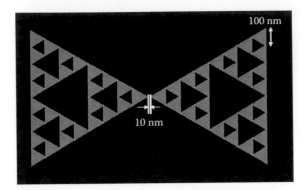

FIGURE 3.4 Numerical finite differences time domain (FDTD) simulation image of the third-order Sierpinski gasket nanoantenna [28], with elementary triangle length of 100 nm and nanogap size of 10 nm. Triangle overlap of 5 nm account for finite bridge size obtained in fabrication, defined by EBL evaporation, and lift-off.

in the nanogap on a very wideband, by exciting plasmon resonance modes in a self-replicating fractal metallic structure [28]. The structure can be extended up to the fifth-order and more, by EBL patterning to engineer nanometric details on a 10-μm scale, controlling size and features of gold on a SiO_2 substrate, of which an example is shown in Figure 3.4. The structures, from the first to the fifth order, have lengths of 400 nm, 780 nm, 1.54 μm, 3.06 μm, and 6.1 μm, with a total area of the solid surfaces being 0.03, 0.1, 0.31, 0.93, and 2.1 μm², respectively.

The cross-sections and field enhancement factors obtainable have been simulated with a finite-difference time-domain (FDTD) numerical solver (Lumerical). Reticule gradation is employed to increase mesh accuracy in the areas of greatest field enhancement, especially in the central nanogap. The nanoantenna is simulated in the band from 400 nm to 1.5 μm, divided in 500 nm bands to achieve a 1.5 h simulation for up to 0.5 ps propagation of the time-domain broadband pulse. Materials are modeled based on their complex permittivity values from literature [40] and their spectra are modeled in the time-domain using a multicoefficient polynomial method.

The fractal structure shows several extinction peaks linked to excited plasmon resonances as shown in Figure 3.5, and the extinction of the incoming radiation is constantly increasing for order beyond third, the best being the fifth order, with the main peak being ~1.4 μm. The patterning process permits one to fine-tune and control the absorption and scattering bands of the nanoantennas.

3.2.4.3 Materials

The most used plasmonic metals are Ag and Au, by virtue of their low loss and ability to support surface plasmon resonances in the visible and UV bands [3]. Cheaper but lossier metals such as Al and Cu can still support visible (Cu) and UV (Al) plasmonic resonances, which make them the right candidates for photovoltaics [41]. The cell and nanoparticle layout needs to be redesigned to account for the different resonant wavelengths to match the absorption spectra, for example, a theoretical study on thin-film GaAs SPP waveguide devices showed that substituting Al for Ag

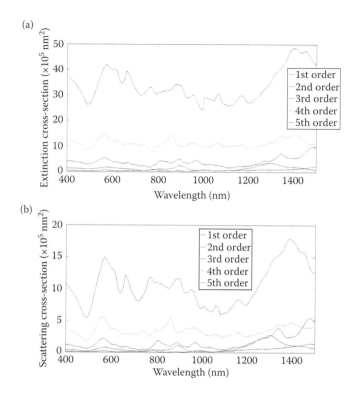

FIGURE 3.5 (**See color insert.**) Numerically simulated (a) extinction and (b) scattering cross-sections of Sierpinski gold nanoantennas, for fractal order between first and fifth and linear polarization excitation.

shifts the SPP resonance to shorter wavelengths, but the absorption enhancement is reduced by the higher Al losses [24].

Thin-film photovoltaic technologies, such as CdTe and CIGS, use elements that are among the rarest materials on earth. Tellurium, for example, is one of the nine rarest metals on Earth, with concentrations between 1 and 5 parts per billion in the crust [3]. Because these elements are scarce, they are usually isolated as by-products of mineral purification for other more common elements. For example, Se and Te are produced largely from Cu mining, and In is produced from Zn mining [6]. Demand for these elements in new technologies is increasing, for example, In for digital displays, driving to high market prices. Improving light trapping can lead to reduction of feedstock material's use through reduction in cell thickness. The current market share of photovoltaics in the total world's energy production is still smaller than 1%, though in a scenario where the demand for renewable energy is constantly increasing. The corresponding market growth will put pressure on feedstock consumption at least for the first two years, as the initial batches of solar cells are produced; after the first period, material recycling can provide relief, though this is strictly linked to the projected lifetime of the modules. Projecting the demand at constant cell thickness from 2005 values, when the US produced 1.5 GW/year of solar power, to a prospective increased power demand of 50 GW/year, the feedstock consumption places impossible demands on materials such as Te, In, and Se [3]. If the cell thickness is reduced to one-tenth of the present figure, the demands on In and Se become reasonable, though still not for Te.

Ultrathin (~10 nm) film devices are also promising for enhancing performance of cells even with poor carrier collection efficiency, such as those that are built with very cheap and low-quality materials such as nanocrystalline or amorphous solids [3]. Their absorbance of normal incident light is typically limited to about 50%, though the addition of a dielectric spacer having a highly reflective backside has been shown in FDTD simulations to suppress not only transmittance of the incident radiation, but also an antireflective behavior from destructive interference front- and back-reflected light. Conversion efficiencies of up to 64% of the Shockley–Queisser limit were shown for composite materials, comprising Ag and thin-film layers of a-Si, CuInSe$_2$, or the organic semiconductor poly(2-methoxy-5-(3'-7'-dimethyloctyloxy)-1,4-phenylenevinylene) (MDMO-PPV) [42].

3.2.5 CALCULATION METHODS

The light–matter interaction is modeled by 3D FDTD calculations [43], aiming for scattering from Ag and Au particles under normal or slanted incidence, using periodic boundary conditions or perfectly matched layers (PMLs). The dielectric functions can be modeled using a Drude model for Ag and a Drude-Lorentz model for Si and Au (to account for dispersion in the visible band), though the best fit is achieved using multicoefficient models with extensive sets of basis functions to better fit complex dispersion profiles.

The wavelengths in use require a mesh size of 0.5–1 nm on the metal particle and in its neighborhood, and the use of conformal mesh technologies [44] can significantly improve accuracy up to one order of magnitude. A total field-scattered field (TFSF) source is placed around the particle to separate the absorption and scattering contributions to the total field extinction [18].

The particle is surrounded by a double monitor box which measures the Poynting vector flow across six surfaces; the contributions are integrated to calculate the total power and scattering can be evaluated separately for the air and substrate domains. A TFSF formulation permits one to remove the incident field contribution from the outer monitor box, which yields the scattered field only. The total field is measured by the inner box and yields the overall extinction power of the particle. By normalization to the incident field intensity, the two quantities result into the scattering cross-section σ_{scat} and the extinction cross-section σ_{ext}. The absorption cross-section is calculated from the difference $\sigma_{\text{abs}} = \sigma_{\text{ext}} - xt_{\text{scat}}$.

For point-like particles (size $\ll \lambda$),the point dipole quasistatic approximation holds [45]:

$$\sigma_{\text{sca}} = \frac{1}{6\pi}\left(\frac{2\pi}{\lambda}\right)|\alpha|^2, \tag{3.1}$$

$$\sigma_{\text{abs}}(\lambda) = \frac{2\pi}{\lambda}\,\text{Im}[\alpha], \tag{3.2}$$

where

$$\alpha = 3V \left[\frac{\varepsilon_p/\varepsilon_m - 1}{\varepsilon_p/\varepsilon_m + 2} \right] \tag{3.3}$$

is the polarizability, V is the volume, ε_p is the metal permittivity, and ε_m is the environment permittivity. When, $\varepsilon_p = -2\varepsilon_m$, polarizability tends to increase (which is limited by losses) and the electrons in the medium oscillate in resonance with the incident wave. This condition defines surface plasmon resonance and is called the Frolich condition.

As the size of patterns with nanoparticles is increased, the electronic response can no longer be considered instantaneous, and it is necessary to evaluate effects such as dynamic depolarization and radiation damping, and the existence of higher-order plasmonic modes (quadrupole, octupole) which are below cut-off in point-like particles [18]. As the electrons experience, with respect to the incoming field, a delay that is variable along the particle, the polarization contributions along the particle are no longer matching, and the surrounding medium can restore neutrality with a lower depolarization field (dynamic depolarization). The restoring force of the oscillator is lower, which corresponds to a red-shift of the resonance. As plasmonic particles in solar cells are designed for high scattering, re-radiation occurs which reduces the polarizability (radiative damping), and thus the energy stored and the quality Q-factor of the oscillator, broadening the plasmon resonance peak. Both effects can be beneficial in solar cell applications as they broaden the light-trapping band toward the IR wavelengths that are poorly absorbed by Si [1].

3.2.5.1 Material Models

Accurate modeling of the noble metals involved in the plasmonic resonance brings forth several issues, as in the visible band where the resonances are usually sought, they experience interband transitions that make the permittivity relationship complex. It is fundamental in time-domain methods such as FDTD since, differently from frequency-domain methods, the complex dielectric permittivity spectrum $\varepsilon(\omega)$ cannot enter the time-stepping scheme, but must be approximated by polynomial function ratios in the complex ω-space.

Far from interband transitions or where interband absorption is low, the Drude model holds, which describes the system as damped free electronic oscillators reacting to an applied field of angular frequency ω [45]:

$$\varepsilon = 1 - \frac{\omega_p^2}{\omega^2 + i\gamma\omega},$$

where ω_p is the bulk plasmon frequency, given by $\omega_p^2 = Ne^2/m\varepsilon_0$, N is the free electron density, e is the electronic charge, m is the effective electronic mass, and ε_0 is the free-space permittivity. The free-space polarizability is thus

$$\alpha = 3V \frac{\omega_p^2}{\omega_p^2 - 3\omega^2 - i\gamma\omega}.$$

For a metal sphere in free space, this results in surface plasmon resonance at $\omega_{sp} = \sqrt{3}\omega_p$, which chiefly depends on N. Al and Ag have the highest densities (UV resonance), while visible wavelengths are obtained from Au and Cu. By changing the environment permittivity, it is possible to influence the polarizability, red-shifting the resonance for increasing permittivity [1].

In the Drude model, the permittivity term ε is modeled in complex ω-space as the combination of an imaginary pole and a pole in the origin. Other models represent the permittivity differently: the Lorentz models with a couple of complex poles and the Debye model with one imaginary pole. The critical point model [46] adds complexity for greater flexibility, by the way of extending the numerator and denominator polynomials to order n. Nevertheless, these models suffer significant inaccuracies when used to model wideband permittivity spectra, making it necessary to model each band (UV, visible, IR) separately.

Recently, a new method has been proposed, which models the material response as a dielectric polarization-dependent both on the field and on its time derivative [47]. It permits remarkably accurate wideband material modeling, for example, permittivity of Si in the 300–1000 nm wavelength range.

3.3 SPECTRAL CONTROL OF LIGHT HARVESTING

Progress of technology is underpinned on the efficiency principle which also governs an evolutionary progress in nature. In solar harvesting by solar cells, search for the schemes which minimize material costs, increase light collection efficiency, and find ways for the highest light-to-electrical conversion efficiency are driving the field.

3.3.1 ENGINEERING OF LIGHT-TO-ELECTRICITY CONVERSION AT IR WAVELENGTHS

Development of efficient solar energy conversion methods has become one of the trends in modern science and technology. Collection and conversion to electricity, a spectrally broad Sun's spectrum, is the main focus of solar cells [48–51]. The most efficient dye-sensitized titania cells achieve close to 100% performance in terms of the incident photon-to-current conversion efficiency (IPCE) even at the wavelengths of 500 ± 50 [50] and 650 ± 50 nm [52], which are considerably longer than the titania's fundamental absorption and are determined by the dye used in the cell.

Challenges are mounting when multiwavelength performance of Grätzel solar cells is attempted by combination of several dyes. So far, there are no practical solutions for harvesting of a sizable 20% part of Sun's IR spectrum [53]. Recently, it was demonstrated that a plasmon-sensitized photoelectric conversion system based on an n-type single crystal of TiO_2-rutile patterned by gold nanoparticles can harvest light and convert it into electricity at IR wavelengths [54]. The photoelectric energy conversion can be controllably extended over a spectral region from sub-bandgap till near-IR energy photons (<1.1 µm) at IPCE level of few percent by engineering an overlap of the absorption and plasmon extinction spectra. The proposed method is expected to find practical applications in other solar energy conversion and light extraction systems, water splitting, artificial photosynthesis systems, catalysis, and *in vivo* IR-sensing applications.

Harvesting of light energy and its conversion is at the focus of practical applications in future power generation. Among a variety of prospective photoelectrical conversion and water splitting schemes, TiO_2-rutile is one of the promising materials [55,56]. Realization of efficient practical solar cells based on titania remains a challenge due to stability of organic or metal complex dyes that act as sensitizers. In contrast, plasmonic structures of Au, Ag, and Cu nanoparticles can successfully perform as nonbleachable photosensitizers. Spectral properties of such nanoparticles and the local light-field enhancement are controlled by their size and shape. Tailored nanoparticles are good candidates for sensitizers in photoelectric conversion systems due to their stability, while the absorption losses and thermal degradation can be controlled. Moreover, the spectral and polarization response of plasmonic nanoparticles can also be flexibly tailored by changing the shape and size.

In order to utilize wide spectral tunability of plasmonic nanoparticles most efficiently, the absorption spectrum of the solar cell substrate should match the extinction spectrum of nanoparticles. Good candidates are the Nb-doped ones and in a reducing hydrogen atmosphere annealed rutile substrates (Figure 3.6a); the later was proposed for creation of photo-active n-type titania as in the first water hydrolysis study [55]. The absorption band spanning from 500 nm toward IR wavelengths is present in the Nb-doped n-type rutile (Figure 3.6a).

3.3.2 Au Nanoparticles as Photo-Sensitizer at IR Wavelengths

Figure 3.6b shows the extinction spectrum of gold nanorod structures which have longitudinal dimensions 2.2 times larger than the lateral on rutile. The corresponding transverse (T-) and longitudinal (L-) modes of plasmon extinction bands appeared at around 750 and 1100 nm, respectively, in water under unpolarized light illumination [54]. Both plasmon bands are at the wavelengths far red-shifted from the fundamental absorption of TiO_2 (0.4 μm) and are determined by dielectric functions of gold, rutile, and solution. The particular spectral positions of the extinction maxima are defined by the size and shape of gold nanoparticles [57] and can be tailored to match the absorption band of the substrate.

When monochromatic light of the wavelength shorter than 450 nm was illuminated onto the semi-insulating TiO_2 without nanorods, photocurrent was observed. This corresponds to the direct excitation of TiO_2, that is, an electronic valence-to-conduction band transition followed by hydrolysis of water. At longer wavelengths, no photocurrent was observed. For the quantitative comparison of photon-to-electron conversion of different samples, the efficiency coefficient, that is, the incident photon-to-current conversion efficiency (IPCE) was calculated at the potential of 0.4 V [saturated calomel electrode (SCE)]. At 400 nm wavelength illumination, the IPCE value is around 100%, while at 450 nm it is around 15 ± 1%.

When gold nanostructured n-type TiO_2 was used as an electrode, the photocurrent appeared even under illumination throughout entire visible spectrum and no current was observed without nanoparticles [54]. Figure 3.6c shows the I–V dependence under different sub-bandgap wavelength illumination and a linear voltage sweep. In all cases, a sharp on–off current response is observed following illumination control. The photocurrent was stable at least for a period of 10 h. In the case of

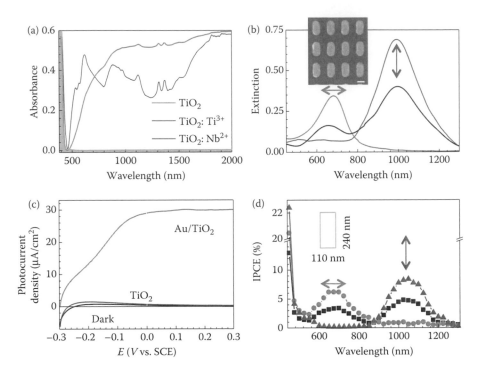

FIGURE 3.6 **(See color insert.)** Plasmonic solar cell. (a) Absorption spectrum of different titania substrates: n-type Nb-doped TiO_2 (Furuuchi Chemical. Co.), reduced TiO_2:Ti^{3+}, and undoped TiO_2; thickness of samples was 0.5 mm. (b) The extinction spectrum of Au nanorod pattern fabricated on the untreated TiO_2 single-crystal measured under unpolarized and polarized illumination in water; polarization is marked by arrows. The inset shows an SEM image of the pattern of gold nanorods of a 240×110 nm^2 footprint. (c) $I–V$ curves measured under different wavelength illumination at linear sweep of 50 mV/s. (d) The incident photon-to-current conversion efficiency (IPCE) spectra of the gold nanorod structure patterned and unpatterned n-type TiO_2:Nd electrode under unpolarized light illumination and for the T and L polarizations, respectively. (Adapted from Y. Nishijima et al., *J. Phys. Chem. Lett.* 10, 2031–2036, 2010.)

gold nanostructures excited by 750 and even 1100 nm illumination, a photocurrent is observed with saturation at 0.2–0.4 V (SCE).

The IPCE values were calculated at 0.4 V and are presented in Figure 3.6d. The IPCE spectra in unpolarized light can be decomposed into two longitudinal and transverse L- and T-modes, correspondingly. The peaks at 700 and 1100 nm wavelengths are matching the extinction maxima for the two polarizations. The light-to-current conversion is more efficient at the wavelengths where the absorption of the substrate and the extinction of the gold nanoparticles is maximized. The pattern was tailored for the most efficient conversion at 1100 nm wavelength and the corresponding performance of the cell was experimentally observed. The obtained IPCE values are at least few times larger than those obtained in nanoporous titania photosensitized by gold nanoparticles [58,59] or titania–gold Schottky junctions [60].

In addition, there were no earlier reports with detectable IPCE values at near-IR spectral range for Grätzel-type solar cells [54].

Transmission of gold nanorods' pattern is approximately $T = 0.63$ at the corresponding extinction of 0.2 which is defined as $-\log(T)$ (Figure 3.6b). Hence, approximately 37% of incidental light is rescattered with negligible absorption in gold at the wavelengths longer than 600 nm [61]. In order to increase extinction and make it not selective to polarization of the incident light, a square footprint pattern of gold nanostructures is made over the surface on n-type rutile [54]. Figure 3.7a shows SEM images of patterns composed of 110×110 nm^2 gold nanoblocks with different block-to-block separations: 90, 190, and 290 nm, respectively. Extinction spectra of the corresponding structures are shown in Figure 3.7b. The maximum extinction value is linearly increasing with the density of gold nanostructures proportionally to the volume fraction of gold. Patterns with extinction of 1.4 (transmission of 3%) efficiently rescatters an impinging light with negligible absorption. The wavelength of plasmon band maximum had a slight blue-shift from 950 to 850 nm with increasing nanostructure density due to dipole-type interaction between the neighboring structures [62].

The photoelectric conversion spectra obtained using the most dense patterns of square nanoblocks of gold are shown in Figure 3.7c with IPCE reaching ~8% [54]. The dependence of IPCE on the volume fraction of gold nanoparticles (their surface

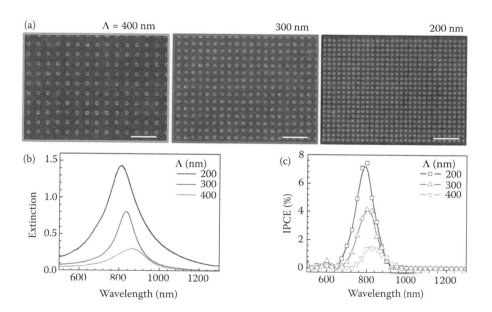

FIGURE 3.7 (**See color insert.**) (a) SEM images of gold nanoblocks of 110×110 nm^2 footprint with different separation of 290, 190, and 90 nm corresponding to the period $A = 400$, 300, and 200, respectively. Scale bar = 1 μm. (b) Extinction spectra of the corresponding gold nanoblock structures. (c) The incident photon-to-current conversion efficiency (IPCE) spectra of the gold nanorod structure patterned n-type TiO$_2$ electrode under unpolarized light illumination. (Adapted from Y. Nishijima et al., *J. Phys. Chem. Lett.* 10, 2031–2036, 2010.)

density) was linear. From the FDTD calculations, the electric field enhancement factor IEI^2 in terms of light-field intensity is around 150 for every nanoparticle and did not depend on the pattern density [63]. The enhancement is mainly caused by the Frolich mechanism, that is, via properties of the dielectric function of gold and its ambient. These results indicate that the photon–electron conversion efficiency is caused by the increased extinction and density of gold nanoparticles at the spectral regions where absorption in the host material, n-type titania, already exist.

3.3.3 Mechanism of Plasmonic Light Harvesting

It is noteworthy that the same design of patterned titania is efficient in the harvesting light at the fundamental absorption band. The IPCE efficiency in the spectrally narrow absorption band at 400–450 nm is approaching 80%; however, this spectrally narrow range contributes equally to the rest of the wide 450–1100 nm window where IPCE is below 10% (for the most dense pattern of nanostructures). The operation mechanism of the photoelectric converter is schematically shown in Figure 3.8. The photocurrent depends linearly on the illumination intensity at the employed wavelengths [54]. This suggests that the one-photon absorption takes place per single photoelectron at the IR spectral range. Illumination intensity was 0.5 W/cm^2 at 800 nm (0.2 W/cm^2 at 1100 nm) and optically nonlinear effects can be neglected. The linear absorption in IR corroborates the mechanism mediated by defects (due to n-type and surface states) at the titania–water–gold interfaces and by direct absorption in gold via sp-to-d band transitions followed by charge separation. Since the defect absorption exists in the depletion zone of titania, the electron transport via electron-hopping and tunneling through a triangular potential barrier is highly probable. Light-field enhancement at the nanoparticle–titania interface is localized in a narrow presurface region of titania (~10 nm as can be estimated by FDTD), where a strong band bending facilitates electron injection into conduction band of titania. There is a strong contribution to light intensity from the field oscillating perpendicular to the interface which facilitates the electron transport and injection into titania [64,65]. The electrons excited into the d-band of gold can tunnel into the conduction band of titania before relaxing into the sp-band and recombining with the hole. Note, for this gold transition to happen, it is not necessary that the photon energy need to be higher than the work function of gold (5.1 eV), which is larger than the bandgap of titania (3.1 eV). Moreover, since light enhancement effects take place, the intra-Au electronic transitions are enhanced and provide an additional channel for light-to-current conversion. This can be considered as a photocatalytic (in terms of enhancing an electronic transport) activity of gold nanoparticles.

Figure 3.8b shows schematically the surface of TiO$_2$ electrode patterned with Au nanoparticles and possible absorption processes. Such presentation, different from the one in panel (a), shows the location of the electron–hole (e–h) processes. The e–h pair generated by photon absorption is then separated by an electron transfer into the conduction band of titania, while the hole can oxidize water on the surface of titania electrode. The IR absorption via defects in titania as well as via the sp-to-d band transitions in gold provide e–h pairs which can be separated before recombination. The holes in gold become delocalized and increase the probability of water oxidation (Figure 3.8b).

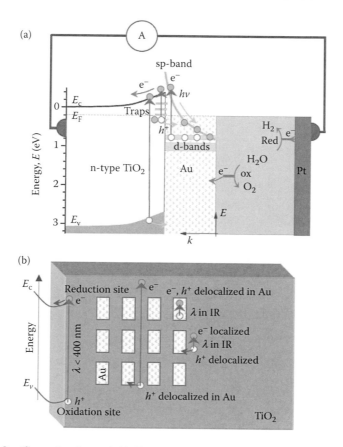

FIGURE 3.8 **(See color insert.)** (a) Energy diagram and schematics of the processes in a Au-nanoparticle sensitized n-type titania solar cell. The absorption takes place via the direct bandgap transitions for photons with energy $hv \geq 3.1$ eV; from the trap defects (which are responsible for the n-type in TiO_2) and d- to sp-band transitions in gold in IR spectral range. Energy position of the water red/ox potentials are shown for discussion of mechanism of charge transfer cycle. The gold particle is shown in the electron dispersion $E(k)$ presentation to reflect the intra-band (sp – d) transitions; k is the wavevector. (b) Schematics of Au-nanoparticles arrangement on the surface of TiO_2 and possible electronic transitions. The electrons e^- are localized on the surface and can flow into bulk of TiO_2; holes h^+ are pinned to the surface at the location of creation or tunnel into Au nanoparticle.

Thermal activation of the photocurrent generation was observed at different wavelengths in the range of increasing temperature from 10 to 60°C [54]. The Arrhenius-type activation of photocurrent follows the rate equation $\ln(k) = \ln(A) - E_a/RT$, where k is the rate constant, A is the preexponential factor, R is the universal gas constant, and T is the absolute temperature. The activation energy was approximately twice smaller, $E_a \sim 12 \pm 2$ kJ/mol (or ~0.124 eV), at the visible–IR spectral range of 650 and 1100 nm wavelengths when compared to that at the close to fundamental transition wavelength of 450 nm with $E_a \sim 28 \pm 2$ kJ/mol (or ~0.29 eV) at the employed range of anodic potentials. This further corroborates an involvement of presurface charge state levels being functional in electron transport to the conduction band of titania. Thermal activation of the light-to-current conversion is a distinct and promising feature of this particular system which contrasts an opposite trend in semiconductor solar cells where increased recombination rate makes cells less efficient at the elevated temperatures.

The photocurrent excitation spectrum closely followed convolution of the absorption of titania and the extinction spectrum of gold nanoparticles according to the T- and L-bands of extinction. This proved that by controlling the plasmon extinction spectrum of gold nanoparticles, photosensitization can be engineered for the required wavelength even at the sub-bandgap and near-IR photon energies. The photon-to-electron conversion is demonstrated for light at the wavelength of 1100 nm.

The photoenergy conversion efficiency is proportional to the density of gold nanoparticles, and the 8% conversion efficiency is obtained with high-density gold nanoparticles. This energy conversion system has the potential for practical applications where sub-bandgap and near-IR wavelengths photons can be harvested from Sun's (or other light source) spectrum. The proposed scheme can also be operated at the fundamental titania absorption at a wavelength shorter than ~450 nm. Degradation problems related to the aging of photosensitizer in Grätzel cells could be alleviated by using Au nanoparticles. Indeed, the nanoparticles of gold show a robust non-photo-bleachable photoluminescence [66]. Such nanotextured surfaces sensitive at the IR wavelengths are promising for light collection *in vivo* at extended depths through the IR transmission window of living tissue.

3.4 ORGANIC SOLAR CELLS

An organic solar cell is a type of photovoltaic cell where the light absorption and charge transport are accomplished by conductive organic molecules or polymers. The molecular structure of these materials features the characteristic system of connected orbitals with delocalized electrons in alternating single and multiple bonds, typical of organic molecules (a conjugated system). A bandgap of 1–4 eV is formed as the degenerate hydrocarbon electrons delocalize, forming bonding–antibonding orbital pairs.

Under these conditions, photonic absorption generates excitons, that is, excited electronic states that, upon migrating to the heterojunction between two different materials, dissociate into free electron and holes to be harvested at the electrodes; the optical absorption length in organic thin films is of the order of 100 nm, which is at least 10 times higher than the exciton diffusion length, limiting efficiency.

Several architectures have been experimented, of which the main ones are summarized in Figure 3.9. The simplest kind comprises a layer of organic polymer sandwiched between two conductors with different work functions (typically indium tinoxide, ITO, and Al/Ca), in order to collect electrons and holes as they are separated from excitons formed within the active region analogous to the p–n junction in semiconductor-based solar cells. As the electric field available is low, the quantum efficiency is very limited; also the film has to be thin due to short diffusion length of carriers; all these make organic solar cells challenging in practical applications. However, several architectures are actively explored since simple roll-to-roll procedures can be used to fabricate large areas at low cost.

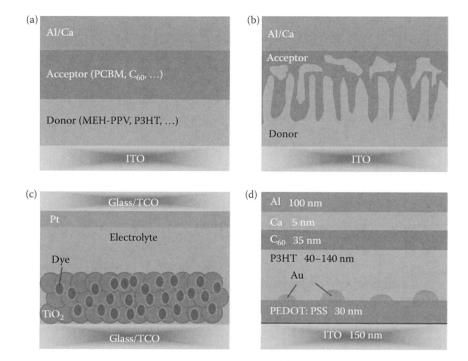

FIGURE 3.9 **(See color insert.)** Some popular geometries of solar cells: examples of (a) organic bilayer and (b) heterojunction, (c) dye-sensitized, (d) plasmonic nanoparticle-enhanced (design from reference [67]). TCO is transparent conductive oxide; see text for details.

The concept is refined in the bilayer organic solar cell in Figure 3.9a, where layer of photo-generation of excitons is composed of two different materials: an acceptor with high electron affinity, in which role fullerene compounds such as phenyl-C_{61}-butyricacidmethyl ester (PCBM) and C_{60} are widely employed, and a donor with lower affinity, where polymers such as poly[2-methoxy-5-(2)ethylhexyloxy)-*p*-phenylene vinylene] (MEH-PPV), and poly(3-hexylthiophene) (P3HT) are used. The potential difference is thus concentrated at the interface, where the high field gradient is more efficient in breaking up excitons. The limit of this configuration is the short diffusion length of excitons, of the order of 10 nm, which is tooth in to achieve enough of light absorbance.

This issue is addressed in bulk heterojunction cells in Figure 3.9b, by blending the donor and the acceptor together, so that the blended zones are of the scale of the exciton diffusion length, which permits one to thicken the layer without compromising the ability of the excitons to break at the interface. The mixing of the polymers must be performed carefully to avoid shorting to the opposite electrode. Heterojunction cells include several of the best-performing organic solar cell designs and also a category of designs where the transparent anode seals a porous transparent material (usually TiO_2) soaked in an organic dye to a liquid electrolyte against a metal cathode. This is known as dye-sensitized solar cell or Grätzel cell shown in Figure 3.9c. The best efficiency of the photo-anodes based on tubular TiO_2 in DSSC (~2.7%) has been recently superseded by nano–microstructured Nb_2O_5 which showed a light to electrical conversion efficiency of ~4.1% [68].

The light absorption and electron–hole pair generation in dye-sensitized cells occur in a molecular dye, normally ruthenium-polypyridine related or I^-/I^{3-} red-ox pair, which sensitizes the host material, a porous TiO_2 layer, which in turn transports the injected carriers, reaching efficiency around 11% [69]. TiO_2 only acts as a conductor, since it has a wide bandgap that makes it transparent to visible and IR radiation, but its porosity contributes to reducing the forward resistance and the recombination probability of minority carriers.

To improve the mechanical stability of highly conductive nanoporous TiO_2 films, the cold isostatic pressing (CIP) compaction technique has been investigated for dye-sensitized cells on ITO-coated polyethylene naphthalate (PEN). Good flexibility was obtained from a room-temperature technique, using P-25 TiO_2 films with a Ru-complex sensitizer, obtaining an incident photon-to-current conversion efficiency (IPCE) of 72% at 530 nm and a maximum power conversion efficiency (PCE) of 7.4% (or the ECE—electric conversion efficiency) [70].

Such structures are attractive for the use of low-cost materials and techniques; the devices themselves can have a high degree of flexibility and transparency, which expand their range of use. The design issues stem from the use of a liquid electrolyte, which requires sealing and is sensitive to temperature changes, and from the chemical stability of the components, which limits long-term reliability. The conversion efficiency is also lower than the best-performing thin-film solar cells, albeit at a much lower cost.

In order to improve higher-wavelength absorption and thermal stability, new donor materials are being researched in substitution of the widely used P3HT. Recently, poly[3-(4-octylphenyl)thiophene)] (POPT) has been investigated in POPT:PCBM heterojunctions, extending absorption up to 750 nm. The obtained V_{oc} was up to 0.67 V, with PCE up to 1.58% and J_{sc} up to 5.5 mA/cm^2 [71]. Templating to control donor crystalline structure had been investigated in lead phthalocyanine (PbPc) cells by introducing a CuI layer, doubling J_{sc} with a 63% fill factor and 2.5% PCE [72].

Several techniques have been explored to reduce thickness without impairing absorption, among which is the stacking of several individual cells. Three-dimensional (3D) fabrication techniques are being pioneered, among which is the folding of a flexible P3HT:PCBM organic solar cell in cylindrical shape with vertical orientation, to increase photon capture and guiding. Guidance of the light beams allows multiple passes through the photoactive layer, doubling J_{sc} with respect to the flat case [73].

A key emerging method is the use of plasmonic nanostructures to enhance light absorption and boost the exciton population through light field enhancement. Low cost, strong light absorption, and the possibility to tune the bandgap through chemistry make these materials attractive for photovoltaics, and research is focused on issues of efficiency and stability improvement.

3.4.1 PLASMONICS IN ORGANIC SOLAR CELLS

Plasmonic enhancement in organic solar cells is usually achieved using Ag particles, either embedded in a polyelectrolyte layer or within the photoactive material.

In the latter case, as the p–n junction is not flat but randomly intermixing, presence of a metal can be detrimental for recombination and possible electrical shorting. A numerical study has been performed regarding the embedding of a regular array of Ag nanoparticles inside the active layer of a P3HT:PCBM heterojunction; it showed a 1.56 times absorption enhancement peaking at ~600 nm for tuned particle diameter and spacing. It was also shown that absorption enhancement mechanism is mainly due to near-field enhancement rather than an enhanced scattering [74]. Introduction of spin-coated Ag nanospheres of 40 nm diameter in a 50-nm-thick MEH-PPV:PCBM bulk heterojunction device resulted in a wideband absorbance increase between 350 and 800 nm, with a maximum of 50% at 500 nm wavelength, due to localized plasmonic effect [75].

Deposition of Au nanorods on the ITO electrode has been accomplished through wet chemical methods, followed by thermal reshaping into nanodots. A polyelectrolyte coating with poly(3,4-ethylenedioxythiophene)poly(styrenesulfonate) (PEDOT:PSS) separates the nanodots from the heterojunction to avoid exciton quenching with nonradiative energy transfer, and the nanodots effect resulted in an increase in the power conversion efficiency from 3.04 to 3.65%, with a short-circuit current density J_{sc} of 11.13 mA/cm^2, and a fill factor of 57% [76]. It is noteworthy that current densities in the order of 10 mA/cm^2 and above are usually considered relevant for practical applications.

In order to avoid device shunting and reduction of plasmonic efficiency by direct deposition of nanoparticles on the transparent front electrode, they can also be deposited on top of the PEDOT:PSS polyelectrolyte layer, which improves reproducibility. This also permits one to use simple electrode position techniques, and in fact pulse-current electrode position was employed to deposit particles of controlled size down to 13 nm close to a P3HT:PCBM heterojunction, obtaining a fill factor of 55% and a power conversion efficiency of 3.69%, increased of 20% with respect to Ref. [35].

The transparent front electrode is usually realized using costly materials such as indium tin oxide, ITO, and therefore it is attractive to replace it with patterned metal, where SPP coupling is exploited to couple light into the substrate and introduce field enhancement to improve performance. An Ag layer on glass patterned with a slit nanocavity array as anode for a copper phthalocyanine (CuPc) cell with a CuPc:C$_{60}$ heterojunction produced a 3.2-fold increase in the power conversion efficiency with respect to an unpatterned Ag anode, with a maximum field enhancement of 65 around 800 nm [77]. In order to reduce costs, an attempt has been made to replace ITO with a Cu-based multilayer transparent electrode, which resulted in a power conversion efficiency up to 87% with respect to ITO [78].

Patterning the counter-electrode to support SPP modes is also effective. In particular, recently the patterning of a very thin (<100 nm) Al counter-electrode with a regular lattice of holes resulted in a I_{sc} enhancement of 47% to 8.9 mA/cm^2 due to the interaction of short-range and long-range SPPs in a device based on a 30-nm-thick P3HT:PC$_{70}$BM heterojunction active layer [79].

It has been shown by silver nanoclusters in tandem organic cells that the field enhancement can have a range of up to 10 nm from the array of nanoparticles of 5-nm size; the process is broadband, persisting even at wavelengths removed from

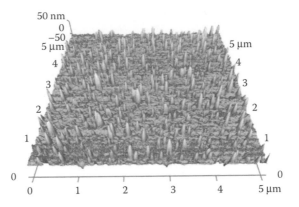

FIGURE 3.10 (**See color insert.**) Atomic force microscopy (AFM) profile of the gold nanoparticle layer. (Adapted from C. Poh, et al., *Opt. Mater. Express* 1, 1326–1331, 2011.)

the surface plasmon resonance. Moreover, the placement of the nanoparticles should be carefully arranged to avoid exciton quenching [34].

A recently proposed structure [67] places gold nanoparticles (chosen for stability and compatibility with the photoactive layer; see Figure 3.10) confined in the p-layer of the P3HT-C_{60} junction to prevent recombination and shorting, but permitting the near-field enhancement due to LSP resonance to affect the depletion zone, without substantial shadowing due to sparse particle placement (Figure 3.9d). Intensity is tuned also by varying the C_{60} layer thickness to adjust the distance from the Ca/Al counter-electrode. A 3D-FDTD numerical model was fitted to experimental results from a thicker P3HT layer, then extended to a thin-film structure to show that the largest LSP enhancement effect is reached when the active layers as deposited so as to make the p–n junction conformal to the plasmonic nanoparticles shape. This also solves a problem on meandering of the electron–hole separation interface and is a promising direction of improvement of organic solar cells.

3.4.2 Effect of Au Nanoparticles

A bilayer organic solar cell was built from ITO-coated glass substrate, spin-coated by PEDOT:PSS to make a film ~30 nm thick. After annealing, a very thin Au layer was deposited by thermal evaporation, which immediately broke forming Au nanoparticles on the PEDOT:PSS surface. This leads to the formation of hemispheric/oblate Au nanoparticles with diameters ranging from 80 to 120 nm and heights from 20 to 40 nm. Over these, a 140 nm P3HT film was deposited, with a further C_{60} layer to form the p–n junction, and a Ca/Al counter-electrode, for a sample experimental area of 3.5×4.0 mm².

The tests were performed under AM1.5 simulated solar illumination with a computer-controlled Keithley SourceMeter 2400. In order to model the electromagnetic behavior of the cell, a 3D-FDTD model was implemented with the Lumerical software, using material refractive indices from literature and linearly polarized illumination. Two power monitors, PA and PB, recorded the power flowing in the z-direction, respectively, above and below the 70-nm wide p–n junction depletion zone. A third monitor, PR, recorded the power reflected from the cell, which is illuminated from the bottom side in Figure 3.9d. The absorbance A was thus calculated as

$$A = \frac{(1 - P_R)(P_B - P_A)}{P_B}.$$ (3.6)

The Au nanoparticles produced an increase in all the performance parameters, in particular, the maximum J_{sc} increased from 0.528 to 0.605 mA/cm^2, while the maximum V_{oc} increased from 0.203 to 0.272 V; this last result possibly due to a Schottky barrier formed by the nanoparticles. The external quantum efficiency difference ΔEQE with respect to the case without nanoparticles showed a resonance peak with an 1.80% increase around 395 nm due to the interband transition absorption of gold, which was correctly modeled by the absorbance difference ΔA simulated by the FDTD simulation, as shown in Figure 3.11, with a slightly higher loss, when the particles were randomly distributed between 30 and 50 nm radius. The proper plasmonic resonance at 675 nm gives a very weak contribution, due to the high thickness of the P3HT layer with respect to the enhancement range of the nanoparticles and the carrier diffusion length, both around 10 nm.

In order to provide better enhancement, it looks necessary to reduce the P3HT thickness to bring the depletion zone closer to the particles, so the P3HT thickness in the model was reduced to 40 nm. The 35-nm depletion zone was allowed to wrap around the particles with a conformal shape, and the best configuration, having a conformal P3HT layer with a flat C$_{60}$ layer above it, achieved a maximum ΔEQE of 3% around 390 nm, as shown by curve B in Figure 3.12. A configuration in which both layers are flat (curve A) takes the depletion zone farther away from the particles, reducing performance, while having both layers conformal (curve C) brings the rear contact too close to the particles, destroying the enhancement.

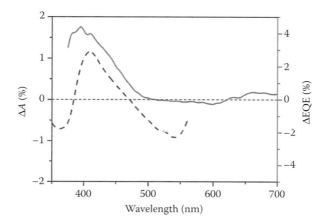

FIGURE 3.11 Experimentally measured changes of electrical conversion efficiency ΔEQE (solid curve) compared with 3D-FDTD simulated absorbance difference ΔA (dashed curve).

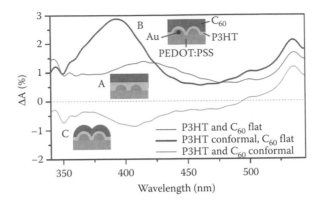

FIGURE 3.12 (**See color insert**.) Simulated absorbance difference A for flat and conformal junction interfaces.

3.5 NANO-SCALE OPTICAL EFFECTS FOR PHOTOCATALYTIC APPLICATIONS

Optical extinction spectra of metallic or dielectric nanoparticles depend strongly on their size and shape, especially when their dissipative losses are small. In this respect, nanoparticles and nanostructured surfaces of noble metals (Ag, Au, Al, Cu) are especially interesting, because their free-carrier plasma behaves as nearly lossless dielectric and can give rise to intense plasmonic resonances at optical wavelengths [80,81]. Resonant response of nanoparticle plasmons is apparent from their extinction spectra, which comprise elastic light scattering and absorptive contributions. The elastic scattering is a consequence of strong resonant localization of optical field in nanoscale regions near the surface of nanoparticles and is often described as "super-focusing" of the optical field into subwavelength-sized spatial regions. Tailoring of plasmonic extinction via nanostructuring of metallic particles and surfaces provides an efficient route to exploitation of linear and nonlinear optical effects in nanoscale "super-focal" regions, such as surface-enhanced Raman scattering/spectroscopy (SERS) for detection and labeling [82,83], light extraction in diode lighting, absorption promotion in solar cells [48], and biological purposes (e.g., cancer diagnostics, DNA detection, optical biosensors, biochemistry [81,82,84–86]). Exploitation of nanoparticle plasmonics depends crucially on fidelity of nanostructuring. Currently, the most precise technique is to define footprint of nanoparticle pattern by electron-beam lithography (EBL), and deposit of a thin metal film on the exposed mask with subsequent lift-off of unwanted areas [57,66,80,87]. EBL enables a flexible choice of the shape and size of nanoparticles which determines spectral and polarization responses [88–90]. However, EBL is too slow when large-area structures with dimensions in the order of centimeters are desired. Also, for a large variety of applications, especially in the fields of photochemistry and catalysis, micro- or nanometer-sized photocatalysts loaded with cocatalysts are the most promising. For example, nanoparticles of noble metals grown (photo-)chemically on smooth surfaces of titania (TiO_2) particles, exhibit strong plasmonic field localization and enhancement in areas where spheres touch the substrate. The field localization may assist photocatalytic action due to transport of metal's electrons into titania [91]. Recent investigations have demonstrated that action spectrum, that is, wavelength dependence of quantum yield for a particular photochemical reaction, is strongly correlated with optical extinction spectrum of the photocatalyst consisting of metal nano-photocatalyst with photocatalytically active microparticles, for example, noble metal deposits on semiconductor particles (titania, titanium-IV oxide) [92,93]. The quantum yield was found to scale proportionally with the size of gold nanoparticles [92]. Such behavior indicates that the transport is promoted at plasmonic resonance by optical near-field concentrated predominantly in a small region at the interface [91] for the required electron transport through the solution–metal, metal–titania, and solution–titania boundaries [93]. The spectral features of action spectrum [93] are also likely related to resonant plasmonic modes and their near-field enhancement, but their origin has not been neither studied in detail nor understood.

Plasmonic optical response of model spherical nanoparticles on TiO_2 substrate using theoretical modeling based on numerical solution of Maxwell's equations by FDTD technique is providing qualitative insights. The polarization orientation plays an important role, since field components normal to the interface are the most efficient in promoting the electronic transport [91]. Numerical simulations of light-field distribution for noble metals [84,94–97] and photoactivities of noble metal/titania composites [59,92,93,97–103] have already been reported as well as the field enhancement and its distribution at the interface between noble metal and titania [64]. One particularly interesting finding of the reported simulations is the plasmonic mode which shows relatively weak optical extinction, but nevertheless exhibits strong field enhancement. Such behavior is indicative of the so-called dark plasmons [104], whose dipolar scattering is inefficient, but field enhancement is strong and is concentrated at the interface, where it can promote photocatalytic activity of the nanoparticles. The electronic transport associated with dark plasmons and polarization effects at the interfaces can open new directions in photocatalytic applications [64].

3.5.1   MODEL AND SIMULATIONS

Theoretical analysis is aimed at structures investigated experimentally in previous studies and follows analysis reported recently [64]. It is relevant to describe the experimental details that were used as an input for the analysis. Geometrical parameters of the model systems can be inferred from the available images of spherical Ag and Au nanoparticles chemically grown (impregnation-reduction method) or photochemically grown (photodeposition) from their solutions on titania substrates (rutile and anatase) [92]. Typical samples are shown in Figure 3.13. For better understanding, for the dependencies of distribution and enhancement of electromagnetic field on sizes and shapes of noble metal nanoparticles, simulations have been carried out for simplified systems of spherical Ag nanoparticles on a smooth titania substrate. This geometry is illustrated in Figure 3.14a.

Optical extinction data are available in reference [105] as diffusive reflectance spectra (DRS). The experimental spectra represent losses due to absorption and

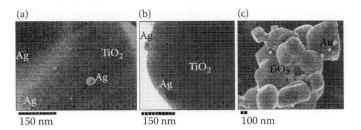

FIGURE 3.13 Typical morphologies of titania microparticles decorated by Ag (a,b) and Au (c) for photocatalytic applications. (Adapted from E. Kowalska, R. Abe, and B. Ohtani, *Chem. Commun.* 2, 241–243, 2009.) (a,c) SEM and (b) combined SEM and transmission electron microscopy (TEM) images. Scale bar in (c) 100 nm. (Courtesy of Dr. E. Kowalska.)

elastic scattering and can be directly compared with calculated extinction spectra, assuming the surrounding environment with refractive index $n = 1$ (air). Since the photoactivities were measured for Ag/TiO_2 powders suspended in acetic acid and 2-propanol (5 vol.%) for its oxidation to carbon dioxide and acetone, respectively, DRS were also taken for suspended powders in water and acetic acid solution (5 vol.%) [93]. The extinction spectra for the environment with refractive index of $n = 1.375$ (2-propanol) is estimated. This modification alters the shape of the extinction bands and red-shifts them [106]. Calculated absorption, scattering, and extinction cross-sections for Au and Ag nanospheres are shown in Figure 3.14b and c.

3.5.2 FDTD SIMULATIONS

Extinction spectra of spherical nanoparticles with radius spin homogeneous environment can be calculated analytically. The expression for extinction cross-section [107] is given as

$$C_{ext} = \frac{24\pi^2 r_p^2 \varepsilon_m^{3/2}}{\lambda} \frac{\varepsilon''}{\left(\varepsilon'' + \left(2 + \frac{12x^2}{5} \right) \varepsilon_m \right)^2 + \varepsilon''^2}, \tag{3.7}$$

where $x = 2\pi N r_p / \lambda$ with N being the refractive index of surrounding medium, ε' and ε'' are the real and imaginary parts of the material on nanoparticle, respectively, and ε_m is the relative permittivity of the surrounding medium.

Equation 3.7 also describes dependence of the extinction cross-section on the nanosphere size. However, analytical analysis is impossible for nanoparticles with more complicated shapes and located in inhomogeneous environments (as is the case for nanospheres on a substrate). Moreover, it does not provide information of the near-field parameters, such as spatial distribution and polarization (which is essential for photocatalytic applications). In these circumstances, the most accessible method for gaining insights into the near-field properties is theoretical modeling based on numerical solution of Maxwell's equations by FDTD technique. This approach has proved to be accurate for metallic nanoparticles having dimensions larger than ~10–15 nm [108].

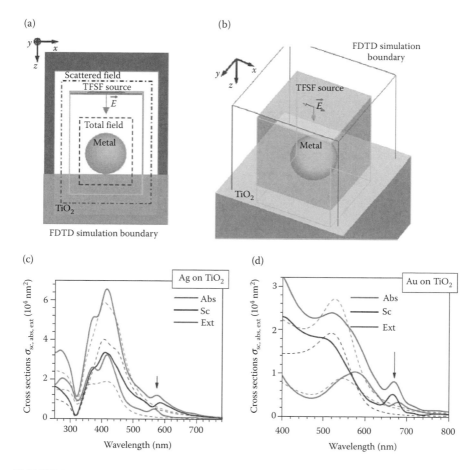

FIGURE 3.14 (**See color insert.**) (a) Geometry of FDTD simulation according to TFSF formulation and (b) structural perspective. The incident wave is linearly polarized along the *x*-axis and is incident on the metallic sphere along the *z*-axis. (c) The extinction, scattering, and absorption cross-sections ($\sigma_{ex} = \sigma_{sc} + \sigma_{abs}$) of a 50-nm-radius silver sphere on rutile; dashed lines mark corresponding cross-sections without rutile substrate. Geometrical cross-section is 0.79×10^4 nm^2. (d) Same as (c), for a 50-nm-radius gold sphere. In (c,d), weak spectral peaks associated with the presence of a substrate are indicated by arrows.

The simulations were performed in 3D space using FDTD Solutions (Lumerical, Inc.) software package. Cross-sectional view of the simulation geometry is illustrated schematically in Figure 3.14a. The model nanoparticle (with size and shape close to those in real samples) is placed on the substrate and is illuminated by a linearly polarized short, broadband optical pulse launched in the TFSF region. Outside the TFSF region, the source intensity is zero, and scattered radiation passing through the boundaries of rectangular area enclosed by scattered-field monitors is recorded. Inside the TFSF region, a similar set of monitors is used for the detection of total radiation. Using this TFSF approximation, the absorbed and scattered field components can be determined from the balance of optical power flowing through the monitors. Frequency dependence (and extinction spectra) can be determined by Fourier-transforming the temporal signals detected by monitors. The spatial domain of FDTD simulations is enclosed by perfectly matched layer (PML) boundaries that act as nearly ideal absorbers for the radiation that has reached the boundaries. Optical near field was monitored on two planes: x–y-plane coincident with surface of the substrate, and x–z-plane coincident with the plane of the drawing. The calculations were performed on a discrete cubic mesh with spacing of 1 nm. Optical properties of gold and silver were described using Lorentz and plasma approximations of the existing experimental data [61]. The substrate was assumed to have a refractive index of $n = 2.5$, close to that of titania; numerical results are not valid at the direct absorption band of titania for the wavelengths shorter than ~400 nm.

3.5.3 PLASMONIC RESONANCES AND NEAR-FIELD PATTERNS

Spheroidal nanoparticles attached to smooth substrates attract wide attention because this geometry is used in many optical applications exploiting field enhancement at the nanoparticle–substrate interface. Such applications include optical sensors, localized laser ablation [109], SERS [110], and photocatalytic reactions [98,102,111]. FDTD simulations demonstrate the relationship between the size and shape of the nanoparticles and their optical scattering spectra as well as optical field enhancement capability [64].

Figure 3.14b shows calculated spectra of scattering, absorption, and extinction cross-sections for Ag sphere with a 50 nm radius on rutile substrate. Geometrical cross-section of a 50 nm radius sphere is 0.79×10^4 nm^2. Reduction of the geometrical cross-section due to penetration of incident waves into silver is negligible, since skin depth in Ag is very small: $\delta = (2\rho/\omega\mu)^{1/2} \sim 2.8$ nm (where $\rho = 1.59 \times 10^{-8}$ Ω m is the resistivity, $\mu = 1.26 \times 10^{-6}$ H/m is the absolute permeability of Ag, and $\omega = 2\pi c/\lambda$ is the cyclic frequency of light at $\lambda = 600$ nm wavelength). As can be seen from Figure 3.14, the extinction cross-section exceeds the geometrical cross-section within the wavelength range 350–600 nm, where it exhibits a major peak centered near the 420 nm wavelength. Distinction between the contributions of a sphere–substrate system (Figure 3.15) and a sphere alone can be seen by examining the reference spectra (calculated for a sphere in air) shown in the same Figure 3.14 by dashed lines. While the main peak has only a minor contribution from the presence of rutile substrate, a new peak arising near the 570 nm wavelength can be observed in Figure 3.14b. Although absolute magnitude of this peak is low, its relative modification is significant (there is no peak in the reference spectrum). Such behavior is a typical signature of structural symmetry breaking in plasmonic structures and is not directly related to the particular dielectric functions of metallic particle and dielectric surface. Such behavior is observed in the cases of Fano resonances in systems of interacting nanoparticles, and in dark plasmons [104]. The latter is known to resemble quadrupole plasmonic resonance whose dipolar character is weakly pronounced and scattering is inefficient, but electrical field enhancement can nevertheless be significant [64].

Extinction spectra of gold spheres (Figure 3.14c) are qualitatively similar to those of silver (b) with main extinction peaks centered at 520 and 420 nm, respectively. Although the main spectral peak is broadened and has no finer features, it is centered

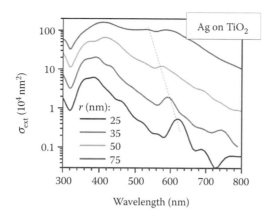

FIGURE 3.15 (**See color insert**.) Dependence of extinction spectra of Ag nanosphere/on TiO$_2$ system on the sphere radius r. The dashed line is guide to the eye to emphasize spectral shift of the characteristic long-wavelength peak (see Figure 3.16). (Adapted from V. Mizeikis, E. Kowalska, and S. Juodkazis, *J. Nanosci. Nanotechnol.* 11, 2814–2822, 2011.)

near the same wavelength. The weaker peak is red-shifted by ~70 nm in comparison to silver spheres and is better-pronounced. Persistence of this peak for both Ag and Au spheres indicates that it is caused predominantly by geometry of the sphere-substrate system.

Practical application of noble-metal nanospheres/rutile systems for photocatalysis reactions usually requires immersion in liquids, such as 2-propanol [92,93]. As can be expected [106], increase in the refractive index to that of 2-propanol ($n = 1.375$) leads to red-shift of the main extinction peaks and suppresses finer features in the spectra [64]. However, at least in the spectra where the long-wavelength peaks can be identified, these peaks exhibit an opposite trend of blue-shift. This trend can be explained using the same arguments as above: when refractive index of environment approaches that of the substrate, the system is transformed to a single sphere (no substrate), and the long-wavelength peak merges with the main peak. It should be noted that in actual experiments with photocatalysts such as titania nanoparticles, the high refractive index of titania and its large volumetric fraction in the suspension creates an environment of an effective higher refractive index causing red-shift of the extinction peaks; this is consistent with experimental observations of red-shifted peaks of the experimental extinction spectra [93]. Due to the augmented refractive index of surrounding, the extinction peaks of metal nanoparticles become red-shifted, and less dependent on the size and refractive index of the immersion solution [88,100,106] as discussed above.

Figure 3.16 shows calculated near-field intensity-enhancement patterns at peak wavelengths of the main extinction peak and the weaker peak. The near-field patterns were obtained from the same simulations as the extinction spectra by detecting temporal signal on a discretized 2D plane. Subsequently, steady-state field values for fixed wavelength were obtained at the discrete spatial positions on the plane by Fourier transform of the temporal signal. The enhancement factor was obtained by normalizing the near-field intensity to that of the incident wave. The field detection

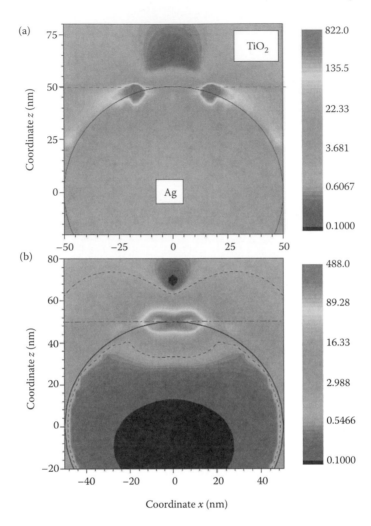

FIGURE 3.16 (**See color insert**.) Cross-sectional pattern of the near-field intensity enhancement factor $|E|_2/E_0|_2$ at ~420 nm (a) and ~570 nm (b) wavelengths. The highest enhancement was 820 at 420 nm, and 490 at 570 nm wavelengths. Outlines of the sphere are emphasized by black solid lines. In (b), black dashed lines emphasize the contour where $E = 1$. (Adapted from V. Mizeikis, E. Kowalska, and S. Juodkazis, *J. Nanosci. Nanotechnol.* 11, 2814–2822, 2011.)

plane coincides with x–z-plane bisecting the center of the sphere. Note that polarization of the incident wave is along the x-axis. Near-field intensity pattern at the 420 nm wavelength (Figure 3.16a) is characterized by two high-intensity spots near the surface of the sphere. The corresponding field patterns in the x–y plane (not shown) form slightly elongated dipolar pattern oriented along the polarization of the incident field. We have verified separately that z-polarized field component (i.e., normal to the substrate surface) E_z is dominant at this wavelength and contributes more than 80% of the total field intensity. This feature is highly sought in photocatalytic applications. However, the spots are displaced by ±20 with respect to the z-axis and a few nanometers away from the substrate, mainly penetrating the metal and its environment (in this case, air with refractive index $n = 1$ is assumed), but not the substrate. Hence, efficiency of this mode at the surface is somewhat limited. Near-field intensity pattern at the 570 nm wavelength (Figure 3.16b) has one high-intensity area at the location where sphere touches the substrate. The corresponding field pattern on x–y-plane (not shown) indicates that this is the only high-intensity spot at the given wavelength. It is quite surprising that field pattern at a longer wavelength (single area) is localized better than at a shorter wavelength (dipolar pattern). However, this observation justifies the assumption about weak dipolar scattering of the new, long-wavelength peak. Its total field intensity enhancement factor is about twice lower compared to that of the previous case, and only about 25% of the total field is z-polarized. Nevertheless, the predominant near-field concentration at the contact point between the sphere and substrate makes the mode centered at 570 nm wavelength particularly interesting for photo-catalytic applications. Qualitatively similar results were obtained for gold. In this case, extinction peaks were red-shifted, and field enhancement factors were lower [106].

The major component of the electrical field at the interface is polarized perpendicular to it and this is even more pronounced in the case of hemispherical particles [64]. This orientation of E-field facilitates charge transport through the interface.

3.6　ENERGY CONVERSION: SOLAR FUEL

Analysis of the light conversion efficiency to electrical and chemical energy is discussed according to Ref. [112] together with possibilities to harness plasmonic effects.

In recent years, series of review articles devoted to solar energy conversion and accumulation have appeared in scientific literature [113–125]. Photolysis or photo-splitting of water into gaseous O_2 and H_2 is described by

$$H_2O \xrightarrow{h\nu} H_2 + \frac{1}{2}O_2$$

$$(3.8)$$

represents one of the most challenging and promising ways of solar energy accumulation, as solar hydrogen can be further used for various purposes, fuel of the future being among them. Since the discovery of photocatalytic water splitting on TiO_2 surface in 1972 [126], the possibilities of practical application of this process remain vague despite of numerous and comprehensive studies [113,115–119,121–124]. According to reference [115], various ways of modification of TiO_2 particles, such as noble metal or noble metal oxide loading on TiO_2 surface, metal ion or anion doping, metal ion implantation, dye sensitization and also addition of sacrificial or other components to the electrolyte lead to rather low rate of water splitting, that is, ~600 $\mu mol\ h^{-1}$ of H_2 and ~300 $\mu mol\ h^{-1}$ of O_2 at best. In the case of unmodified TiO_2 particles, only traces or even no gaseous H_2 and O_2 are found chromatographically [115].

When light-sensitive thin semiconducting layer of TiO_2 is formed on titanium or some other electrically conducting substrate, its potential can be controlled electrochemically using reference electrode, whereas the usage of counter-electrode, usually Pt, and external power source allows to separate anodic and cathodic processes, taking place on the surface of TiO_2 oxide, and also to control their rate. In other words, photosplitting of water can be performed in photoelectrochemical cell (PEC) as photoelectrolysis of water, when evolution of hydrogen and oxygen takes place on different electrodes, that is, O_2 on TiO_2 and H_2 on Pt electrode. Nowotny et al. [117] presented the comprehensive review of most important photoelectrochemical studies performed since 1972 till 2005. In these cases when water splitting on TiO_2 surface was studied in PECs using natural or artificial full-spectrum solar illumination (~100 mW cm^{-2}), the values of energy conversion efficiency, ECE, ranged between ~0.1 and ~1.0%. In the review devoted to vertically oriented TiO_2 nanotubes [120], the highest reported H_2 generation rate under the same conditions of illumination is 1.75 mL (Wh)$^{-1}$, which should correspond to 1.75 l h^{-1} m^{-2} or to photocurrent density of 0.4 mA cm^{-2}.

The rate of hydrogen evolution can be increased significantly using artificial high power UV light sources. For instance, when UV (320–400 nm) illumination of ~100 mW cm^{-2} power is used [120], the highest rate of H_2 generation on TiO_2 nanotubes array is 76 mL (Wh)$^{-1}$, that is 76 L h^{-1}m^{-2} (~3.39 mol H_2), which should correspond to photocurrent density of 18 mA cm^{-2}. According to the authors, the quantum efficiency, QE, in this case is almost 80%. The usage of high-power UV light sources for H_2 production is inexpedient; however, the increase in H_2 generation efficiency demonstrates that concentration of solar light should be very effective.

Thus, in the case of PEC, the values of ECE and H_2 generation rate under natural solar illumination conditions are significantly higher, but still insufficiently high from the technological point of view. According to reference [115], there are three main reasons for low energy conversion efficiency on TiO_2: (i) fast recombination of photogenerated electrons and holes, (ii) fast backward reaction, and (iii) inability to harvest visible and IR light at longer wavelengths than ~400 nm.

Nevertheless, titania and titania-based materials are considered to be the most promising candidates for photoelectrodes for solar hydrogen, since they exhibit outstanding resistance to corrosion and photocorrosion in aqueous environments [115,117], whereas other valence and small bandgap materials are not suitable for this purpose due to very short life time and prohibitive cost [117]. Tandem photoelectrochemical cells, consisting of n- and p-type semiconductor photoelectrodes, or PECs with solid state thin-film multijunctions have been proposed [127] as an option which excludes usage of external current source. In theory, infrared light with photon energy of 1.23 eV could induce reaction Equation 3.8 [117], as the standard potential of water oxidation to molecular oxygen, $E^0_{O_2/H_2O}$ is 1.23 V. Comparison to photon energy with potential $E^0 = 1.23$ V is following. According to $\Delta G = -nFE^0$, where G represents the isobaric–isothermal potential (Gibbs energy) of the reaction or the chemical potential of the substance, n is the number of electrons and F is the Faraday constant, in the case of formation of water molecule $H_2 + O_2 = H_2O$, one would find: $\Delta G = -2 \times 96{,}500 \times 1.23 = -2.374 \times 10^5$ (J mol^{-1}). As there are two H–O bonds in H_2O molecule, the energy corresponding to one bond is 1.187×10^5 J mol^{-1} or 7.409×10^{23} eV mol^{-1}. Division of the latter value by Avogadro number $N_A = 6.02 \times 10^{23}$ gives the energy of one chemical bond, 1.23 eV. This is the minimum photon energy required to break the bond between H and O in H_2O molecule and shows equivalency between thermodynamic and photon energies in terms of chemical bond cleavage.

However, Grätzel [124] has reported that numerous attempts to shift the spectral response of TiO_2 into the visible range or to develop alternative oxides affording water splitting by visible light have been unsuccessful so far, as can be illustrated by references [127–132]. Only UV illumination with much higher photonenergy ($hv \geq 3.2$ eV) is capable of inducing water cleavage [115,117–119,124]. Thorough understanding of the mechanism of oxygen evolution reaction (OER) on TiO_2 electrode surface can help one to find a clue to this problem.

Analysis based on thermodynamical principles and energy conservation reveals the reasons of the low energy conversion efficiency of water photosplitting on TiO_2 surface [112]. The energy conversion efficiencies for the light-to-electrical, electrical-to-chemical, and the light-to-chemical are presented, and general expressions are provided below for the investigated system of water splitting and a possibility to increase light-to-chemical energy conversion efficiency ECE are provided for the current state of the art in solar cell technology [112].

3.6.1 Light-to-Electrical and Chemical Energy Conversion Algebra

Photoelectrochemical properties of light-sensitive TiO_2 layers in the reaction of water splitting were investigated in a three-electrode electrochemical cell supplied with quartz window [113,117,120,133]. The main parameters and characteristics that are measured are as follows:

E_{ph}^{oc}: the open-circuit potential of Ti/TiO_2 electrode under UV illumination; measured electrometrically, that is, with the voltmeter with a very high, ~10^{12} Ω, input resistance;

I_{ph}: the photocurrent, that is, the difference between current values with and without illumination;

I_{ph} versus E: the dependence of photocurrent on the photopotential of Ti/TiO_2 electrode, that is, photovoltammogram (VA).

Schematic representation of water photoelectrolysis cells with external power source and the equipment used for the measurement of the above-indicated parameters is shown in Figure 3.17a. External power source, the voltage of which can be varied within a wide range, is necessary for the realization of water splitting. The scheme, shown in Figure 3.17a, can be realized using potentiostat working in galvanostatic mode. The equivalent circuit of the scheme shown in Figure 3.17a should correspond to that of the scheme shown in Figure 3.17b, where Ti/TiO_2 photoelectrode serves also as photocurrent source. In Figure 3.17b, the difference between the potentials of holes and photoelectrons captured in TiO_2 surface is denoted as ΔE_{ph} (see further), whereas ΔE_{ext} denotes the potential difference between the poles of external power source or, in other words, the potential difference between the anode and the cathode of PEC.

Incident photon-to-electron conversion efficiency IPCE or quantum efficiency QE is calculated according to the following equation [116]:

$$\text{IPCE} = \frac{\text{electrons}}{\text{photons}} = \frac{I_{ph}/q}{P/hv} = \frac{I_{ph}hv}{Pq} = \frac{I_{ph}E_{hv}}{P} \times 100\%,$$

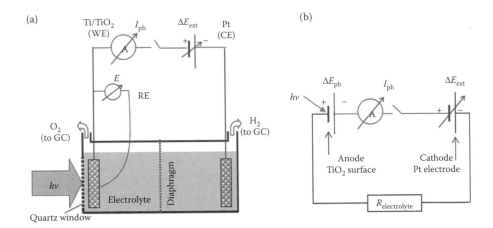

FIGURE 3.17 (a) Schematics of water photoelectrolysis cell with external power source and equipment for the measurement of electric parameters. (Adapted from K. Juodkazis et al., Opt. *Express: Energy Express* 18, A147–A160, 2010.) (b) The equivalent scheme of the electric circuit. WE, RE, CE—working, reference, and counter electrodes, respectively, GC—gas chromatograph, ΔE_{ext}—voltage of external power source, ΔE_{ph}—voltage of photocurrent source, I_{ph}—photocurrent, hv—photon energy.

where hv is the energy of photon, P is the power of incident light in mW cm^{-2}, q is the electron charge 1.602×10^{-19} C, I_{ph} is the photocurrent in mA cm^{-2}, and E_{hv} is the bandgap, in volts, over which an electron excitation takes place by direct absorption of quanta hv. The wavelength dependence of IPCE (λ) is determined via photon energy $hv \equiv hc/\lambda$, here c is the light speed and λ is the wavelength.

Another very important parameter characterizing the efficiency of light-to-electrical energy conversion is ECE, which is understood as a ratio between useful power output P_{out}, and total light power input P:

$$\text{ECE} = \frac{P_{out}}{P} \times 100\%. \tag{3.9}$$

In the case of solar cells [124] (Figure 3.18a), ECE is calculated according to the following equation:

$$\text{ECE} = \frac{I_{ph}^{mpp} V_{mpp}}{P} = \frac{I_{ph}^{sc} V_{ph}^{oc} \text{ FF}}{P} \times 100\%, \tag{3.10}$$

where I_{ph}^{mpp} and I_{ph}^{sc} stand for I_{ph} at the maximum power point and under short-circuit conditions ($R = 0$), respectively; V_{mpp} and I_{ph}^{sc} stand for cell voltage at the maximum power point and under open-circuit conditions ($R = \infty$), respectively; FF is the fill factor which is expressed as FF $= I_{ph}^{mpp} V_{ph}^{mpp}/V_{ph}^{oc} I_{ph}^{sc}$. The meaning of all of the above parameters is explained in Figure 3.18b, which also depicts the response of I_{ph} and V_{ph} as the resistance of the circuit (Figure 3.18b) changes from $R = \infty$ to $R = 0$.

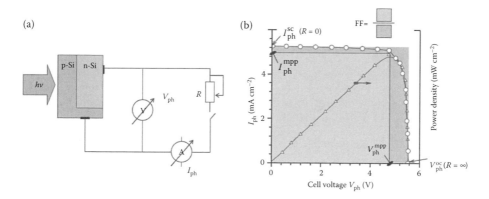

FIGURE 3.18 (a) Schematics of Si solar cell with electric scheme for cell voltage V_{ph} and photocurrent I_{ph} measurement. (Adapted from K. Juodkazis et al., *Opt. Express: Energy Express* 18, A147–A160, 2010.) (b) Typical photocurrent versus photovoltage response when electric circuit resistance R varies from $R = \infty$ (open circuit) to $R = 0$ (short circuit). Definition of the fill factor, FF, is schematically shown.

In the case of photoelectrochemical solar cells with external power source, following equation is used for calculating ECE [120,129,134]:

$$\text{ECE} = \frac{I_{ph}E^0_{O_2/H_2O} - I_{ph}\Delta E_{ext}}{P} \times 100\%, \tag{3.11}$$

where $E^0_{O_2/H_2O} = 1.23$ V. In the latter case, ECE depends on the values of photoelectrode potential, which can be changed with the help of external power source. However, in general case, when $\Delta E_{ext} = 0$, total thermodynamically useful efficiency of light-to-chemical energy conversion should be calculated according to the equation as follows:

$$\text{ECE} = \frac{I_{ph}E^0_{O_2/H_2O}}{P} \times 100\%. \tag{3.12}$$

One can see that according to Equation 3.12, the ECE depends on the value of I_{ph} solely.

The efficiency of electrolysis, that is, the electric energy conversion to chemical energy, can be evaluated according to the following equation:

$$\theta = \frac{E^0_{O_2/H_2O}}{E_a - E_c} \times 100\%, \tag{3.13}$$

where $E_a - E_c$ is the difference between the potentials of the anode and the cathode of the electrolysis cell, that is, the voltage of the cell. It is noteworthy that the set of Equations 3.10, 3.12, and 3.13 represent energy conversion efficiencies of light-to-electrical, light-to-chemical, and electric-to-chemical, respectively. By direct multiplication of Equations 3.10 and 3.13, one would arrive to Equation 3.12, since $I_{ph}^{mpp} = I_{ph}$ and $V_{mpp} = E_a - E_c$.

In the case of water splitting in PEC, $E_a - E_c$ would be equal to $\Delta E_{ph} + \Delta E_{ext}$ (Figure 3.17b). Equation analogous to Equation 3.13 was proposed in reference [135]. The volume of H_2 evolved during electrolysis can be evaluated as described below:

$$V_{H_2} = \frac{22.4It}{53.6} = 0.418It,$$

where V_{H_2} is the volume of hydrogen gas in dm^3, I is the electrolysis current in amperes, t is the time in hours, 22.4 is the volume of mole of gas in dm^3, 53.6 is the charge required to produce one mole of H_2 in A h mol^{-1}.

3.6.2 MECHANISM OF WATER PHOTOELECTROLYSIS

The open-circuit potential of Ti/TiO$_2$ electrode, E_{Ti/TiO_2}^{oc} measured with respect to reference hydrogen electrode, RHE ($E_{RHE} = -0.059$ pH), is about 0.6 ± 0.1 V irrespective of solution pH [136]. Under UV illumination, the potential of Ti/TiO$_2$ electrode in oxygen-free solution of 0.1 M KOH shifts negatively to ~0.2 V, which is insufficiently low for hydrogen evolution to take place on the electrode surface. Therefore, in order to attain the potential of H_2 evolution, that is, $E_{Pt} \leq 0$ V (SHE), minimum voltage of external power source, ΔE_{ext}, should be 0.2–0.25 V [121] and the positive pole of external power source should be connected to Ti/TiO$_2$ electrode (Figure 3.17).

Under UV illumination, oppositely charged zones form in the surface layer of semiconducting n-type titania. The formation of positively and negatively charged zones on TiO$_2$ surface could be envisaged as photochemical reaction:

$$2TiO_2 \xrightarrow{h\nu} TiO_2^+ + TiO_2^- \tag{3.15}$$

where TiO$^+$ and TiO$^-$ represent holes and photoelectrons captured in the TiO$_2$ phase, respectively. The intensity of UV illumination should determine the surface concentration of these particles, the rate of their formation, and recombination and also the value of $E_{Ti/TiO_2}^{oc,h\nu}$. Thus, the semiconducting nature of TiO$_2$ oxide leads to the formation of oppositely charged zones, on the surface of which different electrochemical reactions can take place, that is, the cathodic reaction on negatively charged zones and anodic reaction on positively charged ones. This type of patch formation on the uniformly illuminated surface of titania can be illustrated by Pt deposition from solution [137] where growth of Pt spheroidal nanoparticles forms micropatterns governed by Turing-type instabilities similar to vegetation in semiarid regions.

Positively charged nanoregions participate in the anodic oxidation of H_2O molecules or in the so-called forward reaction, whereas negatively charged ones participate in cathodic reduction of soluble H_2O oxidation products, for example, H_2O_2 and O_2 [138], that is, the backward reaction. Only the potential of negative zones can be measured experimentally (Figure 3.17a), as these zones have the electronic contact with Ti phase. Positively charged zones, which are situated mainly on the same surface of TiO$_2$ electrode, are, in fact, isolated one from another and are galvanically connected with the negative zones and Ti phase only through the solution phase. The potential of positive zones (holes) is more positive than that of negative ones (photo-electrons) by the value of ΔE_{ph}—the voltage of photocurrent source. This potential cannot be measured directly, but this E value can be attained by means of polarizing Ti/TiO$_2$ electrode anodically. Under open-circuit conditions, the value of $E_{Ti/TiO_2}^{oc,hv}$ is determined by the balance of all possible anodic and cathodic reactions on TiO$_2$ surface, similar to mixed potential in the case of metal corrosion processes [139].

It is known from the electrochemistry of titanium [140] that under standard conditions, TiO$_2$ layer forms on the surface of Ti in the anodic range of potentials at $E > 0$ V (SHE), leading to a complete passivation of the electrode.

When the thickness of TiO$_2$ layer is 20–25 nm, the transpassivation of Ti/TiO$_2$ electrode, that is, the oxidation of TiO$_2$ to TiO$_3$ and H$_2$O to O$_2$ begins at 1.8–2.0 V, whereas at a higher TiO$_2$ layer thickness these processes shift toward more positive E values, that is, up to ~3.0 V [141]. No oxygen evolution is observed on TiO$_2$ surface at $E_{O_2/H_2O}^0 = 123$ V. The O$_2$ evolution on TiO$_2$ surface begins at $E \approx 3.0$ V, that is, at E value higher than $E_{OH/H_2O}^0 = 2.8$ V [142], which is the standard potential of water oxidation to OH radicals:

$$H_2O - e^- \Leftrightarrow {}^{\bullet}OH + H^+. \tag{3.16}$$

The latter E value is also close to pE_F^{hv} and E_{VB} of UV light-excited TiO$_2$ (here, pE_F^{hv} and nE_F^{hv}—potential values corresponding to Fermi levels of holes and photoelectrons in UV-illuminated TiO$_2$ surface under open-circuit conditions; E_{CB} and E_{VB}—potential values corresponding to TiO$_2$ conductive and valence bands).

Oxygen evolution on the TiO$_2$ surface under UV illumination has been thoroughly studied [113,117,121,138,143,144]. There is no longer doubt that the process goes through the stages of OH radicals and also forms the peroxide groups O^{2-} and H$_2$O$_2$ [113,138]. Tachikawa ct al. [122] have indicated the following standard potential values of species which play an important role in O$_2$ evolution process: h_{VB}^+ at 2.96 V (pH = 2), h_{tr}^+ (trapped) at (1.6–1.7) V; ${}^{\bullet}OH_{free}$ at 2.72 V, and ${}^{\bullet}OH_{ads}$ over (1.5–1.6) V. In terms of energy, the state of trapped hole h_{tr}^+ and adsorbed OH radical ${}^{\bullet}OH_{ads}$ are close to the state of Ti(6+) oxide—TiO$_3$ [140]. By analogy with results presented in references [145–148], where OER on Ti/RuO$_2$ and Ni/NiO electrodes in the potential range of their transpassivation was studied, the mechanism of oxygen evolution on TiO$_2$ surface under UV illumination or in E range of its transpassivation could be represented by the sequence of the following main steps (reported for the first time in reference [112]):

$$TiO_2 \xrightarrow{hv,E_a} TiO_2^+ + e^- \quad (hv \approx 3.2\,eV) \tag{3.17}$$

$$TiO_2^+ + H_2O \rightarrow TiO_2({}^{\bullet}OH)_{ads} + H^+ \quad (E_{OH/H_2O}^0 = 2.8\,V) \tag{3.18}$$

$$TiO_2({}^{\bullet}OH)_{ads} \rightarrow TiO_3 + H^+ + e^- \quad (E_{O/H_2O}^0 = 2.4\,V) \tag{3.19}$$

$$TiO_3 \leftrightarrow TiOO_2 \quad (Ti^{4+} \text{ peroxide}) \tag{3.20}$$

$$TiOO_2 + H_2O \rightarrow TiO_2 + H_2O_2 \tag{3.21}$$

$$H_2O_2 \rightarrow O_2 + 2H^+ + 2e^- \quad (E^0_{O_2/H_2O_2} = 0.68\,V) \tag{3.22}$$

$$2H_2O_2 \rightarrow 2H_2O + O_2 \quad (\text{disproportionation}) \tag{3.23}$$

The sum of reaction equations 3.12 through 3.17 will give the overall reaction equation 3.10. Step equation 3.13, which, in fact, is oxidation of H_2O molecule to OH radical with $E^0_{OH/H_2O} = 2.8\,V$, determines the potential of OER onset on TiO_2 surface. OH radical, formed in step equation 3.13, is a particle with extremely high oxidizing energy and is capable of oxidizing TiO_2 to TiO_3 (Equation 3.14]. The latter oxide should be stable on TiO_2 surface at $E > 1.8\,V$ [140]. It is highly probable that, similar to electrocatalysis of OER by RuO_2 and $NiOOH$ [145,146,148], oxidation of oxide ion O^{2-} to peroxide ion O_2^{2-} takes place within TiO_3 molecule, which undergoes internal restructuring to Ti(4+) peroxide $TiOO_2$ (Equation 3.15). Hydrogen peroxide formed in chemical reaction equation 3.16 can either be further oxidized electrochemically on TiO_2 surface (Equation 3.17), or disproportionated to O_2 and H_2O in the solution (Equation 3.18). Several alternative mechanisms of O_2 evolution initiated also by OH radicals can be found in literature [138,143,149,150]. The overall equation of photoelectrochemical oxidation of H_2O molecules to O_2 on TiO_2 surface should be expressed by the following reaction, which is more adequate to reality:

$$2H_2O + 4h^+ \xrightarrow{E \geq 2.8\,V,\ h\nu=3.2\,eV} O_2 + 4H^+ \tag{3.24}$$

Since the energy of visible light is too low to induce the formation of OH radicals, all attempts to shift the absorption of TiO_2 into visible region, so that water splitting would be possible under visible light illumination, were not and cannot be successful. Therefore, oxygen evolution on TiO_2 surface is possible either at very high overvoltage with respect to $E^0_{O_2/H_2O} = 1.23\,V$ that is, transpassivation region of Ti/TiO_2 electrode, or under UV illumination ($h\nu \approx 3.2$ eV) only. Due to its chemical and electrochemical stability, TiO_2 does not posses any catalytic properties toward OER. Unmodified titanium anodes are not used for OER due to their passivation. Since in the case of water photoelectrolysis anodic and cathodic processes are interrelated, the low rate of OER determines the low rate of hydrogen evolution reaction.

The way how to overcome this limitation in practical applications of solar hydrogen generation is to use water electrolysis under the most efficient solar cell power supply [112]. At the current development of solar cells, it is possible to improve solar hydrogen generation by at least 17 times when compared to the attempts of direct solar-to-chemical energy conversion on a TiO_2 electrode [112]. It is demonstrated that by separating and optimizing processes of light-to-electric and electric-to-chemical energy conversion, the overall ECE could be increased from ~1% up to ~17% by using solar cells with maximum, that is, ~30%, light-to-electric energy conversion efficiency as well as OER catalyzing anodes for water electrolysis. Then, a solar panel of 20 m^2 area would be capable of producing ~5 m^3 of H$_2$ and ~2.5 m^3 of O$_2$ during 24 h, that is equivalent to accumulation of 12 kA h of electric charge or to ~15 kW h, 54 MJ, 13 Mcal of chemical energy or to ~2.0 L of gasoline. This would consume ~4 L (or ~220 mol) of water.

3.6.3 How Plasmonic Effects Could Help?

Detailed analysis of plasmonic effects [48,80,108] in light-harvesting applications [97,151–154] are actively studied. For the efficient hydrogen and oxygen evolution reactions (HER and OER, respectively) in different water photolysis cells, the overvoltages are used to make reactions more facile. The overvoltage on semiconducting electrode causes the bending of presurface electronic bands, for example, in illuminated regions of TiO$_2$. If the surface of TiO$_2$ anode becomes more positive, the OER rate is enhanced. This corresponds to the pseudo-Fermi level of holes, pE_F^{hv} becoming more positive in respect to the ox-red potentials for the corresponding processes. As a result, the yield of OER becomes large. In terms of quantum and conversion efficiencies, they both scale as $\propto I_{ph}/P$. Since the photocurrent I_{ph} is expected to be proportional to illumination, P, there would be no change in the efficiency. However, if there is a possibility to saturate some of loss channels, for example, surface and presurface recombination can be significantly slowed down by separation of photoinduced electrons and holes, an increase in P is expected to increase QE, ECE. Plasmonic nanoparticles can perform exactly this role of the localized band bending and charge separation. Plasmonic particles redistribute light energy on the surface and create locations where light intensity is enhanced at the spectral region defined via the Frolich mechanism [61] (spectral dependence of the dielectric function of two media created resonance scattering) or the size- and shape-dependent "lightning rod" effect. This causes spatially localized overpotential and charge separation, and hence facilitates the OER. It is noteworthy that by depositing size and shape defined plasmonic nanoparticles as we demonstrated earlier [57,66,90,106,155–157] it is possible to photosensitize the light absorbing substrate and create I_{ph} at the different wavelengths within the absorption band of the substrate (as discussed in Section 3.5 in more details). It is important that photosensitization of titania (or other photoelectrode) at the weak absorption region would not jeopardize the performance of photoanode at the strong band-to-band region where IPCE is the largest. By tailoring the shape and size of plasmonic nanoparticles, it is possible to create strong enhancement of the vertical (to the surface) light field component or to maximize it at the particle–electrode interface for electron transport at the required spectral range [64] (see Section 3.5 for details).

Generic energy storage and conversion formulas suitable for different solar energy-harvesting applications are presented and justified by analysis of the water-splitting mechanism [112]. The used basic principles can be applied to other solar energy conversion systems. Plasmonic effects can be evoked to better control efficiency of energy conversion and this is the most active field of research in solar energy harvesting. The analysis presented in this section demonstrates that it is necessary to search for the most efficient process when several steps of energy conversion are exercised: light-to-electrical and electrical-to-chemical steps can be currently 18 times more efficient when compared to direct light to chemical (example discussed is H$_2$) [112].

3.7   THERMAL EFFECTS IN PLASMONICS: SERS

Thermal effects in plasmonic structures is an actively researched field relevant for nanotweezers, SERS, polymerization, and energy harvesting. The intensity ratio of the anti-Stokes or Stokes modes of molecular vibrations provides access to evaluate temperature at nanoscale. Correlation between the field enhancement by nanoparticles, extinction spectra of nanopatterns, and Raman scattering is discussed next.

3.7.1   SERS USING NANOPARTICLES

The enhancement of light electric field via plasmonic and shape effects of nanoparticles is used for laser nanostructuring and sensing applications. However, engineering, design, and measurement of the E-field enhancement is a challenging task. In SERS applications, detection of low-concentration analyte molecules becomes possible due to the light field enhancement; we have recently demonstrated a 10–12 M sensitivity of pyridine Raman detection on 3D textured surfaces by gold nanoparticles [63]. Langmuir analysis showed a very low surface coverage of the textured surface. The molecules contributing to the recorded signal are occupying the sites of field enhancement, the hot spots, and it is important to assess the thermal effect at the nanovolumes since the analyte molecules can be chemically altered due to temperature rise in the strong electric field. The standard analysis based on the measurement of the ratio of the anti-Stokes to Stokes intensities of Raman scattering intensities could be implemented to estimate the thermal effects [158]. However, the complex extinction spectrum of the nanoparticles should be taken into account to avoid erroneous estimates as shown below.

The polarization effects of nanomaterials in the electric field which has gradients comparable with the molecular size can be accessed via plasmonic light localization effect. The light intensity gradients can be used to effectively polarize and collect molecules in solution with spatial resolution not accessible by other means. Recently, up to 102-fold enhancement in mass transport triggered via Marangoni effect in conventional laser tweezers that created the intensity gradients on subwavelength scale has been demonstrated [159]. Plasmonic laser tweezers [160] can potentially further increase the spatial resolution and sensitivity of molecular agglomeration reaching single-molecular resolution.

The Stokes and much weaker anti-Stokes modes of molecular vibrations on the substrates templated by gold nanoparticles were measured and influence of the extinction spectrum on the measured SERS signals was demonstrated [161]. Simultaneous measurements of both modes are important for thermal analysis of the hot spots with high field enhancements.

3.7.2   SERS SAMPLES

Fabrication of large ~30 × 30 μm² areas of gold square nanoparticles by electron beam lithography, gold sputtering, and lift-off has been described elsewhere [57,156].

Such patterns have narrow size distribution which allowed a high fidelity distinction of the extinction spectra of nanoparticles whose length differed just by 1.2 nm

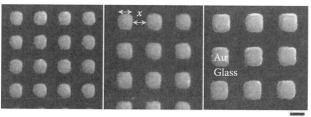

100 nm

FIGURE 3.19 SEM images of gold-on-cover-glass samples used in SERS experiments. The side-length of the square was equal to separation. The side length and separation of the patterns are: $x = 97$, 125, 141 nm, respectively. Polarization of the excitation light in SERS measurements was horizontal. (Adapted from Y. Yokota et al., Influence of plasmonic enhancement on surface-enhanced Raman scattering, in Extended Abstracts of XXIV Int. Conf. on Photochemistry ICP2009, Toledo, Spain, pp. PSII–P82, July 2009; Y. Yokota, Sensing applications of plasmonic nanoparticles, PhD thesis, Hokkaido University, Japan, October 2009.)

(this corresponded to a ~0.3% change in length) [90]. Figure 3.19 shows SEM images of typical patterns of gold nanoparticles [161,162]. The size distribution of the fabricated patterns was statistically analyzed using SEM imaging and it was determined that the central cross-sections of the particles had a ±0.7 nm uncertainty for the particle sizes ranging from 75 to 190 nm (even for the smallest nanosquares the standard deviation was approximately 1%). This is a narrower distribution than the recently reported monodispersity (with standard deviation of 1.6 nm) of chemically synthesized gold nanoparticles [87].

The Raman measurements were carried out under back-scattering conditions in an Olympus BX51 microscope equipped with a water-immersion objective lens of 100× magnification and numerical aperture of NA = 1.0. Excitation source was a solid-state cw-laser at 785 nm. Typical irradiance on the sample was 207 W/cm² (for 2.5 mW laser power) with an acquisition time of 5 s when 10^{-4} M solutions of crystal violet were used. The laser line was blocked by a notch filter with the bandwidth of 9 nm [161,162].

3.7.3 EXTINCTION SPECTRA AND SERS INTENSITY

The extinction spectra of patterns of gold nanoparticles in water are shown in Figure 3.20 for the different particle size. These patterns were used for the SERS experiments. The position of extinction maximum can be set at the required wavelength used for excitation of Raman scattering by choosing the size of nanoparticles. Since the SERS signal depends on the local field enhancement factors at the laser excitation and Raman scattering frequencies, it is important to find the method to maximize overall SERS signals. The power of the Stokes modes in Raman scattering is given as [80]

$$P_s(v_s) = N\sigma_R I(v_1), \tag{3.25}$$

where N is the number of scatters, σ_R is the scattering cross-section ($10^{-31} \leq \sigma_R \leq 10^{-29}$ cm²/molecule), and $I(v_1)$ is the intensity of excitation on the laser frequency. In the

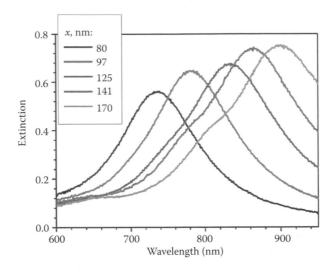

FIGURE 3.20 (**See color insert.**) Extinction spectra of the gold patterned substrates in water. The side length of the squares, x, is shown in the plot. (Adapted from Y. Yokota et al., Influence of plasmonic enhancement on surface-enhanced Raman scattering, in Extended Abstracts of XXIV Int. Conf. on Photochemistry ICP2009, Toledo, Spain, pp. PSII–P82, July 2009; Y. Yokota, Sensing applications of plasmonic nanoparticles, PhD thesis, Hokkaido University, Japan, October 2009; Y. Yokota, K. Ueno, and H. Misawa, *Small* 7, 252–258, 2011.)

case of SERS, the cross-section of scattering, σ_{SERS}, becomes larger due to chemical and electronic contributions of the enhancement which can reach $\sim 10^2$. In addition, the nanoparticles enhance the local electric light field, which is accounted for as a electromagnetic enhancement factor $L(v) = |\mathbf{E}_{loc}(v)|/|\mathbf{E}_0|$, here $\mathbf{E}_0$ is the incident E-field of light. Altogether, the power scattered in the case of SERS is given by [80]

$$P_s(v_s) = N\sigma_{SERS}L(v_1)^2 L(v_s)^2 I(v_1),\qquad(3.26)$$

where the field enhancements at the laser and Raman scattering wavelengths are considered. Since usually the laser and Raman wavelengths are similar, the total enhancement factor $R \equiv L(v_1)^2 L(v_s)^2 \approx |\mathbf{E}_{loc}|^4/|\mathbf{E}_0|^4$ is assumed.

The SERS measurements of aqueous solution of crystal violet was carried out using the substrates templated by gold nanoparticles of different sizes, while the total volume of the gold structures was kept constant [161,162]. Figure 3.21 shows the results for the Stokes and anti-Stokes branches. The observed vibrations can be identified as the ring skeletal vibration of radical orientation at 913 cm^{-1}, in-plane vibrations of ring C–H at 1173 cm^{-1}, and N-phenyl stretching at 1383 cm^{-1} of crystal violet [164].

The anti-Stokes modes were recognizable only on the smallest particles, while the Stokes modes were more intense on the larger particles. The quantitative analysis of the extinction values at the laser and Raman scattering wavelengths along with the SERS intensities of the Stokes and anti-Stokes modes of the 1173 cm^{-1} line is presented in Figure 3.22. The measured extinction spectra of the patterns of nanoparticles in water (Figure 3.20) can be used to normalize the experimental SERS spectra.

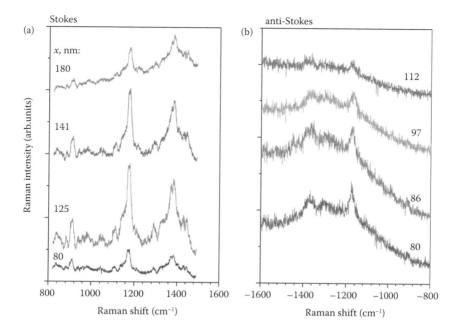

FIGURE 3.21 Spectra of Raman Stokes (a) and anti-Stokes (b) components of the crystal violet on the patterns of Au nanoparticles of different size, x (see, Figure 3.19) plotted with offset for clarity. (Adapted from Y. Yokota et al., Influence of plasmonic enhancement on surface-enhanced Raman scattering, in Extended Abstracts of XXIV Int. Conf. on Photochemistry ICP2009, Toledo, Spain, pp. PSII–P82, July 2009; Y. Yokota, Sensing applications of plasmonic nanoparticles, PhD thesis, Hokkaido University, Japan, October 2009.)

The extinction represents the light intensity losses due to absorption and scattering which occur at the laser excitation, ext_l, as well as at the Raman scattering wavelengths, $ext_{s,as}$, respectively. It is noteworthy that at the used particle size, the scattering is dominating extinction. Figure 3.23 shows the absolute experimental SERS intensity normalized by the extinction at the laser and Raman lines. The extinctions at the excitation and SERS wavelengths were added to account for the cumulative scattering. The Stokes SERS signal becomes almost nanoparticle size-independent, while the anti-Stokes had a tendency to increase as the particle size decreased. The proposed normalization made interpretation of the complex experimental dependencies (Figure 3.22) considerably simpler as discussed in Section 3.7.4. Such normalization using directly measured data is chosen for the following temperature analysis rather than indirect estimations based on FDTD calculations [163]. It is known that the enhancement values obtained by FDTD are dependent on the mesh size and can lead to the overestimated enhancement values [165].

3.7.4 ESTIMATION OF TEMPERATURE AT HOT SPOTS

The experimental SERS spectra of molecules dispersed on well-ordered patterns of gold nanoparticles reflect the local field enhancement which is affected by the interaction of nanoparticles, their size, shape, and volume.

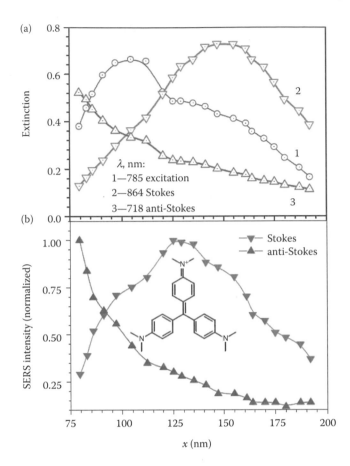

FIGURE 3.22 Extinction and SERS intensity of crystal violet dye on gold nanoparticles. (a) The extinction values of the Au-nanoparticles' patterns at the laser excitation (ext_l), Stokes (ext_s), and anti-Stokes (ext_{as}) wavelengths. (b) The normalized SERS intensity of Stokes and anti-Stokes modes of the 1173 cm^{-1} in-plane vibrations of ring C–H versus the size of Au nanoparticle; the maxima of Stokes and anti-Stokes were at 550 and 43 counts, respectively. Inset in (b) shows the molecular structure of the crystal violet. (Adapted from Y. Yokota et al., Influence of plasmonic enhancement on surface-enhanced Raman scattering, in Extended Abstracts of XXIV Int. Conf. on Photochemistry ICP2009, Toledo, Spain, pp. PSII–P82, July 2009; Y. Yokota, Sensing applications of plasmonic nanoparticles, PhD thesis, Hokkaido University, Japan, October 2009.)

The light scattering and absorption of nanoparticles with dimensions much smaller than the wavelength of light are well described by a dipole model with the corresponding cross-sections [45] (Equations 3.1 through 3.3):

$$\sigma_{sca} = \frac{1}{6\pi}\left(\frac{2\pi}{\lambda}\right)^4 |\alpha|^2, \quad \sigma_{abs}(\lambda) = \frac{2\pi}{\lambda}\,\text{Im}[\alpha], \qquad (3.27)$$

where $\alpha = 3V[(\varepsilon_p/\varepsilon_m - 1)/(\varepsilon_p/\varepsilon_m + 2)]$ is the polarizability, V is the volume, ε_p is the metal permittivity, and ε_m is the environment permittivity.

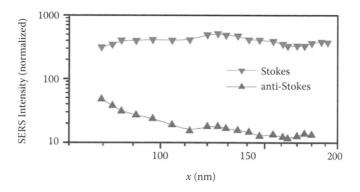

FIGURE 3.23   The normalized SERS intensity for the Stokes $I_s/(\text{ext}_1 + \text{ext}_s)$ and anti-Stokes $I_{as}/(\text{ext}_1 + \text{ext}_s)$ modes at 1173 cm^{-1} (0.145 eV). The extinction values for normalization were taken from Figure 3.22a. Note the log–log presentation.

It is noteworthy that Equation 3.27 accounts for the $1/\lambda^4$ dependence of the scattering. This is important when Raman analysis of the ratio of anti-Stokes and Stokes modes is carried out; the extinction measurements of the fabricated patterns are defined by the cross-sections (Equation 3.27) which already accounts for the $1/\lambda^4$ dependence.

Figure 3.20 shows that the maximum of extinction was increasing with particle size as expected for the polarizability $\alpha \propto V$; the integrated extinction was also following the same trend. The light scattering of the pattern of nanoparticles is reflected in the extinction spectrum (Figure 3.22). The normalized anti-Stokes SERS data (Figure 3.23) show that SERS signal depends on the particle size (equal to separation). Let us estimate the temperature which characterizes the measured Raman modes. Formally, the temperature can be determined from the analysis of intensities of the Raman Stokes and anti-Stokes modes [158]:

$$F(v) = \frac{I_s(v)}{I_{as}(v)} = \frac{n+1}{n} = \exp\left(\frac{hv}{kT}\right), \tag{3.28}$$

where $I_{as,s}$ are the anti-Stokes and Stokes intensities, respectively, v is the absolute value of the frequency difference for the excitation and scattered photons, $n = 1/(\exp(hv/kT) - 1)$ the Bose factor, k is the Boltzmann constant, h is the Plank constant, and T is the absolute temperature.

From the normalized SERS data (Figure 3.23) in the case of the smallest particle size, both weak anti-Stokes and Stokes Raman signals were detected (see Figure 3.21). The temperature defined as the ratio of experimentally measured anti-Stokes and Stokes components is plotted in Figure 3.24. The ratio $F(v)$ was changing from ~7 to 33 over the entire span of nanoparticle sizes. From Equation 3.28, the corresponding temperatures can be found by changing from 600 ± 250°C to 200 ± 70°C, respectively. The high effective temperature defines the population of a particular Raman active vibration mode and can be out of a thermal equilibrium with ambient [166]; note that there is a low number of molecules in the volume of the hot spots which have cross-section of few nanometers. Moreover, emission of the anti-Stokes

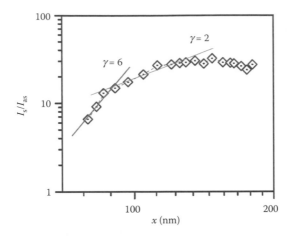

FIGURE 3.24 The SERS intensity ratio of the Stokes and anti-Stokes I_s/I_{as} modes at 1173 cm^{-1} versus nanoparticle size. Note the log–log presentation.

photons cools the molecules (an energy surplus per photon is taken from the environment). The precision of the baseline definition in determination of the amplitudes of Raman signals is another contributing factor to the larger F values. For example, the baseline was found to be strongly dependent on the spectral profile of extinction as can be recognized in the case of anti-Stokes modes on the smallest particles (see Figure 3.21). Future studies focused on the thermal properties in hot spots and molecular dynamics are highly required.

The slope of $\gamma = 2$ in the dependence shown in Figure 3.24 can be interpreted as caused by dipole radiation energy transfer between vibrating molecules. For the larger nanosquares, whose size becomes comparable with the wavelength ($x \gg \lambda$), the radiation is dominant in energy transfer. At the smallest distances comparable to the cross-sections of hot spots of few nanometers in size, the Forster mechanism of energy transfer between molecules is dominating and is characterized with the slope of $\gamma = 6$ (recognizable in the case of smallest nanoparticles in Figure 3.24). On the smallest nanoparticles, the shape was less well defined and the hot spots can be more localized with stronger localization of the light field enhancement. Molecules at the hot spots coupled via the Forster mechanism can explain the slope $1 = 6$. The nano-tweezing effect with strongest light intensity gradient on the smallest nanoparticles can contribute to the Forster mechanism as well. The measured ratio of the SERS signals corroborates such scenario in energy transfer between the molecules distributed over the patterns of gold nanoparticles. The plateau region in the case of the largest particles is most probably determined by the roughness of the gold surface rather than by energy transfer between the particles.

The proposed normalization procedure of SERS intensity by the extinction at the wavelength of excitation and scattering decouples the pure molecular-related contribution in SERS signal from the contribution of the substrate. Such normalization is important for investigation of single molecular properties of Raman vibrations in small volumes of hot spots and for estimation of nonequilibrium population and local temperature. Properties of energy transport within hot spots and well as possibility to facilitate chemical reaction when particular molecular vibrational modes are out of equilibrium is a novel frontier.

3.8 CONCLUSIONS

In this chapter, we discussed some recent results as well as general trends in applications where plasmonic effects can be beneficial in practical terms or for revealing nano–microscale phenomenon. Example of the former would be solar energy harvesting by plasmonics solar cells, while the latter is temperature evaluation at the hot spots and photocatalysis. Those fields are very active, and presenting comprehensive review is neither possible nor practical. Hence, we discussed here basic principles which can reveal underpinning mechanisms and can be possibly used in other fields of science and engineering.

ACKNOWLEDGMENTS

We are grateful to Kestutis Juodkazis and Jurga Juodkazyte for discussions and guidance into plasmonic-related effects in the fields of electrochemistry, hydrogen evolution, and photocatalysis which are presented in Section 3.5.3. Our coauthors Ewa Kowalska and Chung-How Poh made major experimental contributions for the results in Sections 3.5.3 and 3.4.2. PhD project of Yukie Yokota made possible Section 3.7.4. We are grateful for Australian Research Council Discovery grant DP120102980, Linkage grant LP120100161 with Raith-Asia Ltd., and the Joint Development Program for new generation of 3D nanolithography. S.J. is thankful to Swinburne University for his start-up clean room nanolab and dedicated scholarships.

REFERENCES

1. K. R. Catchpole and A. Polman, Plasmonic solar cells, *Opt. Express* 16, 21793–21800, 2008.
2. H. A. Atwater and A. Polman, Plasmonics for improved photovoltaic devices, *Nat. Mater.* 9, 205–213, 2010.
3. V. E. Ferry, J. N. Munday, and H. A. Atwater, Design considerations for plasmonic photovoltaics, *Adv. Mater.* 22, 4794–4808, 2010.
4. N. S. Lewis, Powering the planet (2007 MRS spring meeting plenary address), *MRS Bull.* 32, 808–820, 2007.
5. M. A. Green, K. Emery, Y. Hishikawa, and W. Warta, Solar cell efficiency tables (version 37), *Prog. Photovoltaics* 19, 84–92, 2011.
6. U. S. Geological Survey, 2013, Mineral Commodity Summaries 2013, U. S. Geological Survey, p. 198, ISBN: 978–1–4113–3548–6.
7. M. A. Green, *Solar Cells: Operating Principles, Technology and System Applications*, The University of New South Wales, Sydney, Australia, 1999.
8. E. Yablonovitch and G. Cody, Intensity enhancement in textured optical sheets for solar cells, *IEEE Trans. Electron Devices* 29, 300–305, 1982.
9. E. Yablonovitch, Statistical ray optics, *J. Opt. Soc. Am.* 72(7), 899–907, 1982.
10. J. Müller, B. Rech, J. Springer, and M. Vanecek, TCO and light trapping in silicon thin film solar cells, *Sol. Energy* 77, 917–930, 2004.
11. J. Meier, S. Dubail, S. Golay, U. Kroll, S. Faÿ, E. Vallat-Sauvain, L. Feitknecht, J. Dubail, and A. Shah, Microcrystalline silicon and the impact on micromorph tandem solar cells, *Sol. Energy Mater. Sol. Cells* 74, 457–467, 2002.

12. S. Wiesendanger, S. Fahr, T. Kirchartz, C. Rockstuhl, and F. Lederer, Achieving the Yablonovitch limit in thin-film solar cells with tailored randomly textured interfaces, in *Renewable Energy and the Environment*, OSA Technical Digest (CD) (Optical Society of America, 2011), paper PThB4, 2011.

13. S. E. Han and G. Chen, Toward the Lambertian limit of light trapping in thin nanostructured silicon solar cells, *Nano Lett.* 10(11), 4692–4696, 2010.

14. V. E. Ferry, M. A. Verschuuren, H. B. Li, E. Verhagen, R. J. Walters, R. E. Schropp, H. A. Atwater, and A. Polman, Light trapping in ultrathin plasmonic solar cells, *Opt. Express* 18, A237–A245, 2010.

15. Z. Yu, A. Raman, and S. Fan, Nanophotonic light-trapping theory for solar cells, *Appl. Phys. A* 105(2), 329–339, 2011.

16. C. R. Otey, W. T. Lau, and S. Fan, Thermal rectification through vacuum, *Phys. Rev. Lett.* 104, 154301, 2010.

17. D. K. Gramotnev and S. I. Bozhevolnyi, Plasmonics beyond the diffraction limit, *Nature Photon.* 4, 83–91, 2010.

18. K. R. Catchpole and A. Polman, Design principles for particle plasmon enhanced solar cells, *Appl. Phys. Lett.* 93, 191113, 2008.

19. S. A. Maier, *Plasmonics: Fundamentals and Applications*, Springer, Berlin, Germany, 2007.

20. M. A. Green and M. J. Keevers, Optical properties of intrinsic silicon at 300 K, *Prog. Photovoltaics* 3, 189–192, 1995.

21. S. M. Sze and K. K. Ng, *Physics of Semiconductor Devices*, 3rd ed., John Wiley and Sons, New York, 2007.

22. H. R. Stuart and D. G. Hall, Island size effects in nanoparticle-enhanced photodetectors, *Appl. Phys. Lett.* 73, 3815–3817, 1998.

23. P. N. Saeta, V. E. Ferry, D. Pacifici, J. N. Munday, and H. A. Atwater, How much can guided modes enhance absorption in thin solar cells?, *Opt. Express* 17, 20975–20990, 2009.

24. V. E. Ferry, L. A. Sweatlock, D. Pacifici, and H. A. Atwater, Plasmonic nanostructure design for efficient light coupling into solar cells, *Nano Lett.* 8(12), 4391–4397, 2008.

25. R. A. Pala, J. White, E. Barnard, J. Liu, and M. L. Brongersma, Design of plasmonic thin-film solar cells with broadband absorption enhancements, *Adv. Mater.* 21, 3504–3509, 2009.

26. S. Pillai, K. Catchpole, T. Trupke, and M. Green, Surface plasmon enhanced silicon solar cells, *J. Appl. Phys.* 101, 093105-1–093105-8, 2007.

27. K. Nakayama, K. Tanabe, and H. A. Atwater, Plasmonic nanoparticle enhanced light absorption in GaAs solar cells, *Appl. Phys. Lett.* 93, 121904-1–121904-3, 2008.

28. L. Rosa, K. Sun, and S. Juodkazis, Sierpinski fractal plasmonic nanoantennas, *Phys. Status Solidi Rapid Res. Lett.* 5, 175–177, 2011.

29. E. G. Gamaly, Optical phenomenon on the interface between a conventional dielectric and a uniaxial medium with mixed metal–dielectric properties, *Phys. Rev. E* 51, 3556–3560, 1995.

30. P. Spinelli, M. Hebbink, R. de Waele, L. Black, F. Lenzmann, and A. Polman, Optical impedance matching using coupled plasmonic nanoparticle arrays, *Nano Lett.* 11(4), 1760–1765, 2011.

31. D. Schaadt, B. Feng, and E. Yu, Enhanced semiconductor optical absorption via surface plasmon excitation in metal nanoparticles, *Appl. Phys. Lett.* 86, 063106-1–063106-3, 2005.

32. D. Derkacs, S. Lim, P. Matheu, W. Mar, and E. Yu, Improved performance of amorphous silicon solar cells via scattering from surface plasmon polaritons in nearby metallic nanoparticles, *Appl. Phys. Lett.* 89, 093103-1–093103-3, 2006.

33. X. Chen, B. Jia, J. K. Saha, B. Cai, N. Stokes, Q. Qiao, Y. Wang, Z. Shi, and M. Gu, Broadband enhancement in thin-film amorphous silicon solar cells enabled by nucleated silver nanoparticles, *Nano Lett.* 12, 2187–2192, 2012.

34. B. P. Rand, P. Peumans, and S. R. Forrest, Long-range absorption enhancement in organic tandem thin-film solar cells containing silver nanoclusters, *J. Appl. Phys.* 96, 7519–7526, 2004.

35. S. Kim, S. Na, J. Jo, D. Kim, and Y. Nah, Plasmon enhanced performance of organic solar cells using electrodeposited Ag nanoparticles, *Appl. Phys. Lett.* 93, 073307, 2008.

36. C. Hägglund, M. Zäch, G. Petersson, and B. Kasemo, Electromagnetic coupling of light into a silicon solar cell by nanodisk plasmons, *Appl. Phys. Lett.* 92, 053110-1–053110-3, 2008.

37. I. M. Pryce, D. D. Koleske, A. J. Fischer, and H. A. Atwater, Plasmonic nanoparticle enhanced photocurrent in GaN/InGaN/GaN quantum well solar cells, *Appl. Phys. Lett.* 96, 153501-1–153501-3, 2010.

38. H. Masuda and K. Fukuda, Ordered metal nanohole arrays made by a two-step replication of honeycomb structures of anodic alumina, *Science* 268, 1466–1468, 1995.

39. H. Masuda and M. Satoh, Fabrication of gold nanodot array using anodic porous alumina as an evaporation mask, *Jpn. J. Appl. Phys.* 35, L126–L129, 1996.

40. E. D. Palik, *Handbook of Optical Constants of Solids*, 1st ed., Academic Press, San Diego, CA, 1998.

41. G. H. Chan, J. Zhao, E. M. Hicks, G. C. Schatz, and R. P. Van Duyne, Plasmonic properties of copper nanoparticles fabricated by nanosphere lithography, *Nano Lett.* 7(7), 1947–1952, 2007.

42. C. Hägglund and S. P. Apell, Resource efficient plasmon-based 2D-photovoltaics with reflective support, *Opt. Express* 18(S3), A343–A356, 2010.

43. A. Taflove and S. C. Hagness, *Computational Electrodynamics: The Finite-Difference Time-Domain Method*, 3rd ed., Artech House, 2005.

44. W. Yu and R. Mittra, A conformal finite difference time domain technique for modeling curved dielectric surfaces, *IEEE Microw. Wirel. Compon. Lett.* 11, 25–27, 2001.

45. C. F. Bohren and D. R. Huffman, *Absorption and Scattering of Light by Small Particles*, Wiley-VCH, Weinheim, 2004.

46. J. Leng, J. Opsal, H. Chu, M. Senko, and D. Aspnes, Analytic representations of the dielectric functions of materials for device and structural modeling, *Thin Solid Films* 313–314, 132–136, 1998.

47. A. Deinega and S. John, Effective optical response of silicon to sunlight in the finite-difference time-domain method, *Opt. Lett.* 37, 112–114, 2012.

48. S. Pillipai, K. R. Catchpole, T. Trupke, and M. A. Green, Surface plasmon enhanced silicon solar cells, *J. Appl. Phys.* 101(9), p. 093105, 2007.

49. J. C. Ho, R. Yerushalmi, Z. A. Jacobson, Z. Fan, R. L. Alley, and A. Javey, Controlled nanoscale doping of semiconductors via molecular monolayers, *Nature Mater.* 7, 62–67, 2009.

50. B. O'Regan and M. Gräatzel, A low-cost, high-efficiency solar cell based on dye-sensitized colloidal TiO$_2$ films, *Nature* 353, 737–740, 1991.

51. Y. Bai, Y. M. Cao, J. Zhang, M. Wang, R. Li, P. Wang, S. M. Zakeeruddin, and M. Grätzel, High-performance dye-sensitized solar cells based on solvent-free electrolytes produced from eutectic melts, *Nature Mater.* 7, 626–630, 2008.

52. M. K. Nazeeruddin, A. Kay, I. Rodicio, R. Humphrybaker, E. Muller, P. Liska, N. Vlachopoulos, and M. Grätzel, Conversion of light to electricity by *cis*-X$_2$bis(2,2′-bipyridyl-4,4′-dicarboxylate)ruthenium(II) charge-transfer sensitizers (x = Cl⁻, Br⁻, I⁻, CN⁻, and SCN⁻) on nanocrystalline titanium dioxide electrodes, *J. Am. Chem. Soc.* 115, 6382–6390, 1993.

53. J. Nowotny, T. Bak, M. Nowotny, and L. Sheppard, Titanium dioxide for solar-hydrogen. I. Functional properties, *Int. J. Hydr. Ener.* 32, 2609– 2629, 2007.
54. Y. Nishijima, K. Ueno, Y. Yokota, K. Murakoshi, and H. Misawa, Plasmon-assisted photocurrent generation from visible to near-IR wavelength using a Au-nanorods-TiO₂ electrode, *J. Phys. Chem. Lett.* 10, 2031–2036, 2010.
55. A. Fujishima and K. Honda, Electrochemical photolysis of water at a semiconductor electrode, *Nature* 238, 37–38, 1972.
56. M. Grätzel, Photoelectrochemical cells, *Nature* 414, 338–344, 2001.
57. K. Ueno, V. Mizeikis, S. Juodkazis, K. Sasaki, and H. Misawa, Optical properties of nano-engineered gold blocks, *Opt. Lett.* 30, 2158–2160, 2005.
58. T. Tatsuma and K. Suzuki, Photoelectrochromic cell with a Ag-TiO₂ nanocomposite: Concepts of drawing and display modes, *Electrochem. Commun.* 9, 574–576, 2007.
59. Y. Tian and T. Tatsuma, Plasmon-induced photoelectrochemistry at metal nanoparticles supported on nanoporous TiO₂, *Chem. Commun.* 2004, p. 1810, 2004.
60. E. W. McFarland and J. Tang, A photovoltaic device structure based on internal electron emission, *Nature* 421, 616–618, 2003.
61. P. B. Johnson and R. W. Christy, Optical constants of the noble metals, *Phys. Rev. B* 6, 4370–4379, 1972.
62. Y. Nishijima, L. Rosa, and S. Juodkazis, Surface plasmon resonances in periodic and random patterns of gold nano-disks for broadband light harvesting, *Opt. Express* 20(10), 11466–11477, 2012.
63. Y. Yokota, K. Ueno, S. Juodkazis, V. Mizeikis, N. Murazawa, H. Misawa, H. Kasa, K. Kintaka, and J. Nishii, Nano-textured metallic surfaces for optical sensing and detection applications, *J. Photochem. Photobiol. A* 207, 126–134, 2009.
64. V. Mizeikis, E. Kowalska, and S. Juodkazis, Resonant localization, enhancement, and polarization of optical fields in nano-scale interface regions for photocatalytic applications, *J. Nanosci. Nanotechnol.* 11, 2814–2822, 2011.
65. S. Juodkazis and L. Rosa, Surface defect mediated electron hopping between nanoparticles separated by a nano-gap, *Physica Status Solidi* 4(10), 244–246, 2010.
66. K. Ueno, S. Juodkazis, V. Mizeikis, K. Sasaki, and H. Misawa, Clusters of closely spaced gold nanoparticles as a source of two-photon photoluminescence at visible wavelengths, *Adv. Mater.* 20(1), 26–29, 2008.
67. C. Poh, L. Rosa, S. Juodkazis, and P. Dastoor, FDTD modeling to enhance the performance of an organic solar cell embedded with gold nanoparticles, *Opt. Mater. Express* 1, 1326–1331, 2011.
68. J. Z. Ou, R. A. Rani, M.-H. Ham, M. R. Field, Y. Zhang, H. Zheng, P. Reece et al., Elevated temperature anodized Nb₂O₅: A photoanode material with exceptionally large photoconversion efficiencies, *ACS Nano* 6(5), 4045–4053, 2012.
69. A. Hagfeldt and M. Grätzel, Molecular photovoltaics, *Acc. Chem. Res.* 33(5), 269–277, 2000.
70. H. C. Weerasinghe, P. M. Sirimanne, G. P. Simon, and Y. Cheng, Cold isostatic pressing technique for producing highly efficient flexible dye-sensitised solar cells on plastic substrates, *Prog. Photovoltaics Res. Appl.* 20(3), 321–332, 2012.
71. F. Ouhib, G. Dupuis, R. de Bettignies, S. Bailly, A. Khoukh, H. Martinez, J. Desbrières, R. C. Hiorns, and C. Dagron-Lartigau, Effect of molar mass and regioregularity on the photovoltaic properties of a reduced bandgap phenyl-substituted polythiophene, *J. Polym. Sci. Part A: Polym. Chem.* 50(10), 1953–1966, 2012.
72. H. Shim, H. Kim, J. Kim, S. Kim, W. Jeong, T. Kim, and J. Kim, Enhancement of near-infrared absorption with high fill factor in lead phthalocyanine-based organic solar cells, *J. Mater. Chem.* 22(18), 9077–9081, 2012.
73. S. D. Yambem, K. Liao, and S. A. Curran, Enhancing current density using vertically oriented organic photovoltaics, *Sol. Energy Mater. Sol. Cells* 101, 227–231, June 2012.

74. H. Shen, P. Bienstman, and B. Maes, Plasmonic absorption enhancement in organic solar cells with thin active layers, *J. Appl. Phys.* 106, 073109, 2009.

75. D. Duche, P. Torchio, L. Escoubas, F. Monestier, J. Simon, F. Flory, and G. Mathian, Improving light absorption in organic solar cells by plasmonic contribution, *Sol. Energy Mater. Sol. Cells* 93, 1377–1382, 2009.

76. J.-H. Lee, J.-H. Park, J.-S. Kim, D.-Y. Lee, and K.-W. Cho, High efficiency polymer solar cells with wet deposited plasmonic gold nanodots, *Org. Electron.* 10, 416–420, 2009.

77. N. C. Lindquist, W. A. Luhman, S. Oh, and R. J. Holmes, Plasmonic nanocavity arrays for enhanced efficiency in organic photovoltaic cells, *Appl. Phys. Lett.* 93, 123308, 2008.

78. S. Lim, D. Han, H. Kim, S. Lee, and S. Yoo, Cu-based multilayer transparent electrodes: A low-cost alternative to ITO electrodes in organic solar cells, *Sol. Energy Mater. Sol. Cells* 101, 170–175, 2012.

79. W. Bai, Q. Gan, G. Song, L. Chen, Z. Kafafi, and F. Bartoli, Broadband short-range surface plasmon structures for absorption enhancement in organic photovoltaics, *Opt. Express* 18, A620–A630, 2010.

80. S. A. Maier, *Plasmonics: Fundamentals and Applications*, Springer, New York, 2007.

81. S. Schultz, D. R. Smith, J. J. Mock, and D. A. Schultz, Single-target molecule detection with nonbleaching multicolor optical immunolabels, *Proc. Nat. Acad. Sci.* 97(3), 996–1001, 2000.

82. D. A. Stuart, A. J. Haes, C. R. Yonzon, E. M. Hicks, and R. P. V. Duyne, Biological applications of localized surface plasmon resonance phenomena, *IEE Proc. Nanobiotechnol.* 152(1), 13–32, 2005.

83. A. M. Michaels, M. Nirmal, and L. E. Brus, Surface enhanced Raman spectroscopy of individual rhodamine 6 g molecules on large Ag nanocrystals, *J. Am. Chem. Soc.* 121, 9932–9939, 1999.

84. I. H. El-Sayed, X. Huang, and M. A. El-Sayed, Surface plasmon resonance scattering and absorption of anti-EGFR antibody conjugated gold nanoparticles in cancer diagnostics: Applications in oral cancer, *Nano Lett.* 5(5), 829–834, 2005.

85. J. Homola, S. S. Yee, and G. Gauglitz, Surface plasmon resonance sensors: Review, *Sensor Actuat. B Chem.* 54, 3–15, 1999.

86. I. Lundström, Real-time biospecific interaction analysis, Biosens. *Bioelectron.* 9, 725–736, 1994.

87. P. N. Njoki, I.-I. S. Lim, D. Mott, H.-Y. Park, B. Khan, S. Mishra, R. Sujakumar, J. Luo, and C.-J. Zhong, Size correlation of optical and spectroscopic properties for gold nanoparticles, *J. Phys. Chem.* C 111, 14664–14692, 2007.

88. Y. Xia, Y. Xiong, B. Lim, and S. E. Skrabalak, Shape-controlled synthesis of metal nanocrystals: Simple chemistry meets complex physics?, *Angew. Chem. Int. Ed.* 48, 60–103, 2009.

89. P. W. Barber, R. K. Chang, and H. Massoudi, Electrodynamic calculations of the surface-enhanced electric intensities on large Ag spheroids, *Phys. Rev. B.* 27(12), 7251–7261, 1983.

90. K. Ueno, S. Juodkazis, V. Mizeikis, K. Sasaki, and H. Misawa, Spectrally resolved atomic-scale variations of gold nanorods, *J. Am. Chem. Soc.* 128(44), 14226–14227, 2006.

91. S. Juodkazis, Light-field enhancement at particle–substrate interface (oral), in Australasian Conf. on Optics Lasers and Spectroscopy (ACOLS-09), 29 Nov.–3 Dec., Adelaide, Australia., p. 357, 2009.

92. E. Kowalska, R. Abe, and B. Ohtani, Visible light-induced photocatalytic reaction of gold-modified titanium(IV) oxide particles: Action spectrum analysis, *Chem. Commun.* 2, 241–243, 2009.

93. E. Kowalska, O. O. Prieto-Mahaney, R. Abe, and B. Ohtani, Visible-light-induced photocatalysis through surface plasmon excitation of gold on titania surfaces, *Phys. Chem. Chem. Phys.* 12, 2344–2355, 2010.

94. J. Ctyroky, J. Homola, and M. Skalsky, Modelling of surface plasmon resonance waveguide sensor by complex mode expansion and propagation method, Opt. *Quant. Electron.* 29, 301–311, 1997.

95. M. Haruta, Size- and support-dependency in the catalysis of gold, Catal. Today 36, 153–166, 1997.

96. J. A. Sánchez-Gil, J. V. García-Ramos, and E. R. Méndez, Near-field electromagnetic wave scattering from random self-affine fractal metal surfaces: Spectral dependence of local field enhancements and their statistics in connection with surface-enhanced Raman scattering, *Phys. Rev. B* 62(15), 10515–10525, 2000.

97. Y. Tian and T. Tatsuma, Mechanisms and applications of plasmon-induced charge separation at TiO_2 films loaded with gold nanoparticles, *J. Am. Chem. Soc.* 127, 7632–7637, 2005.

98. V. Rodríguez-González, R. Zanella, G. del Ángel, and R. Gómez, MTBE visible-light photocatalytic decomposition over Au/TiO_2 and $Au/TiO_2–Al_2O_3$ sol–gel prepared catalysts, *J. Mol. Catal. A Chem.* 281, 93–98, 2008.

99. R. S. Sonawane and M. K. Dongare, Sol–gel synthesis of Au/TiO_2 thin films for photocatalytic degradation of phenol in sunlight, *J. Mol. Catal. A Chem.* 243, 68–76, 2007.

100. N. Sakai, Y. Fujiwara, Y. Takahashi, and T. Tatsuma, Plasmon-resonance-based generation of cathodic photocurrent at electrodeposited gold nanoparticles coated with TiO_2 films, *Chem. Phys. Chem.* 10, 766–769, 2009.

101. A. Dawson and P. V. Kamat, Semiconductor–metal nanocomposites: Photoinduced fusion and photocatalysis of gold-capped TiO_2 (TiO_2/gold) nanoparticles, *J. Phys. Chem.* B 105, 960–966, 2001.

102. V. Subramanian, E. E. Wolf, and P. V. Kamat, Catalysis with TiO_2/gold nanocomposites: Effect of metal particle size on the Fermi level equilibration, *J. Am. Chem. Soc.* 126, 4943–4950, 2004.

103. N. Chandrasekharan and P. V. Kamat, Improving the photoelectrochemical performance of nanostructured TiO_2 films by adsorption of gold nanoparticles, *J. Phys. Chem.* B 104, 10851–10857, 2000.

104. T. J. Davis, K. C. Vernon, and D. E. Gómez, Designing plasmonic systems using optical coupling between nanoparticles, *Phys. Rev. B* 79, 155423, 2009.

105. A. A. Verberckmoes, B. M. Weckhuysen, R. A. Schoonheydt, K. Oomsb, and I. Langhans, Chemometric analysis of diffuse reflectance spectra of Co^{2+}-exchanged zeolites: Spectroscopic fingerprinting of coordination environments, *Anal. Chim. Acta* 348, 267–272, 1997.

106. K. Ueno, S. Juodkazis, M. Mino, V. Mizeikis, and H. Misawa, Spectral sensitivity of uniform arrays of gold nanorods to the dielectric environment, *J. Phys. Chem. C* 111(11), 4180–4184, 2007.

107. B. Rogers, S. Pennathur, and J. Adams, *Nanotechnology: Understanding Small Systems*, CRC Press, Taylor & Francis Group, Boca Raton, 2008.

108. E. Hutter and J. H. Fendler, Exploitation of localized surface plasmon resonance, *Adv. Mater.* 16(19), 1686–1708, 2004.

109. S. Juodkazis, Y. Nishi, H. Misawa, V. Mizeikis, O. Schecker, R. Waitz, P. Leiderer, and E. Scheer, Optical transmission and laser structuring of silicon membranes, *Opt. Express* 17(17), 15308–15317, 2009.

110. H. Iwase, S. Kokubo, S. Juodkazis, and H. Misawa, Suppression of ripples on Ni surface via a polarization grating, *Opt. Express* 17(6), 4388–4396, 2009.

111. P. V. Kamat and B. Shanghavi, Interpretation electron transfer in metal/semiconductor composites: Picosecond dynamics of CdS-capped gold nanoclusters, *J. Phys. Chem. B* 101, 7675–7679, 1997.

112. K. Juodkazis, J. Juodkazyte, P. Kalinauskas, E. Jelmakas, and S. Juodkazis, Photoelectrolysis of water: Solar hydrogen achievements and perspectives, *Opt. Express: Energy Express* 18, A147–A160, 2010.

113. A. Fujishima, X. Zhang, and D. A. Tryk, TiO_2 photocatalysis and related surface phenomena, *Surf. Sci. Rep.* 63(12), 515–582, 2008.

114. G. W. Crabtree and N. S. Lewis, Solar energy conversion, *Phys. Today* 60(3), 37–42, 2007.

115. M. Ni, M. K. H. Leung, D. Y. Leung, and K. Sumathy, A review and recent developments in photocatalytic water splitting using TiO_2 for hydrogen production, *Renewable Sustainable Energy Rev.* 11, 401–425, 2007.

116. P. V. Kamat, Meeting the clean energy demand: Nanostructure architectures for solar energy conversion, *J. Phys. Chem. C* 111(7), 2834–2860, 2007.

117. J. Nowotny, T. Bak, M. K. Nowotny, and L. Sheppard, Titanium dioxide for solar-hydrogen. I. Materials requirements, *Int. J. Hydr. Energ.* 32, 2609–2629, 2007.

118. J. Nowotny, T. Bak, M. K. Nowotny, and L. Sheppard, Titanium dioxide for solar-hydrogen. III. Kinetic effects, *Int. J. Hydr. Energ.* 32, 2644–2650, 2007.

119. J. Nowotny, T. Bak, M. K. Nowotny, and L. Sheppard, Titanium dioxide for solar-hydrogen. IV. Collective and local factors in photolysis of water, *Int. J. Hydr. Energ.* 32, 2651–2659, 2007.

120. G. K. Mor, O. K. Varghese, M. Paulose, K. Shankar, and C. A. Grimes, A review on highly-ordered TiO_2 nanotube-arrays: Fabrication, material properties, and solar energy applications, *Sol. Energy Mater. Sol. Cells* 90, 2011–2075, 2006.

121. A. L. Linsebigler, G. Lu, and J.T. Yates, Photocatalysis on TiO_2 surfaces: Principles, mechanisms, and selected results, *Chem. Rev.* 95, 735–758, 1995.

122. T. Tachikawa, M. Fujitsuka, and T. Majima, Mechanistic insight into the TiO_2 photocatalytic reactions: Design of new photocatalysts, *J. Phys. Chem. C* 111, 5259–5275, 2007.

123. V. M. Aroutiounian, V. M. Arakelyan, and G. E. Shahnazaryan, Metal oxide photoelectrodes for hydrogen generation using solar radiation-driven water splitting, *Solar Energy* 78, 581–592, 2005.

124. M. Grätzel, Photoelectrochemical cells, *Nature* 414, 338–344, 2001.

125. T. W. Murphy Jr., Home photovoltaic systems for physicists, *Physics Today*, 42–47, 2008.

126. A. Fujishima and K. Honda, Electrochemical photolysis of water at a semiconductor electrode, *Nature* 238, 37–38, 1972.

127. E. L. Miller, D. Paluselli, B. Marsen, and R. E. Rocheleau, Development of reactively sputtered metal oxide films for hydrogen-producing hybrid multijunction photoelectrodes, *Sol. Energy Mater. Sol. Cells* 88(2), 131–144, 2005.

128. T. Lana-Villarreal and R. Gomez, Interfacial electron transfer at TiO_2 nanostructured electrodes modified with capped gold nanoparticles: The photoelectrochemistry of water oxidation, *Electrochem. Comm.* 7, 1218–1224, 2005.

129. M. Radecka, M. Rekas, A. Trenczek-Zajac, and K. Zakrzewska, Importance of the band gap energy and flat band potential for application of modified TiO_2 photoanodes in water photolysis, *J. Power Sources* 181, 46–55, 2008.

130. R. Beranek and H. Kisch, Surface-modified anodic TiO_2 films for visible light photocurrent response, *Electrochem. Commun.* 9, 761–766, 2007.

131. Y. Tian and T. Tatsuma, Mechanisms and applications of plasmon-induced charge separation at TiO_2 films loaded with gold nanoparticles, *J. Am. Chem. Soc.* 127, 7632–7637, 2005.

132. Z. Zou, J. Ye, K. Sayama, and H. Arakawa, Direct splitting of water under visible light irradiation with an oxide semiconductor photocatalyst, *Nature* 414, 625–627, 2001.

133. A. Survila, P. Kalinauskas, and I. Valsiunas, Photoelektrochemical properties of surface layers formed by anodic oxidation of titanium, *Chemija* 10, 117–121, 1999.

134. B. Parkinson, On the efficiency and stability of photoelectrochemical devices, *Acc. Chem. Res.* 17, 431–437, 1984.

135. U. S. Avachat, A. H. Jahagirdar, and N. G. Dhere, Multiple bandgap combination of thin film photovoltaic cells and a photoanode for efficient hydrogen and oxygen generation by water splitting, *Solar Energ. Mat. Solar Cells* 90, 2464–2470, 2006.

136. J. Juodkazyte, B. Sěbeka, P. Kalinauskas, and K. Juodkazis, Light energy accumulation using Ti/RuO$_2$ electrode as capacitor, *J. Sol. Stat. Electrochem.* 14, 741–746, 2010.

137. S. Juodkazis, A. Yamaguchi, H. Ishii, S. Matsuo, H. Takagi, and H. Misawa, Photo-electrochemical deposition of platinum on TiO$_2$ with resolution of tens of nm by using a mask elaborated with electron-beam lithography, *Jpn. J. Appl. Phys.* 40, 4246–4251, 2001.

138. P. Salvador, Kinetic approach to the photocurrent transients in water photoelectrolysis at n-titanium dioxide electrodes. 1. Analysis of the ratio of the instantaneous to steady-state photocurrent, *J. Phys. Chem. C* 89, 3683–3869, 1985.

139. K. J. Vetter, *Elektrochemische kinetik*, Springer-Verlag, Berlin-Gottingen, 1961.

140. M. Pourbaix, *Atlas d'équilibres électrochimiques*, Gauthier-Villars, Paris, 1963.

141. K. Juodkazis, J. Juodkazyte, Y. Tabuchi, S. Juodkazis, S. Matsuo, and H. Misawa, Deposition of platinum and irridium on Ti surface using femtosecond laser and electro-chemical activation, *Lith. J. Phys.* 43(2), 209–216, 2003.

142. D. Dobos, *Electrochemical Data*, Mir, Moscow, 1980.

143. P. Salvador and C. Gutierrez, The nature of surface states involved in the photo- and electroluminescence spectra of n-titanium dioxide electrodes, *J. Phys. Chem. C* 84(16), 3696–3698, 1984.

144. C. Gutierrez and P. Salvador, Mechanisms of competitive photoelectrochemical oxidation of I– and H$_2$O at n-TiO$_2$ electrodes: A kinetic approach, *J. Electrochem. Soc.* 133, 924–929, 1986.

145. J. Juodkazyte, R. Vilkauskaite, B. Sěbeka, and K. Juodkazis, Difference between surface electrochemistry of ruthenium and RuO$_2$ electrodes, *Trans. Inst Metal Finishing* 85(4), 194–201, 2007.

146. K. Juodkazis, J. Juodkazyte, R. Vilkauskaite, and V. Jasulaitiene, Nickel surface anodic oxidation and electrocatalysis of oxygen evolution, *J. Sol. Stat. Electrochem.* 12(11), 1469–1479, 2008.

147. K. Juodkazis, J. Juodkazyte, V. Sŭkiene, A. Grigucevičiene, and A. Selskis, On the charge storage mechanism at RuO$_2$/0.5 M H$_2$SO$_4$ interface, *J. Sol. Stat. Electrochem.* 12(11), 1399–1404, 2008.

148. K. Juodkazis, J. Juodkazyte, R. Vilkauskaite, B. Sěbeka, and V. Jasulaitiene, Oxygen evolution on mixed ruthenium and nickel oxide electrode, *Chemija* 19, 1–6, 2008.

149. R. Nakamura and Y. Nakato, In situ FTIR studies of primary intermediates of photocatalytic reactions on nanocrystalline TiO$_2$ films in contact with aqueous solutions, *J. Am. Chem. Soc.* 126, 1290–1298, 2004.

150. R. Nakamura, T. Okamura, N. Ohashi, A. Imanishi, and Y. Nakato, Molecular mechanisms of photoinduced oxygen evolution, PL emission, and surface roughening at atomically smooth (110) and (100) n-TiO$_2$ (rutile) surfaces in aqueous acidic solutions, *J. Am. Chem. Soc.* 127, 12975–12983, 2005.

151. J. Lee, J. Park, J. Kim, D. Lee, and K. Cho, High efficiency polymer solar cells with wet deposited plasmonic gold nanodots, *Org. Electron.* 10, 416–420, 2009.

152. C. Chou, R. Yang, C. Yeh, and Y. Lin, Preparation of TiO$_2$/nano-metal composite particles and their applications in dye-sensitized solar cells, *Powder Technol.* 194, 95–105, 2009.

153. T. Hasobe, H. Imahori, S. Fukuzumi, and P. V. Kamat, Nanostructured assembly of porphyrin clusters for light energy conversion, *J. Mater. Chem.* 13, 2515–2520, 2003.

154. H. Imahori and T. Umeyama, Donor–acceptor nanoarchitecture on semiconducting electrodes for solar energy conversion, *J. Phys. Chem. C* 113, 9029–9039, 2009.

155. K. Ueno, S. Juodkazis, V. Mizeikis, D. Ohnishi, K. Sasaki, and H. Misawa, Inhibition of multipolar plasmon excitation in periodic chains of gold nanoblocks, *Opt. Express* 15(25), 16527–16539, 2007.

156. K. Ueno, Y. Yokota, S. Juodkazis, V. Mizeikis, and H. Misawa, Nano-structured materials in plasmonics and photonics, *Curr. Nanosci.* (CNANO) 4(3), 232–235, 2008.
157. Y. Yokota, K. Ueno, V. Mizeikis, S. Juodkazis, K. Sasaki, and H. Misawa, Optical characterization of plasmonic metallic nanostructures fabricated by high-resolution lithography, *J. Nanophotonics* 1(1), p. 011594, 2008.
158. N. V. Surovtsev, On the relation between Stokes and anti-Stokes light scattering in liquids, *Chem. Phys. Lett.* 375, 495–498, 2003.
159. O. A. Louchev, S. Juodkazis, N. Murazawa, S. Wada, and H. Misawa, Coupled laser molecular trapping, cluster assembly, and deposition fed by laser-induced Marangoni convection, *Opt. Express* 16(8), 5673–5680, 2008.
160. A. N. Grigorenko, N. W. Roberts, M. R. Dickinson, and Y. Zhang, Nanometric optical tweezers based on nanostructured substrates, *Nature Photon.* 2, 365–370, 2008.
161. Y. Yokota, K. Ueno, S. Juodkazis, V. Mizeikis, and H. Misawa, Influence of plasmonic enhancement on surface-enhanced Raman scattering, in Extended Abstracts of XXIV Int. Conf. on Photochemistry ICP2009, Toledo, Spain, PSII–P82, July 2009.
162. Y. Yokota, Sensing applications of plasmonic nanoparticles, PhD thesis, Hokkaido University, Japan, October 2009.
163. Y. Yokota, K. Ueno, and H. Misawa, Highly controlled surface-enhanced Raman scattering chips using nanoengineered gold blocks, *Small* 7, 252–258, 2011.
164. E. J. Liang, X. L. Ye, and W. Kiefer, Surface-enhanced Raman spectroscopy of crystal violet in the presence of halide and halate ions with near-infrared wavelength excitation, *J. Phys. Chem.* A 101, 7330–7335, 1997.
165. L. Rosa and S. Juodkazis, Tailoring plasmonic nanoparticles and fractal patterns, in SPIE, vol. 8204, 82041N, 2011.
166. K. Kneipp, Y. Wang, H. Kneipp, I. Itzkan, R. R. Dasari, and M. S. Feld, Population pumping of excited vibrational states by spontaneous surface-enhanced Raman scattering, *Phys. Rev. Lett.* 76(14), 2444–2447, 1996.

4 Polarization-Dependent Plasmonic Chiral Devices

Byoungho Lee and Seung-Yeol Lee

CONTENTS

4.1 INTRODUCTION

Optical activity is a phenomenon related to the birefringence of light that arises from interactions between a material and the spin state of photons. Pioneering research has shown that, when light passes through a bulk medium with optical activity such as quartz [1] or a solution of chiral molecules, the polarization state of the light can be rotated. Such phenomenon was explained by an interaction of light with the chiral unit cell geometry or chiral molecules [2]. In these materials, the asymmetric response of transmission, according to the direction of the incident light, reveals the possibility of producing a polarization rotator and an optical isolator. In the past, investigators attempted to find materials with stronger optical activity due to the high potential for applications of such materials to optical devices [3]. However, this effort does not appear to have been successful, since controlling the optical properties at the molecular level is a difficult task, due to the small scale of molecules and limitations associated with the chemical processes.

On the other hand, rapid advances in nanostructure fabrication techniques have resulted in the production of artificial nanodevices and metamaterial structures. The interaction of an external field with these artificial structures can provide unusual characteristics such as photonic bandgaps [4], a negative refractive index [5,6], polarization conversion in an extremely thin layer [7], and slow light with an ultra-high refractive index [8]. In particular, the effect of polarization-dependent char-acteristics on spiral-shaped or azimuthally periodic structures is one of the most attractive issues due to the fact that their characteristics can be switched by modify-ing the incident polarization.

It has been shown that, when these structures contain metallic portions, the collec-tive oscillation of light and electrons can occur in the metallic structure, often referred to as a surface plasmon polariton (SPP) [9,10], which strongly affects the optical char-acteristics, such as resonant frequency, transmission efficiency, effective refractive index, and the polarization state of the structure. The effects of SPPs on metallic struc-tures vary with the size and geometry, so they cannot be explained by a unique anal-ogy. However, due to the scope of this chapter, we will focus only on analogies that are frequently used in chiral plasmonic devices, such as spin-orbital interactions of SPPs on a metal structure or the concept of topological charge in plasmonic singularity.

The organization of this chapter is as follows. In Section 4.2, we briefly explain the concept of chirality and related physical phenomena, such as spin-orbital inter-actions of SPPs on the metal structure or the concept of topological charge in plas-monic singularity. We then roughly divide chiral plasmonic devices into two groups, based on their relative size and periodicity. Plasmonic devices covered in Section 4.3 are isolated plasmonic structures such as plasmonic lenses [11,12], gratings with nanoholes [13], and chiral metallic nanorod structures [14]. These structures are usu-ally larger than the wavelength scale, so each of them acts as a single nanoscale optical component. When chiral geometry is applied to these structures, each polar-ization component of the light incident on these structures affects the amplitude and phase of induced SPPs differently, due to their chiral geometry. As a result, various phenomena such as plasmonic hot/dark spot switching [12], plasmonic beam switch-ing [15], lateral shifting of plasmonic hot spots through spin-orbital interactions [16], and transmission symmetry breaking in optical nanoapertures [17] are reported. By using these numerous phenomena, polarization-dependent plasmonic structures can be applied to various devices, such as plasmonic switches, data storage devices, and optical tweezing.

On the other hand, in Section 4.4, we focus on the periodic array of subwave-length chiral structures that provide polarization-dependent characteristics with visible, infrared, or THz light incidence. These structures are often referred to as chiral metamaterials, and the principle inside them is quite similar to that in bulk materials (except that atomic or molecular periods are replaced with an artificial structure period) since the unit cell of these materials is much smaller than the scale of incident wavelength. For these cases, only the averaged response from the overall geometry can be observed. Hence, it is much more difficult to explain the working mechanism of SPPs inside a chiral metamaterial than that inside an isolated struc-ture. Nevertheless, we also include a section on chiral metamaterials in this chapter since many studies explain the characteristics of chiral metamaterials with the aid of SPPs [18–23]. With these artificially made periodic chiral structures, numerous interesting results such as asymmetric optical transmission [18,19], extremely strong optical activity in thin metal films [20,21], and negative refractive index materials [22,23] are reported.

As a brief summary, the type and the characteristics of these polarization-dependent isolated and periodic plasmonic devices are categorized in Table 4.1. Although our main object here is to explain the role of SPPs in chiral plasmonic devices, we also include references to operating wavelengths near the THz region in chiral metamaterials, since the design methods are not so different from those used for the optical region.

4.2 PRINCIPLES OF CHIRAL PLASMONIC DEVICES

Since all of polarization-dependent characteristics introduced in this chapter are based on the chiral geometry, we first explain the meaning of "chirality." The word chirality is derived from the Greek word meaning "hand" since the human hand is one of the most familiar examples of chiral geometry. The condition of "chiral geometry" is satisfied when the geometry cannot be the identical shape of its mirror images by rotating or translating the original geometry. Moreover, since the meaning of chirality is not restricted to geometry, it can also be found in electromagnetic waves, by observing the direction of rotation of an electric or magnetic field toward their propagating direction. The physical insight of these chiralities in an optical wave originates from the spin state of the photon. Opposite spin states are "chiral" with respect to each other, since they cannot be identical without mirror imaging. In a macroscopic view, the spin state of a photon can be represented by the optical handedness of an electromagnetic wave, which provides left- and right-handed circular polarization (LCP and RCP) states.

We now discuss the interaction of circularly polarized light and chiral geometry. As mentioned in the introduction, this interaction can be differently explained by the relative scale between the incident wavelength and chiral geometry. When the size of a chiral structure is far smaller than the incident wavelength, the structure can be treated as a chiral molecule that generates optical activity. In addition, structures should be periodically arranged to provide a homogeneous response. The LCP and RCP states have different optical phase velocities, so their superposed polarization can be rotated through the propagation of this medium. The rotation angle per unit distance can be expressed as

$$\chi = \frac{\pi}{\lambda_0}(n_L - n_R),\tag{4.1}$$

where λ_0, n_L, and n_R are the free space wavelength of incident light, and effective refractive indices of the optically active medium with LCP and RCP incidence, respectively. It has been demonstrated that artificially prepared chiral structures can have much stronger optical activity than naturally occurring materials, since they can be optimized by using circuit theory [22] or with the help of SPPs [21]. Some detailed examples related to this topic are discussed in Section 4.4.

On the other hand, chiral geometry cannot be treated as a small chiral molecule with optical activity when the structure size is larger than the wavelength scale. For the case of metallic structures in this region, the interference characteristics of

TABLE 4.1
Configurations and Characteristics of Plasmonic Chiral Devices

Type	Configuration	Characteristics	Operating Wavelength	Reference
Isolated chiral plasmonic devices	Chirality in a nanohole shape	Polarization conversion after transmission	≃4 μm	[24]
		Polarization-dependent EOT effect	≃3 μm	[25]
	Chirality in a patterned slit shape	Hot/dark spot switching	532 nm, 808 nm	[26–28]
		Plasmonic vortex state switching	660 nm	[12]
		Lateral shifting of plasmonic hot spot	532 nm	[16]
	Chirality in a surrounding grating shape	Polarization-dependent enhancement of nanohole transmission	780 nm	[13]
		Evanescent Bessel beam formation by chirality induction	658 nm	[29]
		Polarization-dependent switching of plasmonic beam	650 nm	[15]
	Chirality in a nanorod shape	Plasmonic motor driven by optical torque	810 nm, 1700 nm	[30]
		Symmetry-breaking features in nanospiral	0.8–1.4 μm	[14]
Periodic chiral plasmonic devices (chiral metamaterials)	Single-layered structure	Asymmetric propagation of electromagnetic waves	≃650 nm	[18]
		Optical activity caused by the effect of SPP in gammadion array	700–900 nm	[21]
	Multilayered structure	Negative refractive index material in bridge-shaped structure	≃1 THz (300 μm)	[22]
		Optical activity caused by twisted-cross structure	1–2 μm	[31]

SPPs are quite important for understanding their optical responses. Especially in metal slit patterns or grating structures, the interaction of external light with them can be explained by the spin-orbital interaction of SPPs. A schematic diagram that explains this phenomenon is shown in Figure 4.1. When circularly polarized light is normally incident on the circularly patterned slit or grating structure on a metal film, the SPP excitation efficiency is azimuthally different, since only the electric component perpendicular to slit contributes to SPP excitation. This generates a 2π phase delay of excited SPPs along one cycle of the slit pattern, as shown in Figure 4.1. This can be interpreted differently as follows: the spin state of an incident photon affects the orbital angular momentum of excited SPPs. This phenomenon was introduced in Ref. [26] as the spin-orbital interaction of light in a plasmonic structure.

It is known that this spin-orbital interaction can affect the topological charge number at a singular point generated at the center spot of the metal film with a circular slit pattern. The topological charge number is defined as the number of phase modulations along the circular path near the singular point like the focal point of the optical vortex. The effect of spin-orbital interaction could change the topological charge number of plasmonic vortex or beaming, which can be expressed as

$$l_{\text{total}} = l_{\text{geometry}} + l_{\text{angular}} + l_{\text{spin}}, \qquad (4.2)$$

where l_{geometry}, l_{angular}, and l_{spin} are the angular momentum numbers generated by the chiral geometry, carried by incident light, and generated by the spin-orbital interaction, respectively. Hence, a topological charge can be generated by the spin-orbital interaction without any chirality in the geometry or angular momentum of the incident light. Moreover, it can be shown from Equation 4.2 that an asymmetric spin-orbital

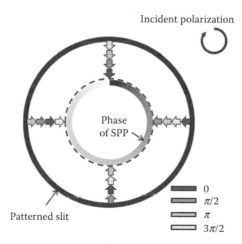

FIGURE 4.1 Schematic diagram for explaining the spin-orbital interaction of light in a circular patterned slit structure. The excitation efficiency of SPPs varies according to the angle between slit-normal direction and electric field direction, which makes an additional 2π phase modulation along the circular slit. Throughout this chapter, RCP is defined as a clockwise field rotating direction.

interaction can be produced when a chiral geometry is involved (nonzero l_{geometry} value). We include a detailed explanation for each plasmonic device using this analogy in Section 4.3, since overall electromagnetic responses are different case by case.

4.3 POLARIZATION-DEPENDENT CHARACTERISTICS IN ISOLATED PLASMONIC STRUCTURES

We now start with isolated chiral plasmonic structures which are generally larger than the operating wavelength scale. In Figure 4.2, we illustrate some representative shapes of isolated chiral plasmonic devices. Depending on the geometry

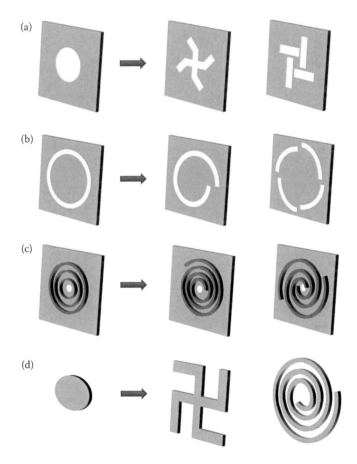

FIGURE 4.2 Schematics of various plasmonic chiral structures and their basic geometries are illustrated. (a) The first group shows geometries modified from a simple metal nanohole aperture. (b) The second group denotes spiral- or whirlpool-shaped plasmonic vortex lens structures that originated from a simple circular slit which is often referred to as a plasmonic lens. (c) The structures shown in the third group are nanoholes surrounded by a spiral or double-spiral dielectric grating structure, which are modifications of the conventional bull's eye structure. (d) The geometries in the last group show gammadion- or spiral-shaped metallic nanorod structures.

and excitation method of the SPPs, these structures can be classified into several groups.

The structures shown in Figure 4.2a were prepared by adding the geometrical chirality to the basic geometry of a nanoaperture. Here, we define the basic geometry as the geometry that contains common characteristics of the group without any chirality. Such types of aperture-geometry modifications are widely used for the enhancement of transmission efficiency or confinement characteristics as well as polarization dependency. Although there are successful examples of producing higher transmission efficiencies, such as C-shaped apertures [32,33] or bowtie-shaped apertures [34,35], we will only cover the case of aperture modifications with chirality, as shown in the right side of Figure 4.2a in this chapter.

The basic geometry of the second group is that of circularly patterned nanoslit structures, as shown in the left side of Figure 4.2b, which are often referred to as a plasmonic lens [11]. The function of a plasmonic lens is to focus the excited surface plasmons at a specific point, so the key parameter in designing a plasmonic lens is the phase characteristics of the excited SPPs by the nanoslit. The addition of a phase is usually accomplished by increasing the radius of the slit monotonically, which can provide a plasmonic vortex at the focal point of a plasmonic lens [36]. Design methods for these chiral plasmonic lens structures are quite different from the cases of the first group and, because of this, they are classified into a different group. The right side of Figure 4.2b shows representative shapes in this group, such as spiral-shaped or whirlpool-shaped slit patterns.

It is well known that the excitation of SPPs from plane wave illumination or the conversion of SPPs back to a radiating far-field wave is possible by using a grating structure which has a period that is matched to the momentum relation [37]. It has been shown that directional beaming or enhanced SPP coupling can be achieved by the help of these gratings. We categorize structures that contain chiral gratings surrounding the metallic nanohole into the third group as shown in Figure 4.2c. The basic geometry of this group is a "bull's eye" structure, which was initially used to enhance transmission efficiency [38] and for generating a highly directional plasmonic beam [39].

Finally, localized surface plasmons, which are generated by the interaction of light and spatially localized electrons in a metallic structure, are another aspect of SPP excitation. These localized SPPs generated in metallic nanorods or nanoparticles could enhance the near-field intensity or optical power flow, and chirality can also be applied to these nanorod structures for polarization-dependent localized SPP excitation. The schematics shown in Figure 4.2d are representative examples for these applications that provide near-field optical activity and switchable optical torque generation.

4.3.1 Polarization-Dependent Interactions in Chiral Metallic Nanoapertures

Since the effect of SPPs on nanoscale metallic structures first came to the fore by the discovery of extraordinary optical transmission (EOT) in nanoholes [40], the first attempt to prepare a chiral nanostructure was related to the attempt to modify metallic nanoholes or nanoapertures in resonance conditions. In nanoapertures without chirality, the condition for strong enhancement of an electric field such as the EOT phenomenon is not related to optical handedness. However, when the chirality of the geometry is included, such symmetrical optical characteristics are broken [41]. One of the representative studies in this area is the gammadion-shaped aperture shown in Figure 4.3a [24,25]. The gammadion geometry has chirality, since the bent wings are arranged in a certain rotating direction. It has four wings with bending angles of 45° and the direction of the twisted clockwise. Parameters such as gammadion width (w) and the length (L) of the gammadion from one end to the other should be carefully designed to provide resonance conditions. Especially, the scale factor between the incident wavelength and the parameter L is of utmost importance for obtaining transmission characteristics that are highly sensitive to optical handedness [24].

In our simulations, the design parameters are set to $w = 150$ nm, $L = 1$ μm, and the incident wavelength is $\lambda_0 = 750$ nm. The scale factor of $L/\lambda_0 \simeq 1.33$ is in the

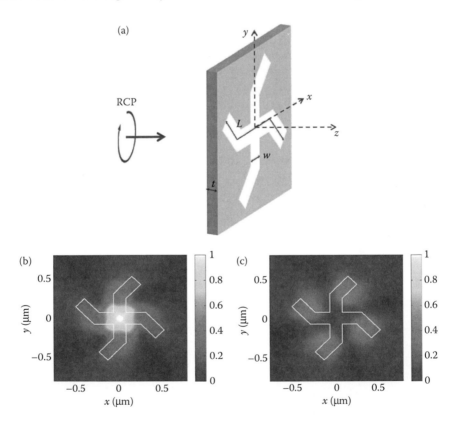

FIGURE 4.3 (a) Geometry of a gammadion-shaped aperture used for asymmetric optical focusing and polarization conversion. Transmitted electric field intensity distributions at 160 nm above the gammadion aperture with (b) LCP and (c) RCP incident waves. For metal substrates, we used a 300-nm-thick ($t = 300$ nm) silver film. The permittivity of silver used in numerical analysis is taken from Ref. [43] throughout this chapter.

reasonable region that provides the polarization-dependent characteristics verified in Ref. [24]. All of the simulation results, except for Figure 4.7, were conducted by rigorous coupled-wave analysis (RCWA) [42], which is a type of Fourier
modal method that is used effectively in mode analysis and polarization-dependent
responses. A field intensity distribution of 160 nm above the gammadion aperture
with RCP and LCP incidences is illustrated in Figure 4.3b and c. As can be seen
from these figures, the transmission from the gammadion structure shows quite
different characteristics with the change in optical handedness. The obvious difference between the two polarization states is shown in the intensity profile: the strong
focusing of light appears only for the case of light with LCP which is an opposite direction compared to the twisted direction of geometry. In contrast, fields are
spread toward the end of the gammadion wings and no focusing is generated when
the RCP light is used for the illumination. Moreover, although not shown here, it
has been demonstrated that the effect of strong polarization conversion is generated from the gammadion structure [24]. Such polarization conversion in a chiral
nanostructure that had been first observed in a gammadion aperture was utilized
in other applications such as in metamaterials, plasmonic beaming, and switching,
which are discussed below.

4.3.2 Switching of the Plasmonic Vortex State in Chiral Plasmonic Lens

As mentioned in the introduction part of this section, the purpose of the plasmonic
lens is to focus SPPs on the focal point of the structure [44]. Before we examine the
chiral plasmonic lens, we briefly review the conventional plasmonic lens. The first
demonstration of a plasmonic lens was done by the illumination of linearly polarized light onto a circular metallic disk structure [11,45]. Although the demonstration
showed that a hot spot was generated at the focal point, it did not exactly match
with the overall intensity of SPPs focused on the focal point, since the tangential
component of an electric field is dominantly measured using a near-field scanning
microscope [11]. Afterward, it turned out that the linear polarization incidence cannot generate a single hot spot due to the opposite phase profile of excited SPPs that
are generated from opposite portion of slits. To solve this problem, some researchers
attempted to excite the SPPs with the same phase profile along the slit position without changing the geometry of plasmonic lens by illuminating the radially polarized
wave [46].

However, owing to the vector beam characteristic of radially polarized light,
aligning the incident beam is too difficult and it can hardly be used for an array of
plasmonic lenses. Therefore, other researchers attempted to resolve the phase mismatch of a plasmonic lens by shifting the slit geometry [26–28,36,47,48]. It was
reported that a half-split ring-shaped plasmonic lens can compensate for the phase
mismatch generated by linearly polarized light [36,47], whereas a spiral-shaped
plasmonic lens is appropriate for circularly polarized light [26–28,48]. The origin of
this phase mismatch is explained by the spin-orbital interaction of the metal nanostructure [27], and it is based on the excitation characteristics of SPPs: In the case
of normal incidence, only the electric field component that is perpendicular to the
direction of slit can induce SPPs [49]. Since the spiral-shaped plasmonic lens has

chirality, it was found that the compensation of phase mismatch can only be achieved for certain types of optical handedness. Otherwise, for the case of opposite handedness, phase mismatch is increased rather than compensated.

By following the design rule introduced in Ref. [12], phase modulation along one cycle of circulation could be designed as an integer number for a given effective wavelength of SPPs. From this, it is possible to generate an arbitrary order of plasmonic vortex, matching the Bessel function. The size of a plasmonic vortex can be determined by the topological charge number, which is defined as the number of phase profile changes over 2π radians along the circular path near the singular point. Owing to the spin-orbital interaction caused by a plasmonic lens, the topological charge number of generated plasmonic vortex can be switched by the optical handedness of the incident beam. Such modulation can change the characteristics of a plasmonic vortex and, because of this, it can function as a polarization-dependent optical component.

Figure 4.4a shows a schematic diagram of a plasmonic vortex lens with three-fold whirlpool-shaped subwavelength slits. Here, the thickness of the metal substrate t should be designed thick enough not to allow tunneling through the metal substrate, but is sufficiently thinner than the propagation length of the waveguide mode inside the metal slit. The width of the slit w should be thinner than the cut-off thickness of the secondary mode. Hence, there is only fundamental mode inside the slit of a plasmonic lens. This condition is important, since the starting phase of excited SPPs should not be mixed with those of higher-order waveguide modes. The radius of a plasmonic lens $r(\phi)$ is determined by

$$r(\phi) = r_0 + \lambda_{SPP} \frac{\mathrm{mod}(l_{geometry}\phi, 2\pi)}{2\pi}, \tag{4.3}$$

where the parameters r_0, λ_{SPP}, and $l_{geometry}$ are the initial radius of the plasmonic lens, the effective wavelength of the surface plasmon mode, and the number of folded slits, respectively. According to Equation 4.3, the total phase retardation amount of $2\pi l_{geometry}$ is generated by the slit geometry along one cycle of circulation. Therefore, by combining the geometrical phase retardation with the spin-orbital interaction effect, mentioned in Section 4.2, the topological charge in the plasmonic vortex can be added or subtracted. Figure 4.4b and c shows the near-field distributions of plasmonic vortex generated on the plasmonic vortex lens with LCP and RCP light incidences, respectively. For these figures, the design parameters are carefully set to $\lambda_0 = 800$ nm, $t = 300$ nm, $w = 250$ nm, and $r_0 = 5$ µm.

As can be expected, plasmonic vortices of different sizes can be achieved by each polarization case. To check whether our results are related to spin-orbital interactions on the subwavelength metallic slit, it is necessary to find the topological charge of a singular point generated at the focal point of the plasmonic vortex lens. A direct method for checking the topological charge is counting the number of phase changes over 2π radians near the singular point. Although this is the simplest method by using the numerical tools, measuring the phase near the singular point is quite a difficult task in experiments. Fortunately, the topological charge can also be measured

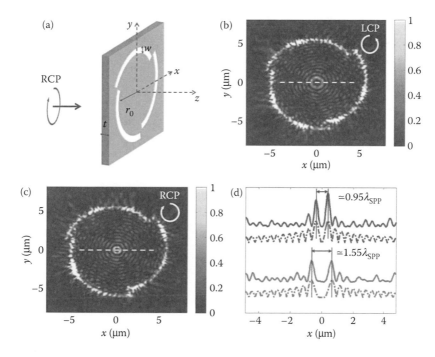

FIGURE 4.4 (a) Geometry of plasmonic vortex lens with three-fold whirlpool-shaped slit. Transmitted electric field intensity distributions of 10 nm above the plasmonic lens structure is shown with the (b) LCP and (c) RCP incident waves. (d) Comparison of the dark spot sizes of plasmonic vortex lens between the LCP and RCP light incidence cases. Upper and lower solid lines show the intensity profiles along the white dotted lines shown in part (b) and part (c), respectively. Dashed-dotted lines illustrate the theoretical sizes of dark spots calculated by Equation 4.2 for both LCP and RCP cases.

by the ring size of dark spot since the near-field distribution follows the Bessel function with the given order of topological charge:

$$E_z(r,\phi,0) \propto E_0 \exp(j(l_{geometry} \pm 1)\phi)J_{(l_{geometry}\pm 1)}(k_{SPP}r), \quad (4.4)$$

where the wavevector of surface plasmon k_{SPP} is given as $k_{SPP} = k_0\sqrt{\varepsilon_{metal}\varepsilon_{air}/\varepsilon_{metal} + \varepsilon_{air}}$. The topological charge of plasmonic vortex l_{total} is given by $l_{total} = l_{geometry} \pm 1$, where the ± 1 after the $l_{geometry}$ term has exactly the same meaning as the l_{spin} shown in Equation 4.2. In Figure 4.4d, we plot the intensity distributions along the white dotted lines shown in Figure 4.4b and c. It can be shown that the sizes of plasmonic vortices are well matched with the first maximum of the Bessel function with the order of topological charge calculated by Equation 4.2 for both polarization cases. These results show that an arbitrary integer order of a Bessel-shaped plasmon vortex could be generated from the specific design of a plasmonic lens and also shows that expanding or shrinking the vortex is possible.

4.3.3 Characteristics of Spiral Bull's Eye Geometry

So far, we discussed the chiral geometry in plasmonic lens structures that are used for polarization-dependent hot or dark spot switching and to control the plasmonic vortex state. The underlying physics for manipulating the plasmonic vortex is laid on the artificial phase shift of focused SPPs which can be obtained by modifying the slit radius. A similar approach can also be applied to the plasmonic bull's eye structure, which is composed of a subwavelength metallic hole surrounded by concentric circular dielectric or metal gratings [50]. In this section, we discuss modifications of chiral geometry in surrounding grating structures.

A conventional bull's eye structure was initially used for enhancing the transmission efficiency of the subwavelength nanohole [38]. In this case, gratings are located at the bottom of the metal substrate, so the incident light directly illuminates the circular gratings. These gratings act as an efficient SPP coupler and focus the SPPs onto the center hole. On the other hand, it has been shown that concentric gratings can also be used for highly directional beam emission from a single nanohole by locating them on top of the metal substrate [39]. After these two demonstrations, various studies have been done in attempting to optimize the bull's eye structure by changing the design parameters [51,52]. For efficient coupling, the most important parameter is the period of the gratings (Λ) that should be matched to the phase matching condition [37]:

$$k_{\text{SPP}} = k_0 \sin\theta + m\frac{2\pi}{\Lambda}, \tag{4.5}$$

where the wavevector of SPP mode is defined as $k_{\text{SPP}} = 2\pi/\lambda_{\text{SPP}}$ and θ is the angle of incident wave. Equation 4.5 is also valid for the directional beaming structure. In this case, θ is the angle of the radiating wave scattered from the gratings. Note that the value λ_{SPP} written in Equation 4.5 is not simply calculated by $\lambda_0/\sqrt{\varepsilon_{\text{air}}\varepsilon_{\text{metal}}/(\varepsilon_{\text{air}} + \varepsilon_{\text{metal}})}$ since λ_{SPP} is the effective wavelength of the SPP mode inside the periodical gratings; so, it is highly dependent on the permittivity, thickness, and width of the gratings. Nevertheless, by using Equation 4.5, not only for bull's eye structures, but also various diffractive optical manipulations can be designed, such as off-axis beaming [53], focusing [54], and bundle beaming [55].

In principle, chiral bull's eye structures are also based on the same mechanism as conventional structures, so the period of the gratings is not modified. However, the phase difference along the azimuthal direction is provided by the monotonic increase in the offset distance from the hole to the grating. The continuous increase in the radius finally provides an Archimedean spiral grating structure, as shown in Figure 4.5a. The radius of the grating can be expressed by the simple relation $r(\phi) = r_1 + \Lambda\phi/2\pi$, where the parameters λ_0, r_0, r_1, and w are set to $\lambda_0 = 800$ nm, $r_0 = 50$ nm, $r_1 = 100$ nm, and $w = \Lambda/2$.

We now show the optical characteristics of the spiral bull's eye structure when it is located at the bottom of the metal substrate. When the spiral bull's eye is used, transmission efficiency enhancement is highly sensitive to the incident polarization state due to the chiral geometry [13,29]. The role of gratings is to couple the incident

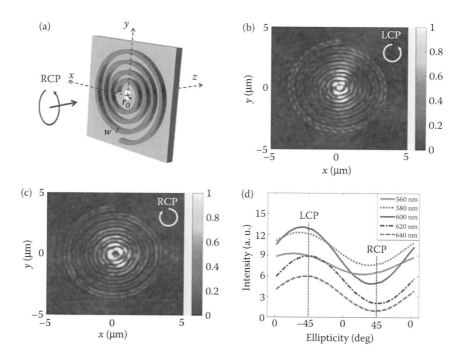

FIGURE 4.5 (**See color insert.**) (a) Schematic diagram of spiral bull's eye structure for the polarization-dependent enhancement of nanohole transmission. Electric field intensity distributions of 10-nm backside of the metal substrate with the (b) LCP and (c) RCP incident waves are shown, respectively, for comparing the field intensity at the center point. (d) Total transmission efficiencies varying with the optical polarization are shown with several grating periods. The results are normalized to the total transmission without the spiral gratings.

wave into the SPPs and merging them to the nanohole by their circular-shaped geometry. However, due to the phase delay caused by the spiral geometry, merged SPPs interfere differently at the center point varying with the incident polarization state [56]. Figures 4.5b and c depict the electric field intensity distribution with RCP and LCP light incidences on the plane 10 nm bottom from the metal substrate, respectively. The field intensity at the center point is dramatically changed by the direction of the optical handedness. As a result, modulation of the field intensity at the center point finally affects the transmission efficiency.

This aspect is clearly shown in Figure 4.5d. Transmission efficiency through the nanohole is strongly dependent on the polarization ellipticity, which has the highest value near an ellipticity of −45°, and is related to the LCP state shown in Figure 4.5b. In addition, the results also show that the effect of the grating period is quite important in terms of enhancing efficiency. Both the polarization sensitivity and overall transmission efficiency could be maximized near the condition of $\Lambda = 600$ nm. Although the detailed calculations are not shown here, this condition is well matched to the highest SPP excitation condition numerically calculated by the RCWA simulation [57].

Moreover, spiral bull's eye structure can also be used for polarization-dependent directional beaming [15]. In an analogy with a conventional bull's eye, the spiral grating is located on top of the metal substrate, as shown in Figure 4.6a. However, when a single spiral grating structure, as illustrated in Figure 4.5a, is used, it is known that the generated beam from the grating does not propagate straightly due to the asymmetric profile of the spiral shape [15]. Therefore, the grating needs to be modified to provide a symmetrical offset distance from the hole to the grating for providing on-axis beaming. The double-spiral grating structure shown in Figure 4.6a can be used for satisfying this condition [15].

It has been shown that the interference of waves scattered from the gratings can generate directional beaming along the z-axis as shown in Figure 4.6b when the incident polarization is opposite to the direction of the increasing grating radius. Here, the full-width at half-maximum of the generated beam at $z = 6.5\ \mu m$ plane is determined to be 976 nm, which is somewhat similar to the value obtained from the beam generated by a conventional bull's eye structure [15]. On the other hand, optical handedness, the same as increasing the direction of the spiral, cannot provide beaming characteristics as shown in Figure 4.6c.

These results can be simply explained by the concept of the addition (or subtraction) of topological charge as was explained in Section 4.2. Compared to the case of a plasmonic lens, a different important point to note is that the dominant electric field component of the z-directional beaming is not E_z but has a tangential component (E_x, E_y). This changes the condition of the directional beaming from $l_{total} = 0$ to $l_{total} = \pm 1$ [15]. Assuming the RCP incidence case of +1, since the double-spiral gratings have the same rotational direction as RCP, they result in a phase retardation of $2\pi \times 2$, so the geometrical angular momentum can be expressed as $l_{geometry} = 2$. The LCP incidence (−1) case shown in Figure 4.6b provides the total topological charge of $l_{total} = 1$.

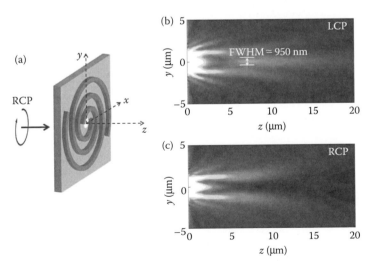

FIGURE 4.6 (a) Schematic diagram of double spiral bull's eye structure for the switching of beaming. Electric field intensity distributions along the y–z-plane from the end of grating layer is shown with the (b) LCP and (c) RCP incident waves, respectively.

Hence, it generates a directional beam with the opposite polarization. However, the RCP case shown in Figure 4.6c gives the total topological charge of $l_{total} = 3$. In this case, the vector sum of the tangential components are canceled out, so the directional beaming cannot be generated.

4.3.4 PLASMONIC RESPONSE OF ISOLATED CHIRAL NANOROD STRUCTURES

Although various applications using the nanorods such as plasmonic sensors with localized SPPs [58] were reported and anomalous spectral response based on electromagnetic-induced transparency [59] was studied recently, only a few studies have been reported for the case where chirality is associated with nanorod structures. Nevertheless, the near-field enhancement characteristics of localized SPP resonance in metallic nanorods and nanoparticles are one of the most anticipated features in plasmonic devices with chiral modifications. For example, isolated chiral nanorod structures have benefits, in that their volumes are small and that they interact strongly with light. It was reported that the optical force transferred from the linear momentum of photons is even strong enough to rotate a nanorod structure formed in a small dielectric fragment [30]. Especially, four-fold windmill-shaped nanorod structures were used for enhancing optical torque and it was demonstrated that the direction of the optical torque could be switched by changing the incident wavelength [30].

In Figure 4.7, we show numerical results obtained from Ref. [30] for a similar gammadion nanorod structure calculated using COMSOL Multiphysics. The specification of the gammadion structure is shown in Figure 4.7a; the width (w), thickness (t), and the overall size of the structure are set at 38, 30, and 200×200 nm, respectively. We first adjusted the incident polarization state to x-directional linear polarization and observed the wavelength-dependent characteristics of the gammadion nanorod structure. In Figure 4.7b and c, the electric field amplitude $\left(|E| - \sqrt{|E_x|^2 + |E_y|^2 + |E_z|^2} \right)$ and time-averaged optical power flow distributions along the plane cutting the middle of the nanorod are shown for two different incident wavelengths of 700 and 1700 nm, respectively. Similar to Ref. [30], the direction of optical power flow is reversed when the incident wavelength is changed, and such reversal of the flow of optical power could result in the reversal of the direction of optical torque, which can be finally used to the light-driven plasmonic motor that is switchable by incident wavelength switching [30].

The switchable plasmonic motor can be realized by the polarization switching as well as by the wavelength switching. Now, we demonstrate the polarization-dependent characteristics of gammadion nanorod structure in fixed wavelength conditions. Figure 4.7d and e shows these results—we calculate electric field amplitude and optical power flow with the RCP and LCP light incidences at the wavelength of 1100 nm. By comparing Figure 4.7d and e with Figure 4.7b and c, although both cases show the conversion of power flow directions, it is more uniformly distributed in cases of circular polarization than the linear polarization. In addition, it is shown that a much stronger and clear optical power flow conversion is obtained by changing the polarization state rather than the wavelength change. Therefore, we expect that a plasmonic

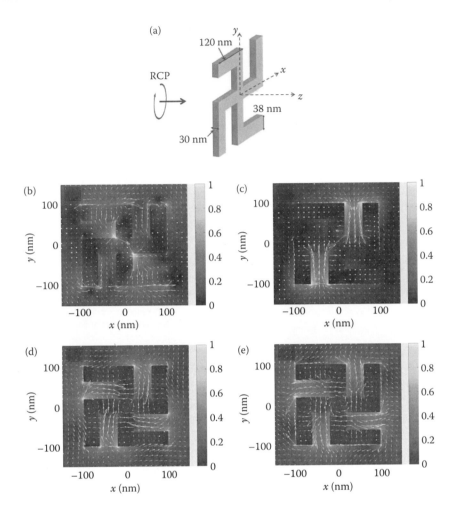

FIGURE 4.7 (a) Schematic diagram of a gammadion-shaped plasmonic nanorod. Wavelength-dependent characteristics of a gammadion nanorod: electric field amplitude (image) and time-averaged power flows (arrows) at incident wavelengths of (b) 700 nm and (c) 1700 nm are depicted, respectively. Polarization-dependent characteristics of gammadion nanorod: results for (d) LCP and (e) RCP light incident with a wavelength of 1100 nm are shown, respectively.

motor can be produced with lower incident power and simpler switching condition by taking advantage of the polarization dependency of chiral nanorod structures.

4.4 OPTICAL CHARACTERISTICS IN CHIRAL PERIODIC STRUCTURE: CHIRALITY IN METAMATERIALS

At this point, we focus on the representative results that demonstrate the existence of chiral metamaterials with polarization-dependent characteristics. Unlike the structures investigated in Section 4.3, the structures in this section are periodically arranged and the size of the unit cell is smaller than the wavelength employed. Therefore, an effect caused by a large geometry such as phase compensation by a spiral geometry or transmission enhancement by a grating is not the main mechanism for how these structures operate. On the other hand, electrons in the metallic portion of a unit cell that are affected by external light could respond in various manners, depending on their geometrical characteristics. Such electron–light interaction in metamaterials is often expressed with the concept of SPPs in the visible and infrared regions. Therefore, these chiral metamaterials can be included in the scope of plasmonic chiral devices. Indeed, much more work has been done concerning chiral metamaterial issues rather than isolated chiral plasmonic devices [18–23,60], and various phenomena such as asymmetric optical transmission [18,19], polarization rotation [17,20], and negative index material [22,23] have been reported. Although some of these works are demonstrated in THz or in longer-wavelength regions, similar analogies could also be adapted in the optical region.

Figure 4.8 shows representative shapes of the chiral metamaterials; each shape provides different optical characteristic due to its own geometry. The connected S-shaped metal nanowire structure shown in Figure 4.8a was reported for asymmetric optical transmission. The asymmetric optical transmission is said to be achieved by the chirality and anisotropy of the unit cell structure [19]. In addition, different resonance characteristics of localized SPPs in nanowires, depending on the optical handedness, lead to different amounts of dissipation, so finally they help asymmetric transmission. Another geometry shown in Figure 4.8b is an array of gammadion structures. It has been shown that the unit cell of a gammadion structure could be used for transmission with polarization rotation [20] or the emission of circularly polarized light from an unpolarized source [21,61]. Moreover, it has been shown

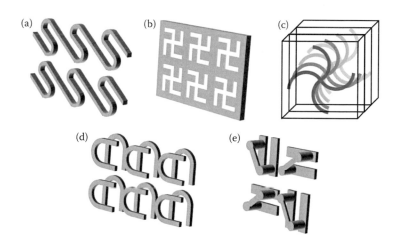

FIGURE 4.8 Schematic diagrams of various metamaterial structures that have chiral geometry. For example, (a) periodically arranged, connected S-shaped nanowire structure, (b) array of gammadion structure, (c) stacked gammadion multiple layer structure with negative refractive index, (d) crossed U-shaped metal structure, and (e) bridge structure based on circuit analogy are illustrated.

that a negative refractive index material can also be made by stacking the periodic gammadion structure with a small twist angle, as shown in Figure 4.8c [23]. Such an approach is an experimental verification of negative refractive index material achieved by single chiral resonance which is different from conventional negative refractive index materials obtained by separate electric and magnetic resonances [62]. In addition to these examples, more complicated structures such as metal screw hole arrays [63], twisted cross-shaped rods [31], chiral bridge structures based on circuit analogy [22], and twisted U-shaped metallic rods [64,65] have been demonstrated for achieving higher optical activity and negative refractive index material in the near-infrared and THz regions.

In Figure 4.9, we reproduce one of the results mentioned above, which is an S-shaped metal nanowire array that is used for asymmetric optical transmission. In order to set the SPP resonance at the near-infrared region, the unit cell length of the proposed metamaterial and the cross-section of the nanowire are set to 550 and 50×50 nm, respectively. Due to the chirality, optical characteristic of the metamaterial with bottom illumination is not same as the case of top illumination, but they are identical when the geometry is flipped along the x- or y-axis. Moreover, for normal incidence case, it can be expected that the flipping of geometry is replaced with that of incident polarization since the same optical characteristic is produced

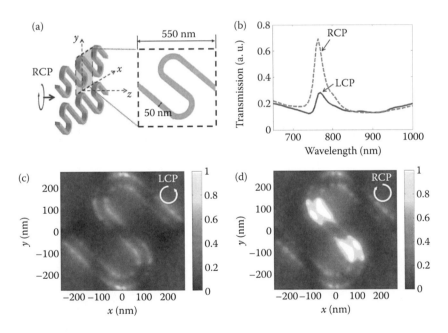

FIGURE 4.9 (a) Overall schematic and unit cell design properties of a connected S-shaped metal nanowire structure. (b) Transmission characteristics of the S-shaped metamaterial structure varying with the incident wavelength from 650 to 1000 nm are shown for both LCP and RCP incidences. Electric field intensity distributions on the unit cell are compared for (c) LCP and (d) RCP light incidences.

when both the geometry and optical handedness are simultaneously reversed. By combining these relations, it would be expected that the optical transmission characteristic of bottom LCP (RCP) illumination is the same as that for top RCP (LCP) illumination. Therefore, comparing RCP and LCP incident cases is sufficient for determining asymmetric optical transmission characteristics [19]. In Figure 4.9b, the spectral response of the S-shaped metamaterial with a wavelength range from 650 to 1000 nm is shown. By comparing the two polarization cases, large amount of transmission difference is observed near the resonance peak located near 770-nm region, which is a resonance peak generated by the effect of SPPs. According to Ref. [19], the origin of such asymmetric transmission is related to the polarization-sensitive excitation of SPPs in the metal. To verify this effect, the field intensity distribution along the plane 10 nm above the metamaterial layer with opposite optical handedness is compared in Figure 4.9c and d. The most strongly enhanced spot in the unit cell geometry is quite different for the two cases. In the case of LCP incidence, the electric field is uniformly enhanced along the metal wire geometry. However, in the case of RCP illumination, the enhancement in electric field is much more heavily concentrated in the middle curve of the S-shaped structure and it provides stronger resonance than the case of LCP illumination. Such a polarization-dependent characteristic of localized SPP resonance in the unit cell structure finally induces the asymmetric optical transmission spectrum [19].

4.5   CONCLUSIONS

In this chapter, we reviewed the principles and optical characteristics of plasmonic chiral structures. Plasmonic chiral structures can be classified into isolated plasmonic chiral devices and chiral metamaterials, which are determined by their periodicity and working mechanisms. In the case of isolated chiral devices, most are obtained by combining the fundamental plasmon-based structure such as a nanohole, nanorod, plasmonic lens, and bull's eye geometry with a chiral geometry such as a gammadion, spiral, and whirlpool shape. Therefore, the characteristics of isolated chiral plasmonic devices have both characteristics that originate from the fundamental plasmonic structure and from chiral geometries. These additions of chirality to the plasmon-based structure could permit the conventional plasmonic effect to be extended into a wide variety of areas, such as polarization-sensitive EOT, hot/dark spot conversion, and plasmonic beam switching. On the other hand, chiral metamaterials, defined by the periodic structures composed of a unit cell with a chiral geometry, have been intensively researched due to their potential for anomalous optical characteristics. Although the operating wavelengths of most chiral metamaterials are currently in the microwave and THz regimes, due to difficulties associated with nanosize unit cell fabrication, some recent studies successfully demonstrated them in the visible and infrared regimes and attempted to explain the optical phenomenon with the help of SPP resonances. It can thus be concluded that the role of SPP in nanosize chiral devices is quite important and reported results for these devices have enormous potential for the application in optical integrated circuit and near-field optics.

ACKNOWLEDGMENTS

This work was supported by the National Research Foundation and the Ministry of Education, Science, and Technology of Korea through the Creative Research Initiatives Program (Active Plasmonics Application Systems).

REFERENCES

1. L. Barron, *Molecular Light Scattering and Optical Activity* (Cambridge University Press, Cambridge, UK, 1982).
2. E. Plum, X.-X. Liu, V. A. Fedotov, Y. Chen, D. P. Tsai, and N. I. Zheludev, Metamaterials: Optical activity without chirality, *Phys. Rev. Lett.* **102**, 113902, 2009.
3. P. V. Lenzo, E. G. Spencer, and A. A. Ballman, Optical activity and electrooptic effect in bismuth germanium oxide ($Bi_{12}GeO_{20}$), *Appl. Optics* **5**, 1688–1689, 1966.
4. S. Y. Lin, J. G. Fleming, D. L. Hetherington, B. K. Smith, R. Biswas, K. M. Ho, M. M. Sigalas, W. Zubrzycki, S. R. Kurtz, and Jim Bur, A three-dimensional photonic crystal operating at infrared wavelengths, *Nature* **394**, 251–253, 1998.
5. R. A. Shelby, D. R. Smith, and S. Schultz, Experimental verification of a negative index of refraction, *Science* **292**, 77–79, 2001.
6. H. J. Lezec, J. A. Dionne, and H. A. Atwater, Negative refraction at visible frequencies, *Science* **316**, 430–432, 2007.
7. N. Kanda, K. Konishi, and M. Kuwata-Gonokami, Terahertz wave polarization rotation with double layered metal grating of complimentary chiral patterns, *Opt. Express* **15**, 11117–11125, 2007.
8. K. L. Tsakmakidis, A. D. Boardman, and O. Hess, 'Trapped rainbow' storage of light in metamaterials, *Nature* **450**, 397–401, 2007.
9. H. Raether, *Surface Plasmons* (Springer-Verlag, Berlin, 1988).
10. W. L. Barnes, A. Dereux, and T. W. Ebbesen, Surface plasmon subwavelength optics, *Nature* **424**, 824, 2003.
11. Z. Liu, J. M. Steele, W. Srituravanich, Y. Pikus, C. Sun, and X. Zhang, Focusing surface plasmons with a plasmonic lens, *Nano Lett.* **5**, 1726–1729, 2005.
12. H. Kim, J. Park, S.-W. Cho, S.-Y. Lee, M. Kang, and B. Lee, Synthesis and dynamic switching of surface plasmon vortices with plasmonic vortex lens, *Nano Lett.* **10**, 529–536, 2010.
13. A. Drezet, C. Genet, J.-Y. Laluet, and T. W. Ebbesen, Optical chirality without optical activity: How surface plasmons give a twist to light, *Opt. Express* **16**, 12559–12570, 2008.
14. J. I. Ziegler and R. F. Haglund, Jr., Plasmonic response of nanoscale spirals, *Nano Lett.* **10**, 3013–3018, 2010.
15. S.-Y. Lee, I.-M. Lee, J. Park, C.-Y. Hwang, and B. Lee, Dynamic switching of the chiral beam on the spiral plasmonic bull's eye structure, *Appl. Optics* **50**, G104–G112, 2011.
16. K. Y. Bliokh, Y. Gorodetski, V. Kleiner, and E. Hasman, Coriolis effect in optics: Unified geometric phase and spin-Hall effect, *Phys. Rev. Lett.* **101**, 030404, 2008.
17. L. D.-C. Tzuang, Y.-W. Jiang, Y.-H. Ye, Y.-T. Chang, Y.-T. Wu, and S.-C. Lee, Polarization rotation of shape resonance in Archimedean spiral slots, *Appl. Phys. Lett.* **94**, 091912, 2009.
18. V. A. Fedotov, P. L. Mladyonov, S. L. Prosvirnin, A. V. Rogacheva, Y. Chen, and N. I. Zheludev, Asymmetric propagation of electromagnetic waves through a planar chiral structure, *Phys. Rev. Lett.* **97**, 167401, 2006.
19. V. A. Fedotov, A. S. Schwanecke, N. I. Zheludev, V. V. Khardikov, and S. L. Prosvirnin, Asymmetric transmission of light and enantiomerically sensitive plasmon resonance in planar chiral nanostructures, *Nano Lett.* **7,** 1996–1999, 2007.

20. E. Plum, J. Zhou, J. Dong, V. A. Fedotov, T. Koschny, C. M. Soukoulis, and N. I. Zheludev, Near-field polarization conversion in planar chiral nanostructures, *Phys. Rev. B* **79**, 035407, 2009.

21. K. Konishi, T. Sugimoto, B. Bai, Y. Svirko, and M. Kuwata-Gonokami, Effect of surface plasmon resonance on the optical activity of chiral metal nanogratings, *Opt. Express* **15**, 9575–9583, 2007.

22. S. Zhang, Y.-S. Park, J. Li, X. Lu, W. Zhang, and X. Zhang, Negative refractive index in chiral metamaterials, *Phys. Rev. Lett.* **102**, 023901, 2009.

23. E. Plum, J. Zhou, J. Dong, V. A. Fedotov, T. Koschny, C. M. Soukoulis, and N. I. Zheludev, Metamaterial with negative index due to chirality, *Phys. Rev. B* **79**, 035407, 2009.

24. A. V. Krasavin, A. S. Schwanecke, N. I. Zheludev, M. Reichelt, T. Stroucken, S. W. Koch, and E. M. Wright, Polarization conversion and 'focusing' of light propagating through a small chiral hole in a metallic screen, *Appl. Phys. Lett.* **86**, 201105, 2005.

25. A. V. Krasavin, A. S. Schwanecke, and N. I. Zheludev, Extraordinary properties of light transmission through a small chiral hole in a metallic screen, *J. Opt. A Pure Appl. Opt.* **8**, S98–S105, 2006.

26. Y. Gorodetski, A. Niv, V. Kleiner, and E. Hasman, Observation of the spin-based plasmonic effect in nanoscale structures, *Phys. Rev. Lett.* **101**, 043903, 2008.

27. S. Yang, W. Chen, R. L. Nelson, and Q. Zhan, Miniature circular polarization analyzer with spiral plasmonic lens, *Opt. Lett.* **34**, 3047–3049, 2009.

28. W. Chen, D. C. Abeysinghe, R. L. Nelson, and Q. Zhan, Experimental confirmation of miniature spiral plasmonic lens as a circular polarization analyzer, *Nano Lett.* **10**, 2075–2079, 2010.

29. T. Ohno and S. Miyanishi, Study of surface plasmon chirality induced by Archimedes' spiral grooves, *Opt. Express* **14**, 6285–6290, 2006.

30. M. Liu, T. Zentgraf, Y. Liu, G. Bartal, and X. Zhang, Light-driven nanoscale plasmonic motors, *Nat. Nanotechnol.* **5**, 570–573, 2010.

31. M. Decker, M. Ruther, C. E. Kriegler, J. Zhou, C. M. Soukoulis, S. Linden, and M. Wegener, Strong optical activity from twisted-cross photonic metamaterials, *Opt. Lett.* **34**, 2501–2503, 2009.

32. X. Shi and L. Hesselink, Ultrahigh light transmission through a C-shaped nanoaperture, *Opt. Lett.* **28**, 1320–1322, 2003.

33. B. Lee, I.-M. Lee, S. Kim, D.-H. Oh, and L. Hesselink, Review on subwavelength confinement of light with plasmonics, *J. Mod. Opt.* **57**, 1479–1497, 2010.

34. L. Wang, S. M. Uppuluri, E. X. Jin, and X. Xu, Nanolithography using high transmission nanoscale bowtie apertures, *Nano Lett.* **6**, 361–364, 2006.

35. L. Wang and X. Xu, High transmission nanoscale bowtie-shaped aperture probe for near-field optical imaging, *Appl. Phys. Lett.* **90**, 261105, 2007.

36. H. Kim and B. Lee, Diffractive slit patterns for focusing surface plasmon polaritons, *Opt. Express*, **16**, 8969–8980, 2008.

37. L. Martín-Moreno, F. J. García-Vidal, H. J. Lezec, A. Degiron, and T. W. Ebbesen, Theory of highly directional emission from a single subwavelength aperture surrounded by surface corrugations, *Phys. Rev. Lett.* **90**, 167401, 2003.

38. T. Thio, K. M. Pellerin, R. A. Linke, H. J. Lezec, and T. W. Ebbesen, Enhanced light transmission through a single subwavelength aperture, *Opt. Lett.* **26**, 1972–1974, 2001.

39. H. J. Lezec, A. Degiron, E. Devaux, R. A. Linke, L. Martín-Moreno, F. J. García-Vidal, and T. W. Ebbesen, Beaming light from a subwavelength aperture, *Science* **297**, 820–822, 2002.

40. T. W. Ebbesen, H. J. Lezec, H. F. Ghaemi, T. Thio, and P. A. Wolff, Extraordinary optical transmission through sub-wavelength hole arrays, *Nature* **391**, 667–669, 1998.

41. L. T. Vuong, A. J. L. Adam, J. M. Brok, P. C. M. Planken, and H. P. Urbach, Electromagnetic spin-orbit interactions via scattering of subwavelength apertures, *Phys. Rev. Lett.* **104**, 083903, 2010.

42. H. Kim, I.-M. Lee, and B. Lee, Extended scattering-matrix method for efficient full parallel implementation of rigorous coupled-wave analysis, *J. Opt. Soc. Am.* A **24**, 2313–2327, 2007.

43. E. D. Palik, *Handbook of Optical Constants of Solids*, 2nd ed. (Academic Press, 1998), *Chap.* 11, 356.

44. L. Yin, V. K. Vlasko-Vlasov, J. Pearson, J. M. Hiller, J. Hua, U. Welp, D. E. Brown, and C. W. Kimball, Subwavelength focusing and guiding of surface plasmons, *Nano Lett.* **5**, 1399–1402, 2005.

45. Z. Liu, J. M. Steele, H. Lee, and X. Zhang, Tuning the focus of a plasmonic lens by the incident angle, *Appl. Phys. Lett.* **88**, 171108, 2006.

46. G. M. Lerman, A. Yanai, and U. Levy, Demonstration of nanofocusing by the use of plasmonic lens illuminated with radially polarized light, *Nano Lett.* **9**, 2139–2143, 2009.

47. Z. Fang, Q. Peng, W. Song, F. Hao, J. Wang, P. Nordlander, and X. Zhu, Plasmonic focusing in symmetry broken nanocorrals, *Nano Lett.* **11**, 893–897, 2011.

48. J. Miao, Y. Wang, C. Guo, Y. Tian, S. Guo, Q. Liu, and Z. Zhou, Plasmonic lens with multiple-turn spiral nano-structures, *Plasmonics* **6**, 235–239, 2011.

49. Z. Liu, Y. Wang, J. Yao, H. Lee, W. Srituravanich, and X. Zhang, Broad band two-dimensional manipulation of surface plasmons, *Nano Lett.* **9**, 462–466, 2009.

50. T. Thio, H. J. Lezec, T. W. Ebbesen, K. M. Pellerin, G. D. Lewen, A. Nahata, and R. A. Linke, Giant optical transmission of sub-wavelength apertures: Physics and applications, *Nanotechnology* **13**, 429–432, 2002.

51. O. Mahboub, S. Carretero-Palacios, C. Genet, F. J. García-Vidal, S. G. Rodrigo, L. Martín-Moreno, and T. W. Ebbesen, Optimization of bull's eye structures for transmission enhancement, *Opt. Express* **18**, 11292–11299, 2010.

52. S. Carretero-Palacios, O. Mahboub, F. J. Garcia-Vidal, L. Martin-Moreno, S. G. Rodrigo, C. Genet, and T. W. Ebbesen, Mechanisms for extraordinary optical transmission through bull's eye structures, *Opt. Express* **19**, 10429–10442, 2011.

53. S. Kim, H. Kim, Y. Lim, and B. Lee, Off-axis directional beaming of optical field diffracted by a single subwavelength metal slit with asymmetric dielectric surface gratings, *Appl. Phys. Lett.* **90**, 051113, 2007.

54. S. Kim, Y. Lim, H. Kim, J. Park, and B. Lee, Optical beam focusing by a single subwavelength metal slit surrounded by chirped dielectric surface gratings, *Appl. Phys. Lett.* **92**, 013103, 2008.

55. S. Kim, Y. Lim, J. Park, and B. Lee, Bundle beaming from multiple subwavelength slits surrounded by dielectric surface gratings, *J. Lightwave Technol.* **28**, 2023–2029, 2010.

56. Y. Gorodetski, N. Shitrit, I. Bretner, V. Kleiner, and E. Hasman, Observation of optical spin symmetry breaking in nanoapertures, *Nano Lett.* **9**, 3016–3019, 2009.

57. H. Kim, J. Park, and B. Lee, Finite-size nondiffracting beam from a subwavelength metallic hole with concentric dielectric gratings, *Appl. Optics* **48**, G68–G72, 2009.

58. T. Chung, S.-Y. Lee, E. Y. Song, H. Chun, and B. Lee, Plasmonic nanostructures for nano-scale bio-sensing, *Sensors*, **11**, 10907–10929, 2011.

59. S. Zhang, D. A. Genov, Y. Wang, M. Liu, and X. Zhang, Plasmon-induced transparency in metamaterials, *Phys. Rev. Lett.* **101**, 047401, 2008.

60. B. Wang, J. Zhou, T. Koschny, M. Kafesaki, and C. M. Soukoulis, Chiral metamaterials: Simulations and experiments, *J. Opt. A Pure Appl. Opt.* **11**, 114003, 2009.

61. K. Konishi, M. Nomura, N. Kumagai, S. Iwamoto, Y. Arakawa, and M. Kuwata-Gonokami, Circularly polarized light emission from semiconductor planar chiral nanostructures, *Phys. Rev. Lett.* **106**, 057402, 2011.

62. J. B. Pendry, A chiral route to negative refraction, *Science* **306**, 1353–1355, 2004.

63. F. Miyamaru and M. Hangyo, Strong optical activity in chiral metamaterials of metal screw hole arrays, *Appl. Phys. Lett.* **89**, 211105, 2006.

64. X. Xiong, W.-H. Sun, Y.-J. Bao, M. Wang, R.-W. Peng, C. Sun, X. Lu, J. Shao, Z.-F. Li, and N.-B. Ming, Construction of a chiral metamaterial with a U-shaped resonator assembly, *Phys. Rev.* B **81**, 075119, 2010.

65. Z. Li, R. Zhao, T. Koschny, M. Kafesaki, K. B. Alici, E. Colak, H. Caglayan, E. Ozbay, and C. M. Soukoulis, Chiral metamaterials with negative refractive index based on four U split ring resonators, *Appl. Phys. Lett.* **97**, 081901, 2010.

5 Optical Metamaterials and Their Fabrication Techniques

Takuo Tanaka

CONTENTS

5.1 INTRODUCTION

Metamaterial is an artificially designed material that consists of nanoscale metal structures. When we designed the structure of metamaterials to be much smaller than the wavelength of light, the light cannot detect each structure owing to its wave nature, and metamaterials act as quasi-homogeneous materials. Hence, they are termed "metamaterials" not "metastructures" (Figure 5.1).

Metamaterial technology covers a wide range of spectra from MHz to several hundreds of THz including the visible light region. We call the metamaterials that work in the high-frequency region such as near-infrared to visible light as "optical metamaterial." Because the wavelength of light is in the micro- to sub-micrometer scale, the structure of optical metamaterials too is in the sub-micrometer or nanometer scale. When the light is illuminated onto such fine metal structures, the free

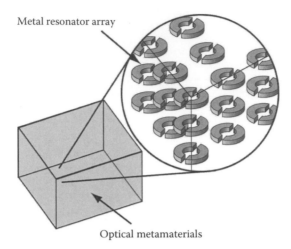

Metal resonator array

Optical metamaterials

FIGURE 5.1 Optical metamaterials.

electrons in the metals oscillate collectively and the so-called "local-mode surface plasmons" are excited in the structures. Therefore, optical metamaterials are sometimes called "plasmonic metamaterials."

The most interesting feature of the metamaterials is that their electromagnetic properties result not only from their composition, but also from their sub-wavelength-engineered metallic structures. By engineering such artificial materials, we can create materials exhibiting desired electromagnetic properties that are normally not attainable with naturally occurring materials. The creation of the magnetically active material is one of the most important and interesting applications of the meta-material because all materials, in nature, lose magnetic response in the visible light frequency region and hence their μ is fixed at unity. The magnetic metamaterial with $\mu \neq 1$ produces a great number of novel materials in the optical frequency region, which enables us to manipulate light freely. For example, "optical cloaking," which renders the object invisible, was proposed [1] and experimentally investigated in the microwave region [2]. The concept of the metamaterial is the introduction of a new paradigm in the research field from microwave to optical region.

In this chapter, first the theoretical background of the optical metamaterials with a discussion about the appropriate materials and structures to gain the magnetism in the visible light region is described, and then some nanofabrication techniques utilized for metamaterials are reviewed. At the end, some applications of the meta-materials for optical device and optical cloaking are introduced.

5.2 STRUCTURE OF OPTICAL METAMATERIALS

In this section, we discuss about the metamaterial technique for modifying magnetic permeability of the materials in the visible frequency region and clarify the suitable materials and structures for metamaterials that work as magnetic metamaterials in the visible region [3–5].

5.2.1 Optical Properties of Metals

First of all, in order to describe the dispersion properties of metals in THz to visible light region, the internal impedance for a unit length and a unit width of the plane conductor ($Z_s(\omega)$) is introduced as

$$Z_s(\omega) = \cfrac{1}{\sigma(\omega)\displaystyle\int_0^T \cfrac{\exp[ik(\omega)z] + \exp[ik(\omega)(T-z)]}{1 + \exp[ik(\omega)T]}\,dz} = R_s(\omega) + iX_s(\omega),$$ (5.1)

$$k(\omega) = \omega\sqrt{\varepsilon_0\mu_0\left[1 + i\frac{\sigma(\omega)}{\omega\varepsilon_0}\right]},$$ (5.2)

where the real and imaginary parts of $Z_s(\omega)$ are the surface resistivity R_s and the internal reactance X_s, respectively. The integration in the denominator of Equation 5.1 indicates the total current flowing through the cross-section of the conductor.

5.2.2 Dispersion Properties of Split Ring Resonators

Figure 5.2 shows the calculation model using a split ring resonator (SRR), which is proposed by Pendry, as a unit element of the metamaterial [6]. In Figure 5. 2, r is the radius of the ring, w is the width of the ring, g is the distance between two rings of SRR, a is the unit-cell dimension in the xy-plane, and l is the distance between adjacent planes of the SRRs along the z-axis.

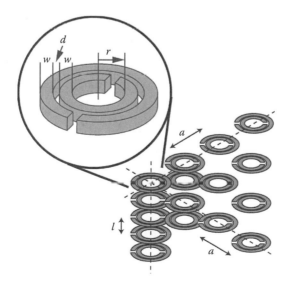

FIGURE 5.2 Sketch of the structure of optical metamaterials and an SRR that is for artificially controlled magnetic permeability.

Based on the dispersive properties of metals described in Equation 5.1, the frequency dependence of the magnetic responses of the metallic SRRs in the optical frequency region was calculated, and the effective permeability (μ_{eff}) of the SRRs is derived as

$$\mu_{eff} = \mu'_{eff} + i\mu''_{eff} = 1 - \frac{F\omega^2}{\omega^2 - 1/CL + i(Z(\omega)\omega/L)}, \tag{5.3}$$

where C and L are the geometrical capacitance and inductance, respectively, and F and $Z(\omega)$ are the filling factor and the ring metal impedance, respectively, defined as

$$F = \frac{\pi r^2}{a^2} \tag{5.4}$$

$$C = \frac{2\pi r}{3}\varepsilon_0\varepsilon_r\frac{K[(1-t^2)^{1/2}]}{K(t)}, \tag{5.5}$$

$$t = \frac{g}{2w + g}, \tag{5.6}$$

$$L = \frac{\mu_0\pi r^2}{l}, \tag{5.7}$$

$$Z(\omega) = \frac{2\pi r Z_s(\omega)}{w}. \tag{5.8}$$

Here, $K[\cdot]$ in Equation 5.5 is the complete elliptic integral of the first kind. For calculating the geometrical capacitance, Gupta et al.'s [7] formula was used to estimate the capacitance coming from the distance between two rings per unit length.

By using Equations 5.1 and 5.3, the empirical values of the plasma frequency (ω_p), and the damping constants (γ) of silver, gold, and copper ($\omega_p = 14.0 \times 10^{15}$ s^{-1} and $\gamma = 32.3 \times 10^{12}$ s^{-1} for silver, $\omega_p = 13.8 \times 10^{15}$ s^{-1} and $\gamma = 107.5 \times 10^{12}$ s^{-1} for gold, and $\omega_p = 13.4 \times 10^{15}$ s^{-1} and $\gamma = 144.9 \times 10^{12}$ s^{-1} for copper), the frequency dispersions of μ_{eff} from 100 to 800 THz were calculated [8].

At the resonant frequency of the SRR, μ_{eff} changes both positively and negatively and takes max μ_{eff} and min μ_{eff} accordingly as shown in Figure 5.3. We defined the change in μ_{eff} as the difference between max μ_{eff} and min μ_{eff}, and we plotted the calculation results in Figure 5.4 for each metal SRRs. From these results, it was clarified that a three-dimensional array of SRRs made of silver can produce a strong magnetic response in the visible light frequency region. As seen from Figure 5.4, silver SRRs exhibit μ_{eff} changes exceeding 2.0 in the entire visible range, which means that μ_{eff} can become a negative value, whereas the responses of gold and copper SRRs are <2.0 in the visible light region.

In Table 5.1, the design strategy of the optical metamaterials that can reveal the magnetic response is summarized. In the lower frequency region, that is, <100 THz, to realize a relatively low-resonant frequency and a low resistance of the metal structure, the structure should possess both a large geometrical capacitance and a wide

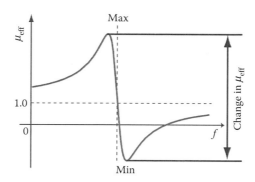

FIGURE 5.3 μ_{eff} changes positively and negatively and it takes max μ_{eff} and min μ_{eff} at the resonant frequency of the SRR.

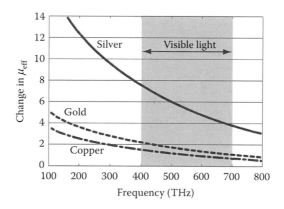

FIGURE 5.4 Frequency dependence of the change of μ_{eff} of the SRRs made of silver, gold, and copper.

TABLE 5.1
Design Strategy of Metamaterial Structure in the Optical Frequency Region Below and Above 100 THz

Frequency	<100 THz	100 THz<
Requirement	Large C and wide ring	Small C and large L
Resonator structure		
Reason of the decrease of magnetic response	Increase of surface resistance	Decrease of L by scaling

width of the ring, and the original shape of SRR, which is concentric double rings with gaps, is appropriate. On the other hand, when the frequency is higher than 100 THz, the effect of the reactance $(X_s(\omega))$ in the ring becomes more dominant than that of the resistance in the ring of the SRR. In such a higher frequency region, to increase the resonant frequency, the resonant structure should have a small geometrical capacitance, and in order to keep the high Q-value and sufficient magnetic responses of the SRRs, a large geometrical inductance of the structure is necessary. To satisfy this requirement, we proposed that a single ring with a number of cuts is more suitable than the original double-ring SRRs, because this is advantageous in preventing the effect of $X_s(\omega)$.

5.3   FABRICATION TECHNIQUES FOR METAMATERIALS

In this section, we will discuss about the several fabrication techniques used for making metamaterials. Owing to the space limitation, only significant results are reviewed.

5.3.1   METAMATERIALS FABRICATED BY PHOTOLITHOGRAPHY TECHNIQUE

Yen et al. [9] reported the experimental result and theoretical verification of the magnetic metamaterials that work at THz region. They fabricated an array of SRRs as shown in Figure 5.5. An SRR consists of two concentric square rings situated opposite to each other with a gap. When a time-varying external magnetic field H is applied to the SRR, an induced current flows around the two rings through the geometrical capacitance of the coplanar strips. The circular current produces the internal magnetic field that results in the effective permeability change.

Yen et al. fabricated the metamaterial structure by using a special photolithography technique termed as "photo-proliferated process" (PPP). The schematic of the process is shown in Figure 5.6.

At the first step, a 5-μm-thick negative photoresist layer is spun onto a quartz substrate, and then the designed SRR pattern was transferred using the contact

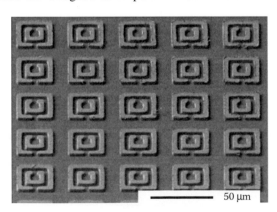

FIGURE 5.5   A secondary ion image of an array of SRR for THz frequency fabricated by PPPs. (Adapted from T. J. Yen et al., Terahertz magnetic response from artificial materials, *Science* **303**, 1494–1496, 2004. Copyright 2004, AAAS.)

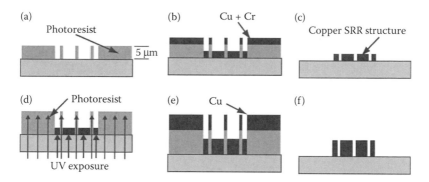

FIGURE 5.6 The process flow of PPP. (a) Resist patterning, (b) deposition of 100-nm-thick Cr and 1-μm-thick Cu layers, (c) lift-off process, (d) photoresist coating and backside UV exposure, (e) evaporation of thick Cu layer, and (f) lift-off process. (Adapted from T. J. Yen et al., Terahertz magnetic response from artificial materials, *Science* **303**, 1494–1496, 2004. Copyright 2004, AAAS.)

photolithographic method (Figure 5.6a). After the photoresist patterning, 100-nm-thick chromium and 1-μm-thick copper layers are deposited by using an electron beam evaporator (Figure 5.6b). Lift-off process using an acetone rinsed in an ultrasonic bath was employed to remove the photoresist to preserve the copper SRR pattern (Figure 5.6c). Photoresist layer is spin-coated again on the SRR pattern for the second lithographic patterning process. In the second lithography process, UV light was exposed from the bottom of the quartz substrate using the copper SRR pattern fabricated in the first lithography as a photomask (Figure 5.6d). After the second UV exposure, evaporation of copper layer and lift-off process are repeated in order to increase the metal thickness of the SRR structures (Figure 5.6e). In the experiment, the magnetic response from the fabricated structure at 1 THz was demonstrated.

5.3.2 METAMATERIALS BY FOCUSED ION-BEAM MILLING TECHNIQUE

Three-dimensional optical metamaterials with a negative index of refraction were reported by Zhang and coworkers [10]. They fabricated the three-dimensional fishnet metamaterials on a multilayer metal–dielectric stack by using focused ion-beam milling (FIB) method. As shown in Figure 5.7a, the 3D fishnet metamaterial consists of a multilayer metal–dielectric film stack. The multilayer stack was deposited by electron beam evaporation of alternating layers of 30-nm silver (Ag) and 50-nm magnesium fluoride (MgF_2) thin films. Twenty-one layers with a total thickness of 830 nm were stacked on a transparent substrate. The fishnet pattern of 22×22 in-plane cells of 860 nm periodicity ($a = 565$ nm and $b = 265$ nm) was fabricated on all the 21 layers by using the FIB techniques. This sample was used for the characterization of the transmittance, and the figure of merit of their sample was 3.5 at $\lambda = 1775$ nm. They made another type of sample consisting of a prism fabricated on the multilayer stack with the number of functional layers ranging from 1 on the one side to 10 on the other side (Figure 5.7b). The prism with a 10×10 fishnet pattern was formed by FIB etching the film. This sample was used to measure the effective

(a)   (b)

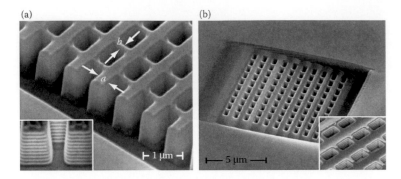

FIGURE 5.7   (a) An SEM image of the 21-layer fishnet structure with the side etched for appearance of the cross-section. The inset shows a cross-section of the pattern taken at 45°. (b) An SEM image of the fabricated 3D fishnet NIM prism. A prism was fabricated on the multi-layer stack with 10 functional layers using FIB. (Adapted from J. Valentine et al., *Nature* **455**, 376–379, 2008. Copyright 2008, Macmillan Publishes Ltd.)

refractive index of the fishnet structure. By measuring the absolute angle of refraction, they concluded that the effective refractive index varies from $n = 0.63 \pm 0.05$ at $\lambda = 1200$ nm to $n = -1.23 \pm 0.34$ at $\lambda = 1775$ nm.

5.3.3   STEREO METAMATERIALS FORMED BY ELECTRON-BEAM LITHOGRAPHY AND ION BEAM ETCHING TECHNIQUES

In 2009, Liu et al. [11] reported the metamaterial structure that consists of an array of a stack of two identical SRRs (Figure 5.8). They fabricated three types of stereo-SRR dimer metamaterials with specific twist angles $\phi = 0°$, 90°, and 180°.

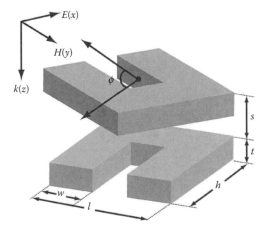

FIGURE 5.8   A schematic diagram of the stereo-SRR dimer metamaterials. The geometrical parameters l, h, w, t, and s are 230, 230, 90, 50, and 100 nm, respectively. The period of SRR dimer in lateral direction is 700 nm. (Adapted from J. Valentine et al., *Nature* **455**, 376–379, 2008. Copyright 2009, Macmillan Publishes Ltd.)

Fabrication process of stereometamaterials is as follows. Three gold alignment marks (size 4×100 µm) with a thickness of 250 nm were fabricated using a lift-off process on a quartz substrate. The substrate was covered with another 50-nm gold film using electron-beam evaporation, and then gold SRR structures were fabricated using electron-beam lithography and Ar^+ beam etching processes. On the SRR patterns, a spacer layer of photopolymer (PC403, 120 nm in thickness) was spin-coated. Subsequently, a second SRR structure was fabricated on the sample using gold film evaporation and electron beam lithography. The total area of the fabricated structures was 200×200 µm.

Transmittance spectra were measured with a Fourier-transform infrared spectrometer combined with an infrared microscope (Figure 5.9). From the transmittance

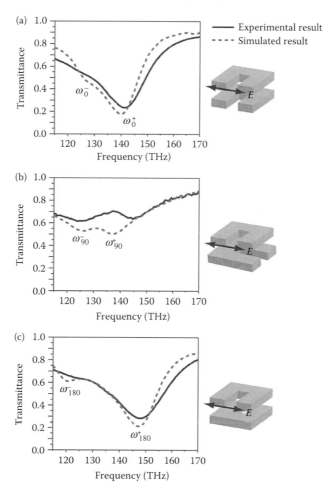

FIGURE 5.9 Transmittance spectra for twisted gold SRR dimer metamaterials: (a–c) 0°, 90°, and 180° twisted gold SRR dimer metamaterials. The gold SRR dimer was fabricated by electron–beam lithography and lift-off process. The SRRs were embedded in a photopolymer (PC403), which served as the dielectric spacer between SRRs.

spectra of three types of stereo-SRR dimer metamaterials, Liu et al. demonstrated that the lower frequency resonant peak (ω^-), which is originated from antiparallel magnetic dipoles, was not distinctly observable than the higher resonant peak (ω^+), which comes from parallel magnetic dipoles, for $0°$ and $180°$ twisted SRR dimer structures. For the $90°$ twisted SRR dimer structure, the splitting of the resonances (ω^- and ω^+) is clearly observed. They concluded that this result can well be explained by the polarization rotation effect arising from the chirality of the $90°$ twisted metamaterial structure.

5.3.4 3D METAMATERIALS FABRICATED BY TWO-PHOTON POLYMERIZATION AND METAL DEPOSITION

Rill et al. [12] demonstrated the three-dimensional metamaterial shown in Figure 5.10 using direct laser writing (DLW), chemical vapor deposition (CVD), and atomic-layer deposition (ALD) techniques. In the first step, a glass substrate is covered with a 2-µm-thick fully polymerized resist film (SU-8) and another SU-8 film is spun-on and exposed using DLW. After the post-baking and developing processes, the SU-8 template is coated with a thin layer of SiO_2 of a few tens of nanometers in thickness using ALD. The SiO_2 surface is exposed to O_2 plasma for 15 min to activate the surface for subsequent silver CVD process. The metal-organic precursor (COD)(hfac)Ag(I) is sublimed at a temperature of $60°C$. A silver film with a thickness of 50 nm was deposited by the CVD process, and 10 CVD cycles are performed to obtain 50-nm-thick silver film.

In order to determine the optical properties of the fabricated metamaterial, transmittance spectra of normal incidence was measured using a Fourier-transform microscope spectrometer. Figure 5.11 shows the experimental results of different heights (d) of elongated SRRs. From the result, the resonant peaks originated from the surface plasmon resonance excited by the incident light being polarized vertically to the grooves of the elongated SRRs are observed.

Formanek et al. [13,14] also reported the 3D fabrication of metallic nanostructures over large areas using a combination of two-photon polymerization and electroless plating techniques. To fabricate numerous metallic 3D structures for

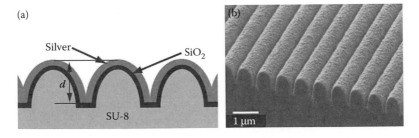

FIGURE 5.10 Photonic metamaterials fabricated by DLW and silver CVD. (a) Schematic diagram of a planar lattice of elongated and all connected SRRs. (b) Electron-beam micrograph of fabricated metamaterial. (Adapted from M. Rill et al., *Nature Mater.* **7**, 543–546, 2008. Copyright 2008, Macmillan Publishes Ltd.)

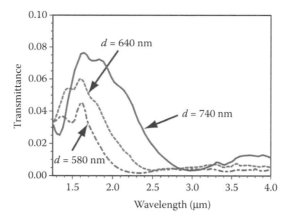

FIGURE 5.11 Normal-incidence optical transmittance spectra. The incident linear polarization is perpendicular to the grooves.

optical metamaterials, multiple laser beam spots created by a micro lens array were introduced to the DLW. Moreover, in order to deposit thin metal films over polymer structures, electroless plating, which is a chemical process and can be effectively realized at ambient conditions, was employed. Electroless plating is suitable for metal deposition onto insulating samples since it allows for a uniform coating over large areas, and even structures with complex shapes and occluded parts can be metal-coated [15–17]. However, polymers are naturally hydrophobic materials and they do not adhere well to metal films due to differences in surface energies [18]. To solve this problem, the chemical modification to the photopolymerizable resin was applied before the fabrication of the 3D polymer structures. Then, a pretreatment using $SnCl_2$ is applied before metal deposition to improve silver nucleation and adhesion on the polymer surface [19]. On the other hand, in order to avoid unwanted metal deposition onto the glass substrate, a hydrophobic coating on the glass slides was applied using dimethyldichlorosilane $((CH_3)_2SiCl_2)$ [20].

Figure 5.12a shows a scanning electron microscope (SEM) image of a large 78×58 μm^2 fabricated area of 3D metalized polymer structures. Figure 5.12b shows an individual structure before coating by electroless plating, composed of a cube (2 μm in size), holding a self-standing spring (2.2 μm in height and 700 nm in inner diameter). To overcoat fine structures, the plating had to be optimized to reduce the thickness of the metal film as much as possible. To do so, they focused on the formation of very small silver particles, with diameters of 20 nm or less, by using a different reductant, changing the concentrations of the reagents, and realizing two consecutive steps. The optimized plating solution was composed of a 0.3 mol/L $AgNO_3$ solution, mixed with a saturated solution of 2,5-dihydroxybenzoic acid $(C_7H_6O_4)$ 300× diluted in water that acts as a reducing agent, in a 1:1 volume ratio. The reaction was performed at 37°C and stopped after 2 min. A second short plating (~10 s) was realized with the same $AgNO_3$ solution, mixed with ammonia (0.2 mol/L) and 10× diluted benzoic acid. Figure 5.12c shows an individual structure after metallization

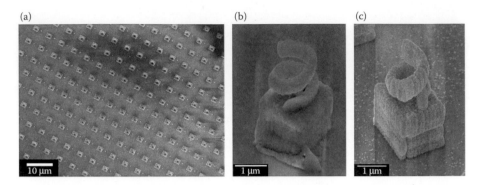

(a)   (b)   (c)

FIGURE 5.12 (a) 78×58 μm^2 SEM image of a 3D periodic structure fabricated on a hydrophobic coated glass surface. (b) Oblique magnified view of an individual uncoated polymer structure composed of a cube (2 μm in size) holding up a spring (2.2 μm in height, 1 μm in inner diameter). (c) SEM image of an individual silver-coated structure after electroless plating.

of the sample. Only a few silver particles adhered on the substrate and the polymer structure was uniformly overcoated by a 50-nm silver film.

5.3.5 Two-Photon-Induced Photoreduction Method

To create an optical metamaterial structure, the fabrication technique is required to have the ability to create arbitrary 3D metallic structures. To satisfy this requirement, Tanaka [21,22] developed a new fabrication technique that used two-photon-induced reduction of metallic complex ions.

Figure 5.13 shows a schematic of this two-photon reduction technique. A mode-locked Ti:Sapphire laser was used as a light source. The laser beam was focused in the material that contains metal ions using an oil-immersion objective lens. When the focused laser beam illuminates the metal-ion solution, metal ions absorb

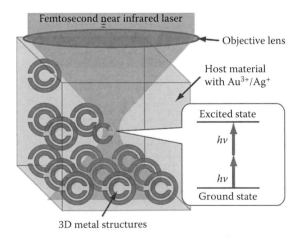

FIGURE 5.13 Schematic of two-photon-induced reduction process.

(a)   (b)

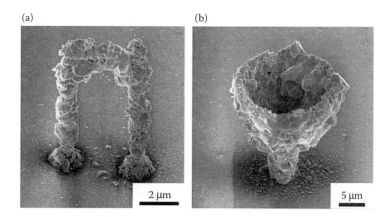

2 μm   5 μm

FIGURE 5.14 SEM images of 3D silver structures. (a) A micro-sized 3D silver gate structure standing on a glass substrate without any support. (b) Top-heavy silver cup on a substrate.

the two photons simultaneously and they are reduced to the metals. Owing to the nonlinear properties of the two-photon absorption process, only at the laser beam spot this metal reduction process occurs and tiny metal particles are created in the 3D space.

The advantage of this technique is that it can create the highly electric conductive metal structures regardless of the micro/nanometer scale. To verify the electrical continuity of the metal structure, the resistivity of the fabricated metal wires was measured and the average of resistivity was determined as 5.30×10^{-8} Ω m. This value is 3.3 times larger than that of bulk silver (1.62×10^{-8} Ω m), indicating the high conductivity of the fabricated silver wires.

Figure 5.14 is the scanning electron micrograph of 3D self-standing silver structures. Figure 5.14a shows silver gate microstructure on a glass substrate, whose width, height, and linewidth of the 3D silver gate are 12, 16, and 1.5 μm, respectively. Figure 5.14b is a top-heavy silver cup. The height and the top and bottom diameters of the silver cup were 26 μm, and 20 and 5 μm, respectively.

In this method, the major problem that inhibited the nanoscale resolution was the unwanted growth of the metal nanocrystal during laser irradiation. Therefore, the main issue to gain the nanometer scale depends on the way of avoiding this unwanted metal particle growth and producing smaller nanoparticles to serve as building blocks. In 2009, Cao et al. [23,24] presented a means to improve the spatial resolution in the fabrication of metallic structures with the aid of a surfactant as a metal growth inhibitor. By using N-decanoylsarcosine sodium as the surfactant, they demonstrated the silver structure the minimum size of which was finer than the diffraction limit of light. Figure 5.15a shows an SEM image of silver line the linewidth of which was 120 nm. Figure 5.15b shows SEM images of the truly freestanding silver pyramids. The close-up view of the silver pyramid shown in the inset of Figure 5.15b reveals that the height of the pyramid was 5 μm and the angle for each edge relative to the substrate was 60°. These silver pyramid structures were strong enough to resist the surface tension in the washing process, which demonstrates that the silver particles were closely combined.

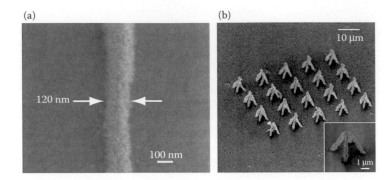

FIGURE 5.15 SEM image of a silver line fabricated using a surfactant-assisted two-photon-induced reduction. (a) The linewidth of the structure was 120 nm. (b) SEM image of the free-standing silver pyramids taken at an observation angle of 45°. Inset is the magnified image showing the detail of the silver pyramid.

5.4 APPLICATIONS OF METAMATERIALS

One of the most popular challenges of the metamaterial application is creating the negative index materials at optical frequency as a way of realizing the sub-wavelength imaging termed "perfect lens" [25]. There are so many articles about negative index materials and perfect lens, the reader can refer to specialized reviews [26]. In this section, we also introduce other applications of the optical metamaterials.

5.4.1 NON-POLARIZED BREWSTER DEVICE

The Brewster effect is widely used as a method to prevent the light reflection occurring at the interface of materials of different indices [27]. The fact that the Brewster effect occurs only for p-polarized light limits its application. In 2006, Tanaka et al. [28] proposed that the magnetically responsive metamaterial enables one to realize the Brewster effect not only for p-polarized light, but also for s-polarized one. The significance of this phenomenon is that the metamaterial can interconnect materials with two different indices while eliminating the reflection arising from the index mismatch, and it can solve the problem of polarization dependence seen in conventional optical components based on the Brewster effect.

Here, the fundamental idea of producing the Brewster effect also for s-polarized light by suitably controlling the magnetic permeability of a material is presented. To simplify the discussion, we consider two isotropic homogeneous materials: material 1 (M_1) and material 2 (M_2) with optical constants ε_1 and μ_1, and ε_2 and μ_2, respectively, as shown in Figure 5.16. The constants ε and μ represent the relative permittivity and the relative permeability, respectively.

According to the Fresnel formula that takes into account both ε and μ, the reflectance for the p- and s-polarization R^p and R^s can be described, respectively, as

$$R^p = \left(\frac{-\mu_2 \sin\theta' \cos\theta' + \mu_1 \sin\theta \cos\theta}{\mu_2 \sin\theta' \cos\theta' + \mu_1 \sin\theta \cos\theta} \right)^2 \tag{5.9}$$

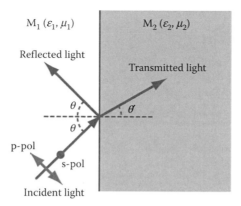

FIGURE 5.16 Calculation model of the Brewster effect for p- and s-polarized light.

and

$$R^s = \left(\frac{\mu_2 \tan \theta' - \mu_1 \tan \theta}{\mu_2 \tan \theta' + \mu_1 \tan \theta} \right)^2, \tag{5.10}$$

where θ and θ' are the incident and refraction angles, respectively. Assuming the numerators of Equations 5.9 and 5.10 to be zero under the condition that the product $\varepsilon_1 \mu_1$ is not equal to $\varepsilon_2 \mu_2$, the Brewster's angles for p- and s-polarized light θ_B^p and θ_B^s are, respectively, found to be

$$\theta_B^p = \tan^{-1} \left[\sqrt{\frac{\varepsilon_2 (\varepsilon_1 \mu_2 - \varepsilon_2 \mu_1)}{\varepsilon_1 (\varepsilon_1 \mu_1 - \varepsilon_2 \mu_2)}} \right] \tag{5.11}$$

and

$$\theta_B^s = \tan^{-1} \left[\sqrt{\frac{\mu_2 (\varepsilon_2 \mu_1 - \varepsilon_1 \mu_2)}{\mu_1 (\varepsilon_1 \mu_1 - \varepsilon_2 \mu_2)}} \right]. \tag{5.12}$$

Figure 5.17a shows the incident angle dependencies of the reflectance R^p and R^s calculated under the condition that M_1 is vacuum ($\varepsilon_1 = \mu_1 = 1.0$) and the M_2 is glass ($\varepsilon_2 = 2.25$ and $\mu_2 = 1.0$). The zero reflectance exists for p-polarized light at $\theta = 56.3°$ and this angle is the standard Brewster's angle. On the other hand, there is no zero-reflectance angle for s-polarized light, thereby indicating that there is no Brewster's angle for s-polarization. Figure 5.17b shows another result calculated under the condition that M_1 is the same as shown in Figure 5.17a ($\varepsilon_1 = \mu_1 = 1.0$), but M_2 is a magnetic material with $\varepsilon_2 = 1.0$ and $\mu_2 = 2.25$. Under this condition, we can see that the reflectance falls to zero for s-polarization at the same angle $\theta = 56.3°$ as in Figure 5.17a. This is also the same Brewster's effect shown in Figure 5.17a, but for

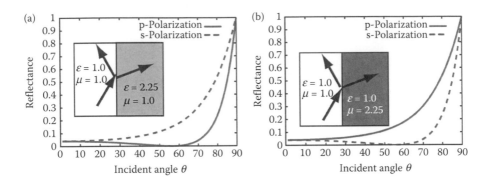

FIGURE 5.17 Incident angle dependence of the reflectance calculated under the condition that (a) M_1 is vacuum ($\varepsilon_1 = \mu_1 = 1.0$) and M_2 is glass ($\varepsilon_2 = 2.25$ and $\mu_2 = 1.0$) and (b) M_1 is vacuum and M_2 is a magnetic material ($\varepsilon_2 = 1.0$ and $\mu_2 = 2.25$).

s-polarization. Previously, it was believed that the Brewster's effect occurs only for p-polarized light because μ of most materials in nature is approximately unity in the optical frequency region. However, these results prove that controlling permeability of the material enables us to realize the Brewster's effect also for s-polarized light. Actually, this Brewster's effect for s-polarization has already been experimentally investigated in the microwave region of 2.6 GHz by employing 2D SRR array [29].

As discussed above, the magnetic material can be used to realize the Brewster's effect for s-polarized light. This new finding also has the significant consequence that if we could produce the Brewster effect for both p- and s-polarized light simultaneously, the light could propagate through the material interface without any reflection at all. The realization of the Brewster effect for p- and s-polarized light simultaneously is the fundamental idea in realizing unattenuated transmission of light across the material boundary. However, Equations 5.11 and 5.12 tell us that the Brewster conditions for each polarization cannot be satisfied simultaneously.

To overcome this conflict, the idea of a uniaxial magnetic metamaterial whose values of ε and μ depend on the direction of the material, which is analogous to a uniaxial crystal, is introduced (Figure 5.18). M_2 is a uniaxial, magnetically active metamaterial that acts as a buffer layer for realizing the perfect light transmission from M_1 to M_3. Since M_2 consists of layer-stacked arrays of the SRRs lying only in the x–y-plane, it responds to a magnetic field oscillating along the z-direction (Hz) and thus changes only μ_2^s. The basic concept of an anisotropic left-handed materials (LHM) was first introduced by Grzegorczyk et al. [30] and they reported inversion of the critical angle and the Brewster's angle in such a material.

We proposed the practical application of the Brewster window for both p- and s-polarization in which light can propagate through the interface between two materials of different refractive indices without any reflection. This metamaterial-based device has a strong impact on optical technologies because the inherent problem of polarization dependence seen in conventional optical component based on the Brewster effect is completely solved.

Figure 5.19 shows an example of the calculation results under the condition that M_1 was a vacuum ($\varepsilon_1 = 1.0$ and $\mu_1 = 1.0$) and M_3 was glass ($\varepsilon_3 = 2.25$ and $\mu_3 = 1.0$)

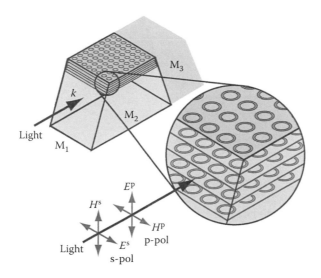

FIGURE 5.18 Uniaxial magnetic metamaterial exhibiting the Brewster effect for both p- and s-polarized light and its calculation model.

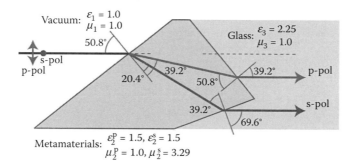

FIGURE 5.19 Design of the uniaxial magnetic metamaterial with the calculated optical constants $\varepsilon_2^p = 1.5$, $\mu_2^p = 1.0$, $\varepsilon_2^s = 1.5$, and $\mu_2^s = 3.29$.

with the additional constraint that the exit angles of p- and s-polarized light to M_3 are zero ($\theta_{ex}^p = \theta_{ex}^s = 0$, i.e., the light was transmitted straight from M_1 to M_3). The solution converged at the points $\varepsilon_2^p = 1.5$, $\mu_2^p = 1.0$, $\varepsilon_2^s = 1.5$, and $\mu_2^s = 3.29$. Under this condition, the Brewster angles for p- and s-polarized light at the interface between M_1 and M_2 were identically 50.8°. This result proved that the light completely passed through both interfaces between two different materials without any loss of reflection.

5.4.2 Optical Cloaking

The freedom of designing of electromagnetic field by the metamaterials provides a new device that conceals an arbitrary object from all the electromagnetic fields; this

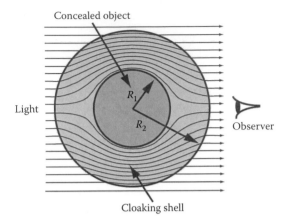

FIGURE 5.20 Optical cloaking using metamaterial shells ($R_1 < r < R_2$). Cloaking is achieved by compressing all field in the region $r < R_2$ into the cloaking shell.

technique is termed "optical cloaking". In 2006, Pendry [1] proposed the basic idea of optical cloaking. By controlling the distribution of permittivity ε and permeability μ, the rays that would have struck the object are deflected, guided around the object, and then returned to their original trajectory. For simplicity, the hidden object is supposed to be a sphere of radius R_1 and the cloaking region to be contained with the annulus $R_1 < r < R_2$. The optical cloaking is achieved by compressing all fields in the region $r < R_2$ into the region $R_1 < r < R_2$ as shown in Figure 5.20. This can be done by making the distribution of the permittivity and permeability as follows.

For region $R_1 < r < R_2$:

$$r' = R_1 + \frac{r(R_2 - R_1)}{R_2}, \tag{5.13}$$

$$\theta' = \theta, \tag{5.14}$$

$$\phi' = \phi, \tag{5.15}$$

$$\varepsilon_r = \mu_r = \frac{R_2}{R_2 - R_1} \frac{(r' - R_1)^2}{r'}, \tag{5.16}$$

$$\varepsilon_\theta = \mu_\theta = \frac{R_2}{R_2 - R_1}, \tag{5.17}$$

$$\varepsilon_\phi = \mu_\phi = \frac{R_2}{R_2 - R_1}. \tag{5.18}$$

For region $r > R_2$:

$$\varepsilon_r = \mu_r = \varepsilon_\theta = \mu_\theta = \varepsilon_\phi = \mu_\phi = 1.0. \tag{5.19}$$

For region $r = R_2$:

$$\varepsilon_\theta = \varepsilon_\phi = \frac{1.0}{\varepsilon_r}, \qquad (5.20)$$

$$\mu_\theta = \mu_\phi = \frac{1.0}{\mu_r}. \qquad (5.21)$$

When the above conditions are satisfied, all fields are excluded from the central region ($r < R_1$) independent of ε and μ in $r < R_1$, and are not reflected at the surface of the cloak.

For the implementation of the cloaking in the microwave region, Smith and coworkers [2] reported the experimental result of 2D cloaking at 8.5 GHz using the layout that consisted of 10 concentric cylinders with SRRs. In the optical frequency region, Shalaev and coworkers [31] proposed the design of a nonmagnetic cloak using radially aligned spheroidal silver wires.

One of the disadvantages of shell-type cloak design is that it can work only over a narrow frequency range, because the curved trajectory of light needs a refractive index n of <1.0, and any metamaterials whose index is <1.0 must be dispersive.

5.5 SUMMARY AND OUTLOOK

The metamaterial concept of creating composites with desired optical properties gives us the opportunity to engineer and specify the electromagnetic properties. Controlling electromagnetic material parameters, such as ε and μ, will open the door for exotic optical phenomena and their applications. From the technical side, the realization of fabricating the metamaterials is still difficult despite the recent progress of the micro/nano fabrication techniques, because 3D metallic micro/nanostructures are necessary to form the isotropic metamaterials. The important issues left to the future are three-dimensionality and mass productivity.

REFERENCES

1. J. Pendry, D. Schuring, and D. R. Smith, Controlling electromagnetic fields, *Science* **312**, 1780–1782, 2006.
2. D. Schurig, J. J. Mock, B. J. Justice, S. A. Cummer, J. B. Pendry, A. F. Starr, and D. R. Smith, Metamaterial electromagnetic cloak at microwave frequencies, *Science* **314**, 977–980, 2006.
3. A. Ishikawa, T. Tanaka, and S. Kawata, Negative magnetic permeability in the visible light region, *Phys. Rev. Lett.* **95**, 237401, 2005.
4. A. Ishikawa and T. Tanaka, Negative magnetic permeability of split ring resonators in the visible light region, *Opt. Commun.* **258**, 300–305, 2006.
5. A. Ishikawa, T. Tanaka, and S. Kawata, Frequency dependence of the magnetic response of split-ring resonators, *J. Opt. Soc. Am. B* **24**, 510–515, 2007.
6. J. B. Pendry, A. J. Holden, D. J. Robbins, and W. J. Stewart, Magnetism from conductors and enhanced nonlinear phenomena, *IEEE Trans. Microwave Theory Tech.* **47**, 2075–2084, 1999.

7. K. C. Gupta, R. Garg, I. Bahl, and P. Bhartia, *Microstrip Lines and Slotlines*, 2nd ed. (Artech House, Norwood, MA, 1996), 375–456.

8. P. B. Johnson and R. W. Christy, Optical constants of the noble metals, *Phys. Rev. B* **6**, 4370–4379, 1972.

9. T. J. Yen, W. J. Padilla, N. Fang, D. C. Vier, D. R. Smith, J. B. Pendry, D. N. Basov, and X. Zhang, Terahertz magnetic response from artificial materials, *Science* **303**, 1494–1496, 2004.

10. J. Valentine, S. Zhang, T. Zentgraf, E. Ulin-Avila, D. A. Genov, G. Bartal, and X. Zhang, Three-dimensional optical metamaterial with a negative refractive index, *Nature* **455**, 376–379, 2008.

11. N. Liu, H. Liu, S. Zhu, and H. Giessen, Stereometamaterials, *Nature Photonics* **3**, 157–162, 2009.

12. M. Rill, C. Plet, M. Thiel, I. Staude, G. vov Freymann, S. Linden, and M. Wegener, Photonic metamaterials by direct laser writing and silver chemical vapour deposition, *Nature Mater.* **7**, 543–546, 2008.

13. F. Formanek, N. Takeyasu, T. Tanaka, K. Chiyoda, A. Ishikawa, and S. Kawata, Three-dimensional fabrication of metallic nanostructures over large areas by two-photon polymerization, *Opt. Express* **14**, 800–809, 2006.

14. F. Formanek, N. Takeyasu, K. Chiyoda, T. Tanaka, A. Ishikawa, and S. Kawata, Selective electroless plating to fabricate complex three-dimensional metallic micro/nanostructures, *Appl. Phys. Lett.* **88**, 83110, 2006.

15. G. O. Mallory and J. B. Hajdu, *Electroless Plating: Fundamentals and Applications,* (American Electroplaters and Surface Finishers Society, Orlando, FL, 1990).

16. A. A. Antipov, G. B. Sukhorukov, Y. A. Fedutik, J. Hartmann, M. Giersig, and H. Mohwald, Fabrication of a novel type of metallized colloids and hollow capsules, *Langmuir* **18**, 6687–6693, 2002.

17. N. Takeyasu, T. Tanaka, and S. Kawata, Metal deposition deep into microstructure by electroless plating, *Jpn. J. Appl. Phys.* **44**, 1134–1137, 2005.

18. L. J. Gerenser, Photoemission investigation of silver/poly(ethylene terephthalate) interfacial chemistry: The effect of oxygen–plasma treatment, *J. Vac. Sci. Technol.* A **8**, 3682–3691, 1990.

19. J. E. Gray, P. R. Norton, and K. Griffiths, Mechanism of adhesion of electroless-deposited silver on poly(ether urethane), *Thin Solid Films* **484**, 196–207, 2005.

20. F. Guan, M. Chen, W. Yang, J. Wang, S. Yong, and Q. Xue, Fabrication of patterned gold microstructure by selective electroless plating, *Appl. Surf. Sci.* **240**, 24–27, 2005.

21. T. Tanaka, A. Ishikawa, and S. Kawata, Two-photon-induced reduction of metal ions for fabricating three-dimensional electrically conductive metallic microstructure, *Appl. Phys. Lett.* **88**, 81107, 2006.

22. A. Ishikawa, T. Tanaka, and S. Kawata, Improvement in the reduction of silver ions in aqueous solution using two-photon sensitive dye, *Appl. Phys. Lett.* **89**, 113102 2006.

23. Y. Cao, N. Takeyasu, T. Tanaka, X. Duan, and S. Kawata, 3D metallic nano-structure fabrication by surfactant-assisted multi-photon-induced reduction, *Small* **5**, 1144–1148, 2009.

24. Y. Cao, X. Dong, N. Takeyasu, T. Tanaka, Z. Zhao, X. Duan, and S. Kawata, Morphology and size dependences of silver microstructures on fatty salts-assisted multiphoton photoreduction microfabrication, *Appl. Phys. A* **96**, 453–458, 2009.

25. J. Pendry, Negative refraction makes a perfect lens, *Phys. Rev. Lett.* **85**, 3966–3969, 2000.

26. D. Smith, J. Pendry, and M. Wilshire, Metamaterials and negative refractive index, *Science* **305**, 788–792, 2004.

27. M. Born and E. Wolf, *Principles of Optics*, 6th ed. (Pergamon Press, Oxford, 1980).

28. T. Tanaka, A. Ishikawa, and S. Kawata, Unattenuated light transmission through the interface between two materials with different indices of refraction using magnetic metamaterials, *Phys. Rev. B* **73**, 125423, 2006.
29. Y. Tamayama, T. Nakanishi, K. Sugiyama, and M. Kitano, Observation of Brewster's effect for transverse-electric electromagnetic waves in metamaterials: Experiment and theory, *Phys. Rev. B* **73**, 193104, 2006.
30. T. M. Grzegorczyk, Z. M. Thomas, and J. A. Kong, Inversion of critical angle and Brewster angle in anisotropic left-handed metamaterials, *Appl. Phys. Lett.* **86**, 251909, 2005.
31. W. Cai, U. K. Chettiar, A. V. Kildishev, and V. M. Shalaev, Optical cloaking with meta-materials, *Nature Photonics* **1**, 224–227, 2007.

6 Resonant Nanometric Apertures in Metallic Films

Ann Roberts and Xiao Ming Goh

CONTENTS

6.1 INTRODUCTION

The propagation of electromagnetic waves through apertures in metal screens (or films) has attracted considerable attention since the development of the theory of electromagnetism. Significant interest was initially motivated by applications in radar and by understanding the penetration of conducting surfaces by electromagnetic pulses. A large number of theories with varying degrees of approximation were developed during the middle part of the twentieth century (for reviews of early work in this area, see Refs. [1–4]). In the late twentieth century, interest in this topic was reinvigorated by the increasing availability of nanofabrication tools such as focused ion beam milling and electron beam lithography and led to the first observation of "extraordinary optical transmission" (EOT) [5]. This subsequently sparked a significant interest in understanding the propagation of visible and near-infrared light through subwavelength apertures [6,7]. Furthermore, with increasing interest in the development of optical antennas [8], it is timely to consider apertures as a category of "slot" antennas that can play a key role in a range of applications. This chapter is aimed at investigating a class of apertures that exhibit distinct localized resonances. We focus on coaxial (annular) and cross-shaped apertures and discuss the influence of the geometry, optical properties of the film and environment, aperture arrangement, and phase shifts on resonance.

6.2 TRANSMISSION OF ELECTROMAGNETIC WAVES THROUGH APERTURES

Circular apertures (see review [2]) and infinitely long slots [9–13] have been the subject of many investigations into transmission of electromagnetic waves through apertures. In the absence of computers, a range of approximate techniques were developed, most notably to investigate the extreme limits of apertures that are either very small compared to the wavelength of the electromagnetic waves with which they are interacting [14], or very large [15]. A number of these investigations were a result of interest shown in radar and microwave communications and, as a consequence, the approximation that the metals involved were perfectly conducting was widely used. Andrejewski [16] developed a rigorous model describing diffraction by a circular aperture in an infinitesimally thin, perfectly electrically conducting (PEC) metallic screen. Bethe [14] was the first to examine transmission of light through a small aperture in the walls of a perfectly conducting waveguide. The wall in which the aperture was located was assumed to be infinitesimally thin (in addition to being perfectly conducting). Bethe showed that the transmitted far-field can be considered as being produced by crossed electric and magnetic dipoles and also that the transmitted power normalized to the area of the aperture with radius a varied as $(k_0 a)^4$. He also provided expressions for the near-zone fields that were subsequently corrected and the results extended by Bouwkamp [17].

6.2.1 TRANSMISSION THROUGH CIRCULAR APERTURES IN THICK, PERFECTLY CONDUCTING SCREENS

Neither Bethe nor Bouwkamp looked at the influence of the thickness of the screen on transmission through circular apertures. Initially, investigations in this area probed at the influence of screen thickness on transmission through infinitely long slits [18]. Although of interest, since parallel plate waveguides can support a TEM mode and simply connected waveguides, such as circular waveguides, cannot, slits can potentially transmit electromagnetic waves with arbitrarily long wavelengths. Circular apertures, on the other hand, have a cut-off wavelength above which the fields in the aperture are all evanescent and the transmission decreases strongly with increasing thickness. There were also extensions of Bethe's "small-hole" theory to the inclusion of the effects of screen thickness [19], but the first rigorous theoretical investigation of the transmission of electromagnetic waves through circular apertures in thick PEC screens was undertaken by Roberts [20]. The theoretical approach involved a modal method [21] wherein the fields within the cavity formed by the circular aperture in a PEC film were expanded in circular waveguide modes and the fields above and below the screen were expressed as integrals over plane waves. By applying appropriate boundary conditions at the upper and lower interfaces of the screen, an infinite set of linear equations for the amplitudes of the waveguide modes can be derived and solved computationally by truncating the sums and inverting the resulting matrices. In this way, the influence of screen thickness can be examined.

Figure 6.1 shows the transmissivity calculated using the modal method (described above) through a circular aperture of radius a in PEC films with different thicknesses h, as a function of normalized wavenumber, $k_0 = 2\pi/\lambda$, where λ is the free-space wavelength. The incident field is a normal plane wave and the transmission is normalized to that in the geometric optics limit to yield the transmissivity. For zero thickness, the result replicates that developed using other rigorous methods [16]. It is evident that as soon as the finite thickness of a metal screen is taken into account, issues that are more frequently associated with waveguide propagation become apparent. In particular, as made transparent, using the modal formulation, only certain waveguide modes will propagate within the apertures. Although evanescent modes are required in the model for completeness, they do not transport energy. Since circular apertures do not support a TEM mode, there is a wavelength above which all modes within the waveguide are evanescent and, even in the case of thin screens, it is expected that the total power transmitted through the aperture would decrease rapidly with either increasing wavelength or screen thickness.

Another aspect of including the screen thickness in calculations (and one that also emerges naturally by considering the modal method) is the concept of resonances

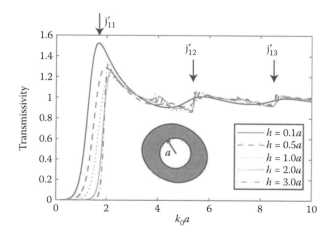

FIGURE 6.1 (**See color insert.**) Transmission through a circular aperture of radius, a, calculated using the modal method for different thicknesses, h, of the screen. Transmission is normalized to that through an aperture in the geometric optics limit. Cut-off frequencies for the first three TE circular waveguide modes (i.e., the first three roots of J_1') are shown.

within the cavity formed by the aperture and its upper and lower boundaries. As the group velocity of a given mode approaches zero, its phase velocity approaches infinity and results in a zeroth-order Fabry–Perot resonance. Hence, near the cut-off wavelengths of a given mode, there will be a significant increase in transmission.

From Figure 6.1, it is evident that there are wavelengths where the transmissivity (the total transmission normalized to that in the geometric optics limit) is greater than unity. The wavelengths at which these resonances appear correspond to the cut-off wavenumbers of different circular aperture waveguide modes or higher-order Fabry–Perot resonances that depend on the thickness of the metal. In the case of normal incidence, symmetry considerations show that an incident plane wave can couple only to the $n = 1$ waveguide modes. The (normalized) cut-off wavenumber for various circular aperture waveguide modes is also shown. These correspond to the roots of the derivatives of Bessel functions, $J_1'(x)$, noted as j_{1n}' are shown. It is also apparent that for the thicker screens, the $(k_0 a)^4$ dependence of the Bethe small-hole model as $k_0 a \rightarrow 0$ is no longer satisfied since the transmissivity approaches zero much more rapidly.

6.2.2 Coaxial Apertures

Note that it is possible to achieve transmissivities of ~1.5 in a circular aperture in a thin screen, so super-unity transmissivities can be seen in conventional electromagnetic treatments of simple apertures in perfectly conducting screens. More striking effects, however, can be found when investigating transmission through more complex apertures with more degrees of freedom. If the equivalent modal method is now applied to a coaxial (annular) aperture in a thick conducting screen, where the circular waveguide modes of the expansion described above are replaced by the modes of coaxial waveguides, even higher transmissivities and cavity Q factors can be achieved. A coaxial aperture of radius a and central stop radius b in a perfectly conducting film of thickness h is considered here. In Figure 6.2, the transmission through a circular aperture for a fixed aperture radius a and thickness of $h = 1.1a$ is considered for different values of the radius of the central stop b. The aperture is illuminated with a normally incident plane wave and the transmission normalized to that in the geometric optics limit. It is apparent that very high transmissivities, an order of magnitude greater than that achievable for a circular aperture and that obtainable in the geometric optics limit, can be obtained. As b decreases, that is, the width of the gap increases, both the maximum transmissivity and the Q of the resonance decrease, and in the limit where $b \rightarrow 0$, the transmissivity approaches that in the case of a circular aperture.

Note that for normally incident plane waves, it is not possible to excite the TEM mode of the coaxial aperture, so in the example presented here, it has no role in the resulting transmission. As was the case with the circular aperture, resonances are located at the cut-off wavelengths of the dominant TE modes. As will be discussed further in the next section, the high Q is a result of the increased capacitance of the structure compared to a simple circular aperture, which also leads to a longer wavelength (lower wavenumber) cut-off compared to the circular aperture as can be seen

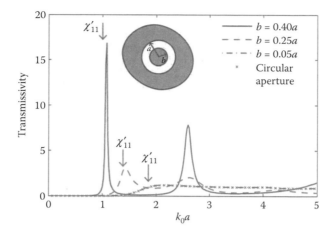

FIGURE 6.2 (**See color insert.**) Transmission through a coaxial aperture with varying inner radii, b, and a circular aperture in a screen of thickness $h = 1.1a$ as a function of normalized wavenumber.

from a comparison of Figures 6.1 and 6.2. Furthermore, the existence of higher-order Fabry–Pérot resonances, dependent on the thickness of the screen, in an aperture are apparent—note the resonances at values of k_0a near 2.6. Hence, apertures in thick, perfectly conducting screens possess complex transmission spectra that depend on their transverse geometry and screen thickness and can also exhibit a transmissivity higher than unity and higher Qs than exhibited by simple circular apertures.

6.3 FREQUENCY-SELECTIVE SURFACES

Apertures (and their complementary shapes) also play a significant role in the performance of frequency-selective surfaces (FSSs) designed to behave as filters (low-pass, high-pass, or bandpass) or beam-splitters in the far-infrared and millimeter-wave regions of the electromagnetic spectrum [22–26]. FSSs consist of regular two-dimensional arrays of metal patches (a capacitive grid or mesh) or a complementary array of apertures in a metallic film (an inductive mesh). At long wavelengths, compared to the periodicity of the array, inductive meshes are perfect reflectors (long-pass filters), whereas the transmission through capacitive meshes approaches 100% (short-pass filters) [26]. One notable feature of these meshes is the availability of equivalent transmission line models [25,26] to describe their performance that provides a useful framework for structure design and applications. Specifically, the grid can be assigned an impedance corresponding to a lumped element with a characteristic capacitance and inductance. Since, at long wavelengths, the materials involved are assumed lossless, the resistance is taken to be zero.

Being two-dimensional diffraction gratings, meshes exhibit a range of anomalies and surface excitations associated with the existence of multiple diffracted orders. In general, if a freestanding square mesh (i.e., one not supported by a substrate) has a period d in both transverse directions, then the (p,q)th diffracted order propagating at an angle θ to the normal to the mesh surface is given by

$$\sin\theta_{pq} = \sin\theta_0 + \frac{\lambda}{d}\sqrt{p^2 + q^2},$$

where θ_0 is the angle of incidence and λ is the free-space wavelength. It is apparent that if $\lambda > d$, then at normal incidence there will be only one propagating transmitted and one reflected order—all other orders being evanescent. If, however, this condition is not met, there will be multiple propagating transmitted and reflected orders.

Early theoretical investigations focussed on infinitesimally thin meshes, although the modal method [21] was developed to study thick screens and was applied to the investigation of arrays of rectangular [21] and circular [22,27] apertures in thick screens. As is the case with single apertures, the modal method can be used to evaluate diffraction by these structures, but now only a discrete set of propagating and evanescent plane waves are required to model the reflected and transmitted fields. By combining the features of both capacitive and inductive grids into a single mesh, it is possible to develop a bandpass filter [24,28]. Two grids of this nature proposed were arrays of cross-shaped [24] and annular (coaxial) [28,29] apertures in a thick, perfectly conducting film (Figure 6.3). These structures possess characteristic resonances at which distinct transmission peaks occur in the transmission spectrum. The wavelength at which resonance occurs is (as is the case for single apertures) located near the cut-off wavelength for the longest wavelength propagating mode that the aperture can support.

Figure 6.4 shows the transmission through a square array of coaxial apertures in a PEC film calculated using the modal method. Two screen thicknesses are considered, $0.5d$ and d, where d is the array period, and two different coaxial geometries

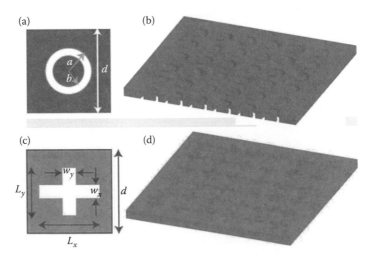

FIGURE 6.3 (**See color insert.**) Arrays of annular (a, b) and cross-shaped (c, d) apertures in thick metallic films.

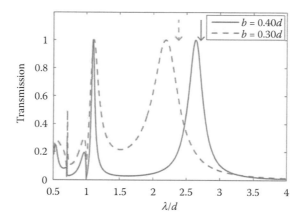

FIGURE 6.4 Transmission through an array of coaxial apertures with period, d, in a PEC film of thickness $0.50d$, outer radius $0.45d$, and inner radii of either $0.40d$ or $0.30d$. The location of the cut-off wavelength of the corresponding TE_{11} waveguide mode is shown.

are shown in the figure. In both cases, the outer radius of the apertures is $0.45d$, and examples where the inner radius, b, is $0.40d$ and $0.30d$, are also shown. The corresponding cut-off wavelengths of the TE_{11} coaxial waveguide modes are also shown in the figure. As was the case with single apertures, a clear resonance associated with the cutoff of the TE_{11} coaxial waveguide mode is apparent and this resonance shifts to shorter wavelengths as the gap size in the coaxial apertures increases. The transition of an order from propagating to evanescent when θ_0 or λ is changed leads to the appearance of anomalies associated with the excitation of surface waves. Examples in Figure 6.4 are at $\lambda/d = 1$ and $1/\sqrt{2}$, where the $(\pm 1, 0)$, $(0, \pm 1)$, and $(\pm 1, \pm 1)$ diffracted orders transition from propagating to evanescent.

In the case of both cross-shaped and annular apertures, the resonance can be tuned to have a much higher Q than an array of simple circular or square apertures. As is seen in the case of isolated annular apertures, the primary factors influencing the Q are: (i) the linewidth of the apertures (narrower apertures produce a higher Q) and (ii) (as shown in Figure 6.5) the thickness of the screen in which they are located (thicker screens produce a higher Q) [28,30]. Both the cross-aperture and coaxial FSS also possess the useful properties of exhibiting polarization-independent performance. One attractive feature of these bandpass grids (and particularly the coaxial geometry) is the fact that the dominant long-wavelength transmission resonance can be tuned to be at a wavelength longer than any Wood anomalies. For example, a narrow coaxial waveguide has a TE_{11} mode cut-off wavelength approaching the circumference of the aperture. Hence, the apertures can be arranged in an array so that the longest wavelength resonance is clearly separated from Wood anomalies associated with diffracted orders "passing off."

FSS can also be cascaded to produce a more complex transmission spectrum. In this case, the equivalent circuit model is invaluable since it provides a straightforward modeling tool for assessing their performance when used in tandem [25].

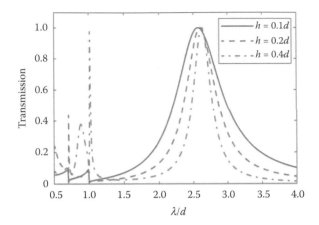

FIGURE 6.5 Transmission through an array of annular apertures with outer radius $a = 0.45d$ and inner radius $b = 0.40d$ in PEC screens of varying thickness. Normal incidence plane wave illumination.

6.4   APERTURE PROPERTIES

6.4.1   EFFECTS OF SCREEN CONDUCTIVITY

Although the modal method is useful for eliciting broad conclusions about the transmission of electromagnetic waves transmitted through isolated and periodic arrangements of apertures and their performance at long wavelengths where metals have a very high conductivity, the optical properties of the metal film play an increasingly important role as the wavelength decreases into the infrared, near-infrared, and visible regions of the electromagnetic spectrum.

Studies based on analytical expressions developed for describing the fundamental optical modes in slot and rectangular apertures [31,32] have revealed the impact of a real metal (compared to a PEC) on the cut-off wavelengths in these structures. A number of studies have explored the influence of finite conductivity on arrays of coaxial apertures in the visible and near-infrared regions of the electromagnetic spectrum [33–37]. By taking into account the finite conductivity of a real metal, the resonance of the fundamental guided mode of the aperture is shown to shift to longer wavelengths than that of a perfect conductor.

Figure 6.6 shows the computed transmission through freestanding arrays of annular apertures with identical dimensions (period 500 nm, outer radius 150 nm, inner radius 100 nm, and thickness 140 nm), but with a metal composed of different materials. The finite element method (FEM) as implemented in COMSOL Multiphysics [38] was used to perform these calculations. Periodic boundary conditions were used on transverse boundaries, and perfectly matched layers (PMLs) introduced at the upper and lower surfaces. The incident field was excited by a uniform surface current, and the transmission (relative to the incident power) is determined by integrating the z-component of the Poynting vector over a surface above the lower PML. The optical constants of silver and gold were taken from Johnson and Christy [39] and aluminum from Palik [40]. These are compared with

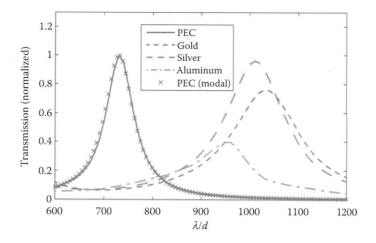

FIGURE 6.6 Calculated transmission through an array, with a period of 500 nm, of coaxial apertures with an outer radius of 150 nm and an inner radius of 100 nm, in various freestanding metallic films of thickness 140 nm.

that obtained using both the modal method for a PEC and introducing PEC boundaries into the FEM calculation.

From Figure 6.6, it is apparent that for all metals, a resonance still exists and that the strongest transmission through apertures in real metals occurs at a longer wavelength than for PEC. It is also apparent that the longest resonant wavelength occurs for gold, whereas the highest transmission is for silver, is consistent with the behavior of the relative permittivities. It is also worth noting that the agreement between the rigorous modal method and the FEM calculation in the PEC case is excellent. Note that these plots are for freestanding films, and any surface wave resonances would occur at shorter wavelengths than the shortest wavelength shown here. These resonances are, therefore, independent of the excitation of surface plasmon polariton (SPP) or other surface waves.

Although there are differences in the location, shape, and strength of resonances when apertures are located in real metals in the visible and near-infrared regions of the electromagnetic spectrum, the concept of resonance in these apertures is carried over to this regime. Furthermore, the red-shifting associated with the transition from a PEC to a real metal provides a means of moving the localized aperture resonance away from regions where the excitation of surface waves strongly affects transmission and into the region where there is only one transmitted and one reflected diffracted order. As a consequence, structures and devices based on these apertures can be tailored to have properties in common with metamaterials.

6.4.2 Effects of Aperture Size

The overall size of the aperture also plays a critical role in determining its resonant wavelength independent of whether the aperture is isolated or in an array. This is readily apparent when considering apertures in PEC films since the resonant

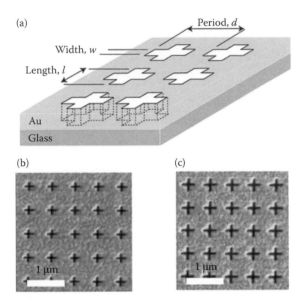

FIGURE 6.7 Periodic arrangement of cross-shaped apertures (a), with period, d, arm-length, l, and width, w, in a metallic film of thickness h. SEM images of arrays A (b) and C (c). (From Lin, L., L.B. Hande, and A. Roberts, *Applied Physics Letters*, 2009. **95**: 201116. With permission.)

wavelengths scale with the overall dimensions of the aperture. Here, we focus on cross-apertures which, in the case of PEC films, have similar properties to coaxial apertures. The lack of a TEM mode in the cross-apertures is not significant since, as discussed previously, it cannot be excited in coaxial apertures by normally incident plane waves. Cross-apertures can be thought of as a superposition of two, orthogonal and rectangular, slots and, if symmetric, can be designed to be insensitive to the polarization of the incident field. Conversely, it is straightforward to tune the geometry to introduce a controlled birefringence. Also, in contrast to coaxial apertures, metal films containing cross-apertures can be self-supporting.

The influence of aperture size on transmission in the near-infrared has been considered in the case of cross-shaped apertures where structures with a nominally fixed period (~600 nm) and arm-widths (~40 nm), but varying arm-lengths were fabricated and characterized [41]. Figure 6.7 shows the scanning electron microscope images of typical structures used in this research and Figure 6.8 shows the transmission calculated using the FEM (a) and experimentally determined transmission spectra (b). Geometric parameters for these structures are given in Table 6.1. It is apparent that increasing the overall sizes of the apertures (without substantially changing the gap size) red-shifts the aperture resonances.

6.4.3 SINGLE APERTURES VERSUS ARRAYS OF APERTURES

It is instructive to compare the transmission through a single aperture with a periodic arrangement of identical apertures in PEC screens of the same thickness. Given

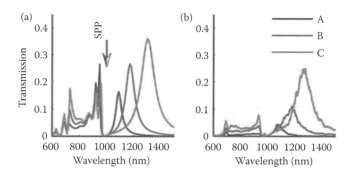

FIGURE 6.8 **(See color insert.)** Transmission spectra for arrays A, B, and C. Calculated (a) and measured (b) results. The approximate location of the (0, ±) Au/glass SPP excitation is shown in (a). (From Lin, L., L.B. Hande, and A. Roberts, *Applied Physics Letters*, 2009. **95**: 201116. With permission.)

that the mechanism underlying bandpass FSS relies on resonances of the individual apertures and the high transmission through apertures relies on localized resonances of those same apertures, it would be expected that high transmission would occur at the same wavelengths in both arrays and isolated apertures. Figure 6.9 shows the transmission, calculated using the modal method, through a square array of coaxial apertures in a PEC film compared with the transmissivity through a single aperture in a film of the same thickness. The resonance at $2.6d$ is apparent in both structures as is a higher-order, Fabry–Pérot resonance near $1.1d$. The most noticeable difference is the appearance of the Wood anomaly in the case of the periodic array at $\lambda = d$, where the (±1, 0) and (0, ±1) diffracted orders undergo the transition from propagating to evanescent fields. An anomaly near $\lambda = 0.7d$ also arises from the transition of the (±1, ±1) diffracted orders from propagating to evanescent. These anomalies distort the single-aperture transmission spectrum and can also lead to the appearance of Fano resonances [42] as the resonance from the Wood anomaly couples to the localized aperture resonance.

Similarly, when the effects of finite conductivity are taken into account, we see distinct aperture resonances, whether or not the apertures are located in an array. Figure 6.10 shows FEM-computed transmission through cross-shaped apertures with identical geometries either isolated or in a periodic array [41]. It is apparent that, despite changes in the overall shape of the transmission spectrum, resonances exist in the presence and absence of order. Furthermore, Orbons et al. [43] (Figure 6.11)

TABLE 6.1

Physical Dimensions of Fabricated Devices with Data Shown in Figure 6.7

Label	d_x (nm)	L_x (nm)	w_x (nm)	d_y (nm)	L_y (nm)	w_y (nm)
A	609	253 ± 9	33 ± 3	628	231 ± 5	28 ± 4
B	604	299 ± 7	38 ± 5	629	283 ± 10	37 ± 4
C	611	351 ± 5	38 ± 5	625	341 ± 6	40 ± 4

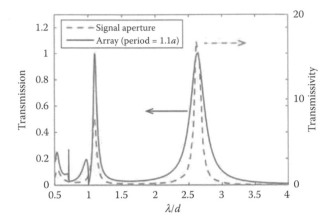

FIGURE 6.9 Transmission through isolated coaxial apertures (dashed line) and identical apertures arranged in an array with period d. Apertures have outer radius $0.45d$, inner radius $0.40d$, and are located in a PEC film with thickness $0.5d$.

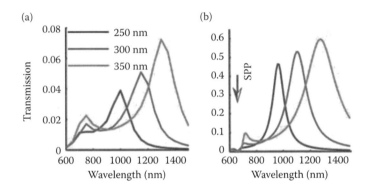

FIGURE 6.10 (**See color insert.**) Simulated transmission spectra for (a) single apertures and (b) an arrangement of apertures with the same dimensions with periodicity 400 nm. The location of the Au/glass $(0, \pm1)$ SPP is shown in (b). (From Lin, L., L.B. Hande, and A. Roberts, *Applied Physics Letters*, 2009. **95**: 201116. With permission.)

considered regular and random arrangements of identical coaxial apertures and saw localized resonances [termed cylindrical surface plasmons (CSPs)] that persisted in the absence of arrangement order.

6.4.4 ENHANCED OPTICAL TRANSMISSION

In 1998, Ebbesen et al. [5] observed what was originally termed EOT where relatively high (10%) transmission through periodic arrangements of subwavelength (150 nm diameter) apertures in a thin (200 nm) silver film on a quartz substrate was observed. The key features of this phenomenon [subsequently more commonly referred to as "enhanced optical transmission" (EOT)] are the fact that the apertures are so small that they do not support a propagating mode at the wavelength at which peaks in

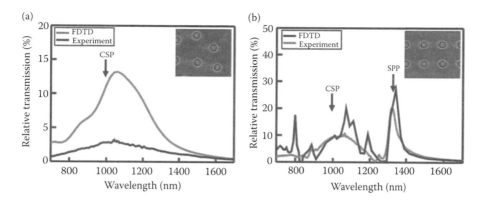

FIGURE 6.11 **(See color insert.)** Transmission (experimental and calculated) through random and regular arrangements of coaxial apertures with outer radius 160 nm and inner radius 88 nm in a silver film of thickness 140 nm on a silica substrate. (a) Transmission through a random arrangement of apertures; (b) transmission through an array with periods 780 nm (in the horizontal direction) and 820 nm (vertical). Locations of localized CSP and SPP excitation wavelengths are shown. Insets show SEM images of the samples. (From Orbons, S.M. et al., *Optics Letters*, 2008. **33**(8): 821–823. With permission.)

transmission occur, the transmission is higher than would be expected through an equivalent density of randomly arranged apertures and the relationship between the periodicity of the arrays and the enhanced transmission. There has been considerable debate about the fundamental origin of EOT with attention being primarily focussed on the role played by SPPs and other diffractive effects. For a review, see Ref. [7].

Although some attention has been given to the role played by the specific geometry of the apertures themselves, it is interesting to consider EOT in the context of apertures that are known to exhibit clear resonances whether or not they are arranged in some order, are randomly organized, or are isolated. Looking at these structures, it is possible to locate, with some sensitivity, the location of the resonance and investigate the behavior of the structure with changes in the geometry of the apertures or their arrangement.

If these apertures are placed in an array and the geometry of the apertures is permitted to vary, then the complex interplay between the array dynamics and the aperture can be investigated. For example, by tuning the localized resonance to wavelengths either greater than or less than the wavelength at which SPP excitation occurs, and for periodic and random arrangements of apertures, it is straigtforward to demonstrate EOT [43]. In the results shown in Figure 6.11, it is evident that the localized aperture resonance has been tuned to be at a wavelength less than the (0, ±1) SPP. It is apparent that in the case of the regular array, the transmission near 1320 nm (~20%) is significantly higher than that for the random arrangement (<5%).

6.4.5 Transition from Simple Apertures to Resonant Apertures

From Figure 6.2, it is apparent that as the radius of the inner stop in an annular aperture in a PEC film decreases, the transmission through the aperture approaches that

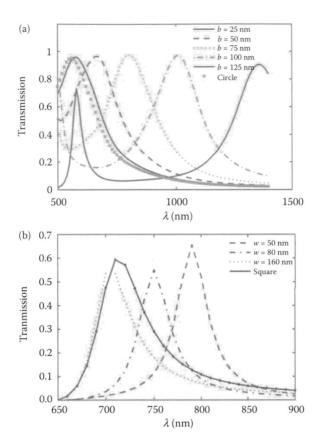

FIGURE 6.12 Normalized calculated transmission through periodic ($d = 400$ nm) arrays of apertures in 140-nm-thick silver films. In (a), transmission through coaxial apertures with a fixed at 150 nm and b varying is shown, while in (b) cross-shaped apertures with fixed arm-lengths ($L = 200$ nm) and different arm-widths w are shown. There is no substrate in (a), while in (b) the substrate has an index of 1.52.

of a circular aperture with the same radius as the annular aperture. It is instructive to consider the transformation from simple circular or square apertures to highly resonant coaxial or cross-shaped apertures in the case of apertures in metallic films with finite conductivity [44]. Figure 6.12 shows the transmission spectra, calculated using the FEM, as a function of wavelength for periodic arrays of (a) coaxial apertures and (b) symmetric crosses. In both cases, the arrays had a period of 400 nm and were located in a silver film of thickness 140 nm. The coaxial apertures had an outer radius of 150 nm, but the radius of the inner "stop" varied from 25 to 125 nm. It is apparent that as the size of the stop increases and the width of the gap decreases, the resonance is red-shifted in line with theoretical predictions [36,37], although the increase in Q seen in the PEC case is less apparent. Similarly, as the width of the cross-arms in the case of the cross-apertures decreases, the aperture becomes less "square-like" and resonances are red-shifted. Also shown in these plots is the transmission through the equivalent circular and square apertures. It is apparent that as the central stop disappears or the arm-widths approach their half-lengths, the behavior of the coaxial aperture or cross-shaped aperture approaches that of a circular or square aperture.

6.4.6 INFLUENCE OF THE SUBSTRATE AND THICKNESS OF METAL

In the fabrication of thin-film devices, substrates are commonly employed to lend robustness to the structure as well as to provide support to the isolated metallic elements in coaxial apertures. Beyond these structural enhancements, however, the presence of a substrate can also strongly influence the transmission through apertures in metallic films [45–49] and their nanoparticle counterparts [50, 51]. Structures relying on surface wave excitation [45,47,48] for facilitating high transmission notably exhibit a red-shift in the peak wavelength as a result of the presence of a substrate. This emerges naturally from a consideration of the relevant dispersion relations. In the case of high transmission arising from excitation of localized resonances within the apertures, however, the strongly confined fields occurring within the apertures at resonance do not extend for any appreciable distance into the substrate and, as a consequence, any influence of the substrate tend to be less apparent [33,34,36,37,41,43,44,52,53].

Studies into the impact of a dielectric substrate on transmission have been performed for FSS [25,54], where the performance of the device is regarded as the analog of an electrical circuit, with an effective capacitance, C_0, and inductance, L_0, in the case of lossless media. The capacitance of an infinitesimally thin, perfectly conducting capacitive grid formed on a semi-infinite substrate with refractive index, n, is shown to be [25]

$$C_{sub} = C_0 \left(\frac{1 + n^2}{2} \right),$$

where C_0 denotes the capacitance in the absence of a substrate. This results in a shift in the resonant free-space wavelength, λ_{res}, to

$$\lambda_{res} = \lambda_0 \sqrt{\frac{1 + n^2}{2}},$$

where λ_0 is the resonant wavelength when the grid is surrounded by vacuum. Resonances arising from metamaterials likewise exhibit similar effects that stem from capacitance changes within the structure, including sensitivity to the thickness of the metal film [55,56], as well as the presence of a substrate or dielectric overlayer [55,56]. An increase in the film thickness, however, results in a shift in the resonant frequency [55,56] and inherently reduces the relative contribution of the substrate to the total capacitance in the structure. This leads to a corresponding decrease in sensitivity to the presence of the substrate [55].

The interplay between the thickness of the PEC, the substrate index, and the location of the resonant wavelength in the case of an array of coaxial apertures is investigated in Figure 6.13. In Figure 6.13a, it can be seen that the resonant wavelength, calculated using the modal method, increases monotonically with increasing substrate index n for different screen thicknesses. It is apparent that as the metal thickness $h \to 0$, the

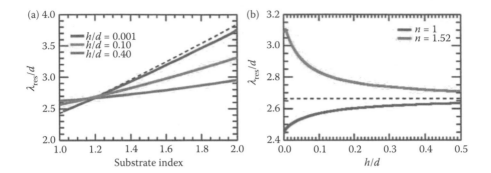

FIGURE 6.13 (**See color insert.**) The location of the transmission maxima associated with the longest localized aperture resonance wavelength as a function of (a) substrate index for structures with films with different thickness, h, and (b) film thickness for structures on substrates of refractive index $n = 1$ and $n = 1.52$. The periodicity of the array is d, and the outer and inner radii of the rings are $0.45d$ and $0.40d$. The resonant wavelength predicted by the relevant equation is shown as a dotted line in (a). The cut-off wavelength for the TE_{11} coaxial waveguide mode is indicated by the dotted line in (b). (From Roberts, A. and L. Lin, *Optical Materials Express*, 2011. **1**(3): 480–488. With permission.)

wavelength dependence approaches that given for λ_{res} in the above equationand that the sensitivity of the resonant wavelength decreases as the screen thickness increases. Figure 6.13b shows the location of the resonance as a function of film thickness for freestanding PEC films and films supported by a substrate of refractive index 1.52. In the case of very thick films, it can be observed that the resonant wavelength ($2.66d$) tends toward that of the TE_{11} coaxial waveguide mode.

It should be noted that the trends discussed here show consistency with similar research involving THz metamaterials [55,56] and have implications for a range of applications including sensing.

The FEM is used to extend this work to the study of coaxial apertures in silver films. The location of the resonant wavelength as a function of Ag thickness is shown in Figure 6.14 for two different substrate indices. Again, the sensitivity to substrate index decreases as the film thickness increases. Roberts and Lin [30] also showed that since the fields around the structure associated with these aperture resonances are strongly localized, both in the transverse and vertical directions, the location of the resonance can be tuned by selective removal of the substrate in the vicinity (approximately tens of nanometers) of the aperture.

6.4.7 INFLUENCE OF THE OPTICAL PROPERTIES OF THE APERTURE FILLING

It is also instructive to consider the role that the optical properties of the material filling the aperture has on its transmission. This is important as this provides another means of tuning the performance of devices and also provides an insight into their potential as sensing elements. Figure 6.15 shows the location of the long-wavelength maximum in the calculated transmission through an array of annular apertures in a silver film either freestanding or supported by a substrate with a refractive index of

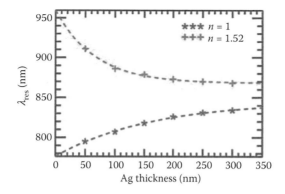

FIGURE 6.14 Location of transmission maxima in the transmission through arrays of either freestanding or substrate-supported arrays of coaxial apertures in silver films as a function of film thickness. The outer and inner radii of the apertures are 125 and 75 nm, respectively. (From Roberts, A. and L. Lin, *Optical Materials Express*, 2011. **1**(3): 480–488. With permission.)

1.52. The thickness of the film is 140 nm and the outer and inner radii of the apertures are 125 and 75 nm, respectively. It is apparent that the location of the resonance increases linearly with the refractive index. As discussed in the previous section, the resonance occurs at a longer wavelength for the case with the higher-index substrate. The difference in the resonant wavelengths decreases, however, as the refractive index of the filling increases. The sensitivity of the resonant wavelength in the case of the freestanding structure is 670 nm/RIU and in the presence of the substrate 630 nm/RIU. Hence, the presence of the substrate slightly reduces the sensitivity of the resonance to the aperture filling. Although the sensitivity is not as high as some

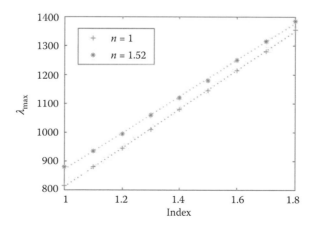

FIGURE 6.15 Locations of the long-wavelength transmission maximum in the transmission spectrum of an array of annular apertures in a silver film of thickness 140 nm as the refractive index of the material filling the apertures is varied. The outer radius of the apertures is 125 nm and the inner radius is 75 nm. Cases where the array is freestanding and supported by a substrate with refractive index 1.52 are shown. Straight-line fits are also shown.

other sensing elements, the robustness of the resonances to illumination conditions (as discussed in the next section) and the aperture arrangement could underpin new sensing elements for use in adverse conditions.

6.4.8 POLARIZATION AND ANGLE OF INCIDENCE

Another important feature of the localized resonances within apertures is the fact that these can persist when illuminated off-normal and at different polarizations. In general, the bulk of theoretical and computational studies into the electromagnetic behavior of aperture array devices employ configurations involving plane wave illumination. It should, however, be noted that plane wave illumination is generally only an approximation in experimental research, and a superposition of plane waves with various angles of incidence and polarizations can be utilized for expressing focused beams or beams with finite-spot sizes using a plane wave decomposition [53,57]. It is, furthermore, noteworthy to consider the case of off-normal illumination given that some features in the transmission spectra; in particular, those whose dominant mechanism underpinning their high transmission stem from the excitation of surface waves exhibit strong dependence not only on the optical properties of the metal and surrounding media, but also on the angle of incidence and polarization [58–60]. In contrast, the excitation of localized resonances within apertures has been shown to be relatively insensitive to off-normal incidence in the case of annular [53,61] and cross-shaped [62] aperture arrays, as well as variations in polarization for annular aperture arrays [53].

The robustness to changes in incidence parameters for the specific case of symmetric cross-shaped aperture arrays in a silver film formed on a glass substrate ($n = 1.52$) can be observed in Figure 6.16 which shows the experimentally determined and computed transmission as a function of angle of incidence for both TE and TM polarization. The broad transmission peak at a wavelength of ~800 nm shown in Figure 6.16a and b stems from LSP resonances in the apertures.

It is apparent that the location of spectral features associated with the excitation of the LSP resonances is robust to changes in angle of incidence, where the transmission efficiency and LSP peak bandwidth are largely unchanged for angle of incidence values less than $\theta = 30°$. In the case of TM illumination, the spectral location of the $(-1, 0)$ SPP mode is red-shifted for larger angles of incidence. For values of $\theta \geq 20°$, interference from the SPP mode results in the LSP peak shifting to a longer wavelength. This influence on location or shape of the LSP peak can be moderated by ensuring that the $(-1, 0)$ SPP and the LSP modes are well separated. As discussed in the previous section, cross-shaped apertures can easily be tailored to tune the characteristics of the LSPs occurring within it. By increasing the arm-length and/or decreasing the arm-width of the crosses, the localized resonances can be tuned to occur far from SPP resonances [32,41] to reduce perturbation from the SPPs.

The study of the behavior of metallic structures illuminated with more complex beams may also generate a useful platform for the development of robust plasmonic devices. The rigorous modeling of scattering of a focused spot by a grating [63], the excitation of surface plasmons on metallic film surfaces by focussed beams [64] and optical vortex beams [65], and the focusing of surface plasmons by illuminating

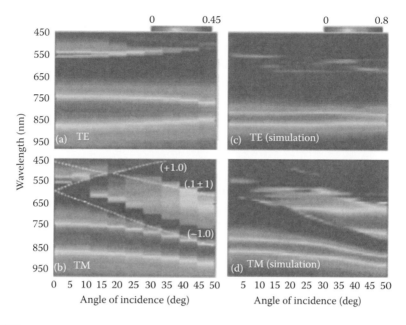

FIGURE 6.16 (**See color insert.**) Angle-dependent light transmission through arrays of cross-shaped apertures in a Ag film formed on a glass substrate ($n = 1.52$). Measured spectra for (a) TE and (b) TM illumination. The white dotted lines shown in (b) represent the calculated dispersion curves of SPPs on Ag/glass interface. Simulated results for (c) TE and (d) TM illumination. The array periodicities are: $p_x = 360$ nm and $p_y = 358$ nm; cross-arm-lengths: $L_x = 240 \pm 6$ nm and $L_y = 241 \pm 6$ nm; width of arm-length $L_x{:}w_x = 64 \pm 7$ nm, width of arm-length $L_y{:}w_y = 68 \pm 7$ nm. Subscripts x and y denote different dimensions along the two axes of the array. The total area of the entire array is 85×68 μm^2.

a coaxial aperture with radially polarized light [66] are some examples of recent research into this area. It has also been shown that aperture arrays can act as a spatial filter, preferentially transmitting certain spatial frequencies [53]. For example, the TEM mode can be excited by TM-polarized incident light at higher angles of incidence at certain resonant frequencies or illumination with a radially polarized beam [53] dictated by the thickness of the metal.

The studies of the interaction of off-normal incidence plane waves with the aperture array discussed here show the feasibility of filtering the angular spectrum of certain beams, as well as utilizing the incident beam to excite different electromagnetic modes of the structure as a means of modifying the beam transmission. Such investigations may provide useful insights into application in areas such as wavefield modification, spatial filtering, and interrogation of plasmonic devices.

6.5 RESONANCE AND PHASE

It is well known that resonances in RLC circuits are accompanied by significant changes in phase. In this section, we discuss phase variations in the field transmitted through arrays of cross-shaped apertures. As discussed previously, a major advantage

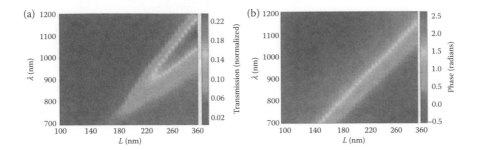

FIGURE 6.17   (**See color insert.**) Simulated results for the (a) normalized transmission and (b) phase variation of light passing through a single cross-shaped aperture formed in a 140-nm-thick silver film as function of wavelength λ and arm-length L. The arm width is fixed at 40 nm.

of the cross-shaped aperture lies in the simplicity of tuning the characteristics of the LSPs occurring within it. It has been shown previously that LSP resonances existing in cross-shaped apertures, whether isolated or in periodic arrays, can be tuned by simply altering the arm-length of the crosses [41]. The transmission efficiency and phase variation of light passing through a single, isolated cross-shaped aperture as a function of wavelength and arm-length, L, is shown in Figure 6.17.

It is apparent that, as the transmission passes through a maximum, the phase variation of the transmitted waves undergoes a rapid modulation. The calculated normalized transmission and phase of the transmitted field for a single cross-shaped aperture as a function of aperture arm-length at a wavelength of 850 nm are shown in Figure 6.18. It is clear that near LSP resonances at a specific wavelength, the amplitude and phase of the transmitted field can be tailored by simply tuning the arm-length L of the aperture.

From Figure 6.18, it can be observed that near the resonant dimension of the aperture, the variation in amplitude and phase of the transmitted field shows similar trends to that observed in an electrical RLC resonant circuit, with a total phase modulation of about π. The total available phase modulation can be further extended by varying the dimensions of other structure parameters, such as the cross-arm-width

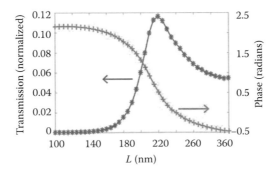

FIGURE 6.18   Calculated normalized transmission (*) and phase (+) of the transmitted field for a single cross-shaped aperture as function of arm-length at $\lambda = 850$ nm.

or thickness of the metal film. In particular, by increasing the thickness, so that the first-order Fabry–Pérot resonance is close to the zeroth order observed here, the phase excursion could be increased to closer to 2π.

This aperture-geometry-dependent phase property can be employed to introduce a spatial phase modulation onto an incident optical wavefield by means of varying aperture (or other structural parameters) sizes across an aperture array device around resonant dimensions for a particular design wavelength. Recent examples of applications that utilize this approach include wavefront control devices such as 2D plasmonic lenses [67,68]. Earlier studies into other plasmonic devices involving arrays of nanoholes with fixed diameters [69], nanoslits with varying depths [70], widths [71], and modulated slit positions [72] in metallic films have also been demonstrated. These, however, possessed the disadvantages of being either polarization-dependent and/or producing only a cylindrical focus. Furthermore, by introducing a tailored asymmetry into the aperture geometry, it is possible to produce plasmonic waveplates [73] and other birefringent materials.

6.6 CONCLUSION

Here we have discussed properties of coaxial and cross-shaped apertures in metallic films. Similarities and differences between isolated apertures and holes arranged in regular arrays were discussed. Early computational research into structures in PEC films was presented and the extent to which insights that could be extended from this work to apertures in real metals were discussed. The influence of aperture geometry and also the optical properties of surrounding media were discussed. Apertures with subwavelength dimensions play a key role as components in new plasmonic systems with potential applications such as optical antennas, sensing elements, and devices for optical computing.

ACKNOWLEDGMENTS

The authors would like to thank colleagues and former staff and students who have contributed to the research discussed in this chapter. They are particularly indebted to Shannon Orbons and Ling Lin. The financial support of the University of Melbourne and the Australian Research Council's Discovery Projects funding scheme (project number DP0878268) are also gratefully acknowledged.

REFERENCES

1. Bouwkamp, C.J., Diffraction theory. *Reports on Progress in Physics*, 1945. **17**: 35–99.
2. Butler, C.M., Y. Rahmat-Samii, and R. Mittra, Electromagnetic penetration through apertures in conducting surfaces. *IEEE Transactions on Antennas and Propagation*, 1978. **26**(1): 82–93.
3. Karczewki, B. and E. Wolf, Comparison of three theories of electromagnetic diffraction at an aperture. Part I: Coherence matrices. *Journal of the Optical Society of America*, 1966. **56**(9): 1207–1214.
4. Karczewki, B. and E. Wolf, Comparison of three theories of electromagnetic diffraction at an aperture. Part II: The far field. *Journal of the Optical Society of America*, 1966. **56**(9): 1214–1219.

5. Ebbesen, T.W. et al., Extraordinary optical transmission through sub-wavelength hole arrays. *Nature*, 1998. **391**(6668): 667–669.
6. Garcia de Abajo, F.J., Light scattering by particle and hole arrays. *Reviews of Modern Physics*, 2007. **79**(October–December): 1267–1290.
7. Weiner, J., The physics of light transmission through subwavelength apertures and aperture arrays. *Reports on Progress in Physics*, 2009. **72**: 064401-1–064401-19.
8. Bharadwaj, P., B. Deutsch, and L. Novotny, Optical antennas. *Advances in Optics and Photonics*, 2009. **1**: 438–483.
9. Mata Mendez, O., M. Cadilhac, and R. Petit, Diffraction of a two-dimensional electromagnetic beam wave by a thick slit pierced in a perfectly conducting screen. *Journal of Optical Society America*, 1983. **73**(3): 328–331.
10. Roumiguieres, D. et al., etude de la diffraction par une fente pratiquée dans un écran infiniment conducteur d'épaisseur quelconque. *Optics Communications*, 1973. **9**(4): 368–373.
11. Hongo, K. and G. Ishii, Diffraction of an electromagnetic plane wave by a thick slit. *IEEE Transactions on Antennas and Propagation*, 1978. **26**(3): 494–499.
12. Neerhoff, F.L. and G. Mur, Diffraction of a plane electromagnetic wave by a slit in a thick screen placed between two different media. *Applied Scientific Research*, 1973. **28**(1): 73–88.
13. Wirgin, M., Influence de l'épaisseur de l'écran sur la diffraction par une fente. *C. R. Acad. Sci. Paris*, 1970. **270**: 1457–1460.
14. Bethe, H.A., Theory of diffraction by small holes. *Physical Review* 1944. **66**(7–8): 163–182.
15. Millar, R.F., The diffraction of an electromagnetic wave by a large aperture. *Proceedings of the IEE, Part C: Monographs*, 1957. **104**(6): 240–250.
16. Andrejewski, W., Strenge Theorie der Beugung ebener elektromagnetischer Wellen an der vollkommen leitenden Kreisscheibe und an der kreisförmigen Öffnung im vollkommen leitenden ebenen Schirm. *Numerische Ergebnisse. Naturwissenschaften*, 1951. **38**(17): 406–407.
17. Bouwkamp, C.J., On the diffraction of electromagnetic waves by small circular disks and holes. *Philips Research Reports*, 1950. **5**: 401–402.
18. Leviatan, Y., Electromagnetic coupling between two half-space regions separated by two slot-perforated parallel conducting screens. *IEEE Transactions on Microwave Theory and Techniques*, 1988. **36**(1): 44–52.
19. Leviatan, Y., Study of nearzone fields of a small aperture. *Journal of Applied Physics*, 1986. **60**(5): 1577–1583.
20. Roberts, A., Electromagnetic theory of diffraction by a circular aperture in a thick, perfectly conducting screen. *Journal of the Optical Society of America A Optics Image Science and Vision*, 1987. **4**(10): 1970–1983.
21. Chen, C.-C., Transmission through a conducting screen perforated periodically with apertures. *IEEE Transaction on Microwave Theory and Techniques*, 1970. **18**(9): 627–632.
22. Chen, C.-C., Diffraction of electromagnetic waves by a conducting screen perforated periodically with circular holes. *IEEE Transaction on Microwave Theory and Techniques*, 1971. **19**(5): 475–481.
23. Chen, C.-C., Transmission of microwave through perforated flat plates of finite thickness. *IEEE Transaction on Microwave Theory and Techniques*, 1973. **21**(1): 1–6.
24. Compton, R.C. et al., Diffraction properties of a bandpass grid. *Infrared Physics*, 1983. **23**(5): 239–245.
25. Whitbourn, L.B. and R.C. Compton, Equivalent-circuit formulas for metal grid reflectors at a dielectric boundary. *Applied Optics*, 1985. **24**(2): 217–220.
26. Ulrich, R., Far-infrared properties of metallic mesh and its complementary structure. *Infrared Physics*, 1967. **7**: 37–55.

27. Bliek, P. et al., Inductive grids in the region of diffraction anomalies: Theory, experiment and applications. *IEEE Transaction on Microwave Theory and Techniques*, 1980. **28**(10): 1119–1125.

28. Roberts, A. and R.C. McPhedran, Bandpass grids with annular apertures. *IEEE Transactions on Antennas and Propagation*, 1988. **36**(5): 607–611.

29. Roberts, A. and R.C. Compton, A vector measurement scheme for testing quasi-optical components. *International Journal of Infrared and Millimeter Waves*, 1990. **11**(2): 165–174.

30. Roberts, A. and L. Lin, Substrate and aspect-ratio effects in resonant nanoaperture arrays. *Optical Materials Express*, 2011. **1**(3): 480–488.

31. Collin, S., F. Pardo, and J.L. Pelouard, Waveguiding in nanoscale metallic apertures. *Optics Express*, 2007. **15**(7): 4310–4320.

32. Gordon, R. and A.G. Brolo, Increased cut-off wavelength for a subwavelength hole in a real metal. *Optics Express*, 2005. **13**(6): 1933–1938.

33. Baida, F.I., Enhanced transmission through subwavelength metallic coaxial apertures by excitation of the TEM mode. *Applied Physics B Lasers and Optics*, 2007. **89**(2–3): 145–149.

34. Baida, F.I. and D. Van Labeke, Light transmission by subwavelength annular aperture arrays in metallic films. *Optics Communications*, 2002. **209**(1–3): 17–22.

35. Baida, F.I., D. Van Labeke, and B. Guizal, Enhanced confined light transmission by single subwavelength apertures in metallic films. *Applied Optics*, 2003. **42**(34): 6811–6815.

36. Haftel, M.I., C. Schlockermann, and G. Blumberg, Role of cylindrical surface plasmons in enhanced transmission. *Applied Physics Letters*, 2006. **88**(19): 193104.

37. Haftel, M.I., C. Schlockermann, and G. Blumberg, Enhanced transmission with coaxial nanoapertures: Role of cylindrical surface plasmons. *Physical Review B*, 2006. **74**(23): 235405-1–235405-11.

38. COMSOL *Multiphysics*. Available from: http://www.comsol.com/.

39. Johnson, P.B. and R.W. Christy, Optical constants of the noble metals. *Physical Review B*, 1972. **6**(12): 4370–4379.

40. Palik, E.D., *Handbook of Optical Constants of Solids*, 1985: Academic Press, Boston.

41. Lin, L., L.B. Hande, and A. Roberts, Resonant nanometric cross-shaped apertures: Single apertures versus periodic arrays. *Applied Physics Letters,* 2009. **95**: 201116.

42. Luk'yanchuk, B. et al., The Fano resonance in plasmonic nanostructures and metamaterials. *Nature Materials*, 2010. **9**(September): 707–715.

43. Orbons, S.M. et al., Dual resonance mechanisms facilitating enhanced optical transmission in coaxial waveguide arrays. *Optics Letters*, 2008. **33**(8): 821–823.

44. Orbons, S.M. and A. Roberts, Resonance and extraordinary transmission in annular aperture arrays. *Optics Express*, 2006. **14**(26): 12623–12628.

45. Brolo, A.G. et al., Surface plasmon sensor based on the enhanced light transmission through arrays of nanoholes in gold films. *Langmuir*, 2004. **20**: 4813–4815.

46. Kang, J.H. et al., Substrate effect on aperture resonances in a thin metal film. *Optics Express*, 2009. **17**: 15652–15658.

47. Krishnan, A. et al., Evanescently coupled resonance in surface plasmon enhanced transmission. *Optics Communications*, 2001. **200**: 1–7.

48. Pang, Y., C. Genet, and T. Ebbesen, Optical transmission through subwavelength slit apertures in metallic films. *Optics Communications*, 2007. **280**: 10–15.

49. Shuford, K.L. et al., Substrate effects on surface plasmons in single nanoholes. *Chemical Physics Letters*, 2007. **435**: 123–126.

50. Vernon, K.C. et al., Influence of particle–substrate interaction on localized plasmon resonances. *Nano Letters*, 2010. **10**(6): 2080–2086.

51. Zhang, S. et al., Substrate-induced Fano resonances of a plasmonic nanocube: A route to increased-sensitivity localized surface plasmon resonance sensors revealed. *Nano Letters*, 2011. **11**: 1657–1663.

52. Baida, F.I. et al., Origin of the super-enhanced light transmission through a 2-D metallic annular aperture array: A study of photonic bands. *Applied Physics B Lasers and Optics*, 2004. **79**(1): 1–8.

53. Roberts, A., Beam transmission through hole arrays. *Optics Express*, 2010. **18**(3): 2528–2533.

54. Compton, R.C., L.B. Whitbourn, and R.C. McPhedran, Strip gratings at a dielectric interface and application of Babinet's principle. *Applied Optics* 1984. **23**(18): 3236–3242.

55. Chiam, S.Y. et al., Controlling metamaterial resonances via dielectric and aspect ratio effects. *Applied Physics Letters* 2010. **97**: 191906.

56. O'Hara, J.F. et al., Thin-film sensing with planar terahertz metamaterials: Sensitivity and limitations. *Optics Express*, 2008. **16**: 1786–1795.

57. Wolf, E., Electromagnetic diffraction in optical systems. I. An integral representation of the image field. *Proceedings of the Royal Society of London, Series A*, 1959. **253**(1274): 349–357.

58. Altewischer, E., M.P. van Exter, and J.P. Woerdman, Polarization analysis of propagating surface plasmons in a subwavelength hole array. *Journal of the Optical Society of America B Optical Physics*, 2003. **20**(9): 1927–1931.

59. Grupp, D.E. et al., Crucial role of metal surface in enhanced transmission through subwavelength apertures. *Applied Physical Letters*, 2000. **77**(11): 1569–1571.

60. Ghaemi, H.F. et al., Surface plasmons enhance optical transmission through subwavelength holes. *Physical Review B*, 1998. **58**(11): 6779–6782.

61. Van Labeke, D. et al., An angle-independent frequency selective surface in the optical range. *Optics Express*, 2006. **14**(25): 11945–11951.

62. Lin, L. and A. Roberts, Angle-robust resonances in cross-shaped aperture arrays. *Applied Physics Letters,* 2010. **97**(6): 061109-1–061109-3.

63. Brok, J.M. and H.P. Urbach, Rigorous model of the scattering of a focused spot by a grating and its application in optical recording. *Journal of the Optical Society of America A: Optics, Image Science, and Vision*, 2003. **20**(2): 256–272.

64. Bouhelier, A. et al., Surface plasmon interference excited by tightly focused laser beams. *Optics Letters*, 2007. **32**(17): 2535–2537.

65. Tan, P.S. et al., Surface plasmon polaritons generated by optical vortex beams. *Applied Physics Letters,* 2008. **92**: 111108-1–111108-3.

66. Lerman, G.M., A. Yanai, and U. Levy, Demonstration of nanofocusing by the use of plasmonic lens illuminated with radially polarized light. *Nano Letters*, 2009. **9**(5): 2139–2143.

67. Goh, X.M., L. Lin, and A. Roberts, Plasmonic lenses for wavefront control applications using two-dimensional nanometric cross-shaped aperture arrays. *Journal of the Optical Society of America B: Optical Physics*, 2011. **28**(3): 547–553.

68. Lin, L. et al., Plasmonic lenses formed by two-dimensional nanometric cross-shaped aperture arrays for Fresnel-region focusing. *Nano Letters*, 2010. **10**: 1936–1940.

69. Gao, H. et al., Broadband plasmonic microlenses based on patches of nanoholes. *Nano Letters*, 2010. **10**(10): 4111–4116.

70. Sun, Z. and H.K. Kim, Refractive transmission of light and beam shaping with metallic nano-optic lenses. *Applied Physics Letters*, 2004. **85**(4): 642–644.

71. Verslegers, L. et al., Planar lenses based on nanoscale slit arrays in a metallic film. *Nano Letters*, 2009. **9**: 235–238.

72. Chen, Q. and D.R.S. Cumming, Visible light focusing demonstrated by plasmonic lenses based on nano-slits in an aluminum film. *Optics Express*, 2010. **18**(14): 14788–14793.

73. Roberts, A. and L. Lin, Plasmonic quarter-wave plate. *Optics Letters*, 2012. **37**(11): 1820–1822.

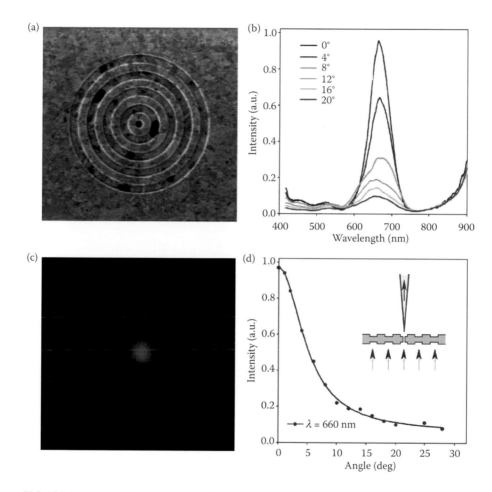

FIGURE 2.1 (a) FIB image of a bull's eye structure in a suspended silverfilm and (b) its transmission spectra for different collection angles. The tail beyond 800 nm is an artifact of the measurement. The structure is illuminated under normal incidence with unpolarized light; the period of the grating is 500 nm, the groove depth is 60 nm, the hole diameter is 250 nm, and the film thickness is 300 nm. (c) Optical image and (d) angular intensity distribution at the wavelength of maximum transmission. (After H. J. Lezec et al., Beaming light from a subwavelength aperture, *Science 297*, 820–822, 2002. © AAAS 2002, with permission.)

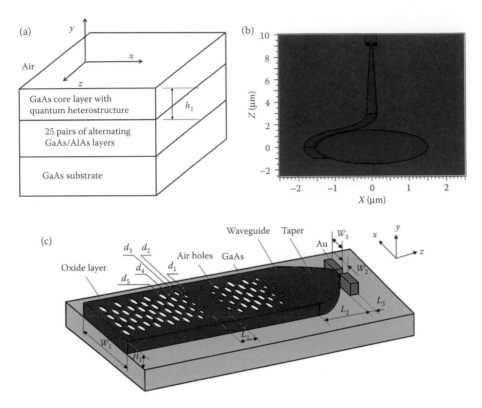

FIGURE 2.3 (a) Schematic of the epitaxially layered structure and (b) tapered coupling of light from a microdisk laser to a nanoantenna. (After H. T. Hattori et al., Coupling of light from microdisk lasers into plasmonic nano-antennas, *Opt. Express* **17**, 20878–20884, 2009. © OSA 2009, with permission.) (c) Schematic of light coupling from a photonic crystal laser into a nanoantenna by using a parabolic taper. (After Z. Li et al., Merging photonic wire lasers and nanoantennas, *J. Lightw. Technol.* **29**, 2690–2697, 2011. © IEEE 2011, with permission.)

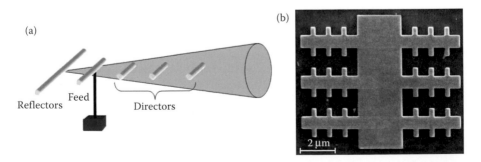

FIGURE 2.5 (a) Typical geometry of a five-element RF Yagi–Uda antenna and (b) SEM image of Charnia-like structure.

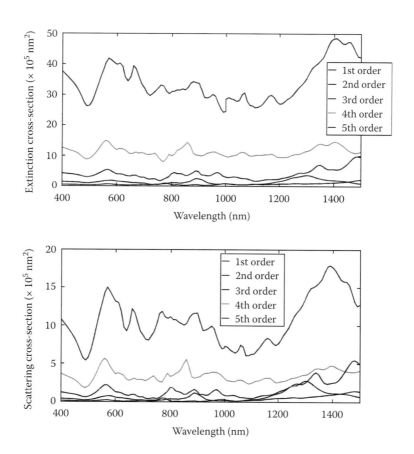

FIGURE 3.5 Numerically simulated (a) extinction and (b) scattering cross-sections of Sierpinski gold nanoantennas, for fractal order between first and fifth and linear polarization excitation.

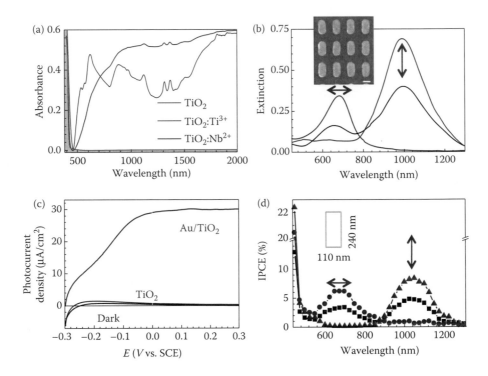

FIGURE 3.6 Plasmonic solar cell. (a) Absorption spectrum of different titania substrates: n-type Nb-doped TiO_2 (Furuuchi Chemical. Co.), reduced TiO_2:Ti^{3+}, and undoped TiO_2; thickness of samples was 0.5 mm. (b) The extinction spectrum of Au nanorod pattern fabricated on the untreated TiO_2 single-crystal measured under unpolarized and polarized illumination in water; polarization is marked by arrows. The inset shows an SEM image of the pattern of gold nanorods of a 240×110 nm^2 footprint. (c) *I–V* curves measured under different wavelength illumination at linear sweep of 50 mV/s. (d) The incident photon-to-current conversion efficiency (IPCE) spectra of the gold nanorod structure patterned and unpatterned n-type TiO_2:Nd electrode under unpolarized light illumination and for the T and L polarizations, respectively. (Adapted from Y. Nishijima et al., *J. Phys. Chem. Lett.* 10, 2031–2036, 2010.)

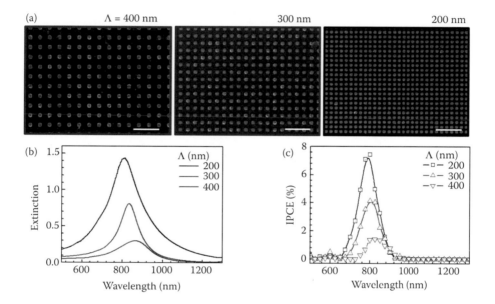

FIGURE 3.7 (a) SEM images of gold nanoblocks of $110 \times 110 \ nm^2$ footprint with different separation of 290, 190, and 90 nm corresponding to the period $A = 400$, 300, and 200, respectively. Scale bar = 1 μm. (b) Extinction spectra of the corresponding gold nanoblock structures. (c) The incident photon-to-current conversion efficiency (IPCE) spectra of the gold nanorod structure patterned n-type TiO_2 electrode under unpolarized light illumination. (Adapted from Y. Nishijima et al., *J. Phys. Chem. Lett.* 10, 2031–2036, 2010.)

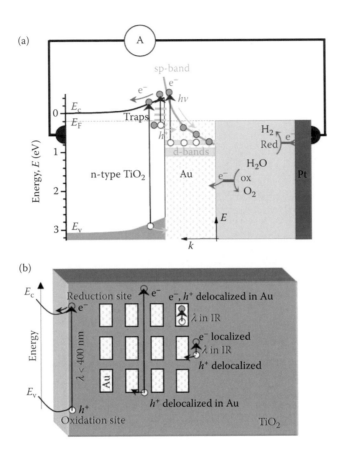

FIGURE 3.8 (a) Energy diagram and schematics of the processes in a Au-nanoparticle sensitized n-type titania solar cell. The absorption takes place via the direct bandgap transitions for photons with energy $h\nu \geq 3.1$ eV; from the trap defects (which are responsible for the n-type in TiO_2) and d- to sp-band transitions in gold in IR spectral range. Energy position of the water red/ox potentials are shown for discussion of mechanism of charge transfer cycle. The gold particle is shown in the electron dispersion $E(k)$ presentation to reflect the intra-band (sp − d) transitions; k is the wavevector. (b) Schematics of Au-nanoparticles arrangement on the surface of TiO_2 and possible electronic transitions. The electrons e⁻ are localized on the surface and can flow into bulk of TiO_2; holes h⁺ are pinned to the surface at the location of creation or tunnel into Au nanoparticle.

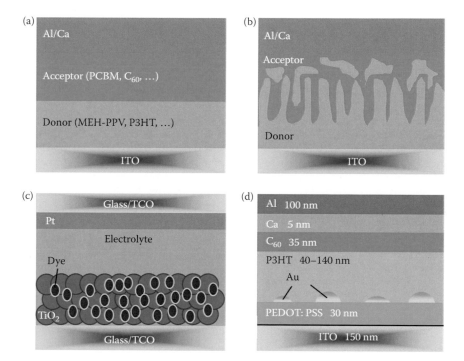

FIGURE 3.9 Some popular geometries of solar cells: examples of (a) organic bilayer and (b) heterojunction, (c) dye-sensitized, (d) plasmonic nanoparticle-enhanced (design from reference [67]). TCO is transparent conductive oxide; see text for details.

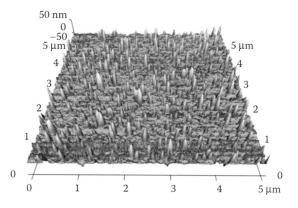

FIGURE 3.10 Atomic force microscopy (AFM) profile of the gold nanoparticle layer. (Adapted from C. Poh, et al., *Opt. Mater. Express* 1, 1326–1331, 2011.)

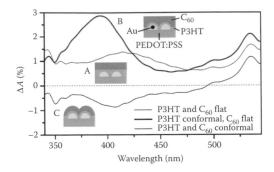

FIGURE 3.12 Simulated absorbance difference *A* for flat and conformal junction interfaces.

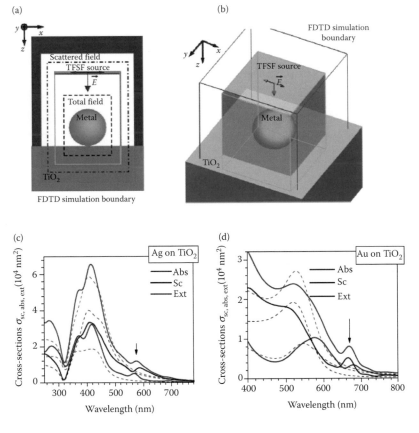

FIGURE 3.14 (a) Geometry of FDTD simulation according to TFSF formulation and (b) structural perspective. The incident wave is linearly polarized along the *x*-axis and is incident on the metallic sphere along the *z*-axis. (c) The extinction, scattering, and absorption cross-sections ($\sigma_{ex} = \sigma_{sc} + \sigma_{abs}$) of a 50-nm-radius silver sphere on rutile; dashed lines mark corresponding cross-sections without rutile substrate. Geometrical cross-section is 0.79×10^4 nm^2. (d) Same as (c), for a 50-nm-radius gold sphere. In (c,d), weak spectral peaks associated with the presence of a substrate are indicated by arrows.

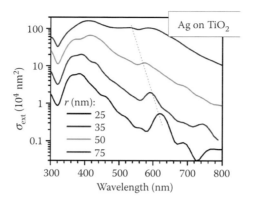

FIGURE 3.15 Dependence of extinction spectra of Ag nanosphere/on TiO$_2$ system on the sphere radius r. The dashed line is guide to the eye to emphasize spectral shift of the characteristic long-wavelength peak (see Figure 3.16). (Adapted from V. Mizeikis, E. Kowalska, and S. Juodkazis, *J. Nanosci. Nanotechnol.* 11, 2814–2822, 2011.)

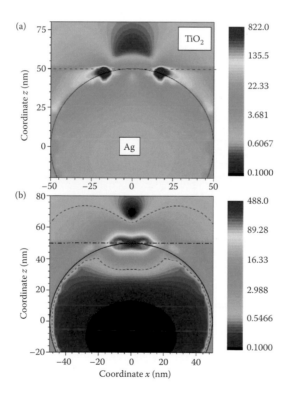

FIGURE 3.16 Cross-sectional pattern of the near-field intensity enhancement factor $|E|_2/E_0|_2$ at ~420 nm (a) and ~570 nm (b) wavelengths. The highest enhancement was 820 at 420 nm, and 490 at 570 nm wavelengths. Outlines of the sphere are emphasized by black solid lines. In (b), black dashed lines emphasize the contour where $E = 1$. (Adapted from V. Mizeikis, E. Kowalska, and S. Juodkazis, *J. Nanosci. Nanotechnol.* 11, 2814–2822, 2011.)

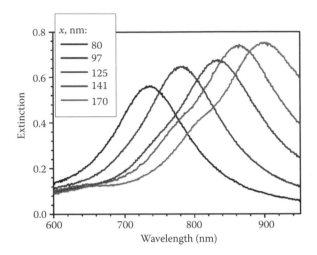

FIGURE 3.20 Extinction spectra of the gold patterned substrates in water. The side length of the squares, x, is shown in the plot. (Adapted from Y. Yokota et al., Influence of plasmonic enhancement on surface-enhanced Raman scattering, in Extended Abstracts of XXIV Int. Conf. on Photochemistry ICP2009, Toledo, Spain, pp. PSII–P82, July 2009; Y. Yokota, Sensing applications of plasmonic nanoparticles, PhD thesis, Hokkaido University, Japan, October 2009; Y. Yokota, K. Ueno, and H. Misawa, *Small* 7, 252–258, 2011.)

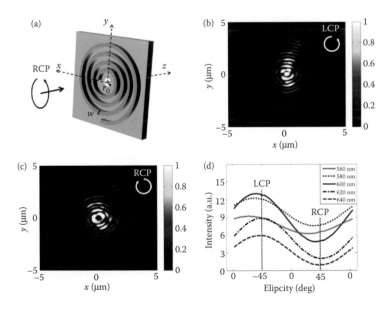

FIGURE 4.5 (a) Schematic diagram of spiral bull's eye structure for the polarization-dependent enhancement of nanohole transmission. Electric field intensity distributions of 10-nm backside of the metal substrate with the (b) LCP and (c) RCP incident waves are shown, respectively, for comparing the field intensity at the center point. (d) Total transmission efficiencies varying with the optical polarization are shown with several grating periods. The results are normalized to the total transmission without the spiral gratings.

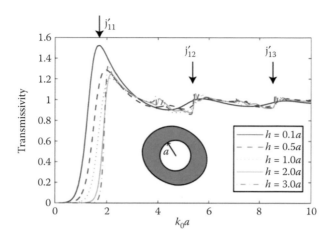

FIGURE 6.1 Transmission through a circular aperture of radius, a, calculated using the modal method for different thicknesses, h, of the screen. Transmission is normalized to that through an aperture in the geometric optics limit. Cut-off frequencies for the first three TE circular waveguide modes (i.e., the first three roots of J_1') are shown.

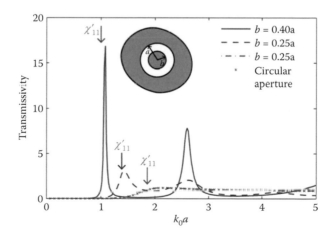

FIGURE 6.2 Transmission through a coaxial aperture with varying inner radii, b, and a circular aperture in a screen of thickness $h = 1.1a$ as a function of normalized wavenumber.

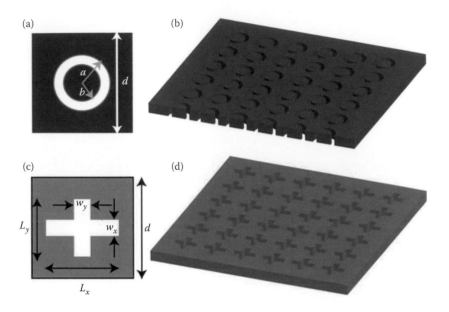

FIGURE 6.3 Arrays of annular (a, b) and cross-shaped (c, d) apertures in thick metallic films.

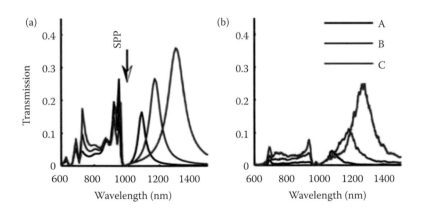

FIGURE 6.8 Transmission spectra for arrays A, B, and C. Calculated (a) and measured (b) results. The approximate location of the $(0, \pm)$ Au/glass SPP excitation is shown in (a). (From Lin, L., L.B. Hande, and A. Roberts, *Applied Physics Letters*, 2009. 95: 201116. With permission.)

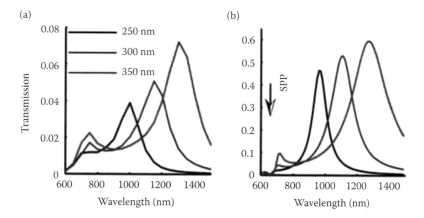

FIGURE 6.10 Simulated transmission spectra for (a) single apertures and (b) an arrangement of apertures with the same dimensions with periodicity 400 nm. The location of the Au/glass (0, ±1) SPP is shown in (b). (From Lin, L., L.B. Hande, and A. Roberts, *Applied Physics Letters*, 2009. 95: 201116. With permission.)

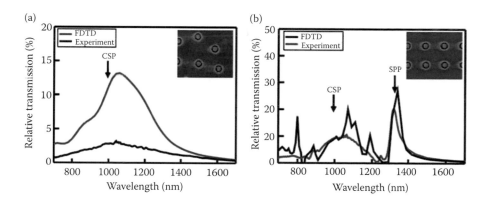

FIGURE 6.11 Transmission (experimental and calculated) through random and regular arrangements of coaxial apertures with outer radius 160 nm and inner radius 88 nm in a silver film of thickness 140 nm on a silica substrate. (a) Transmission through a random arrangement of apertures; (b) transmission through an array with periods 780 nm (in the horizontal direction) and 820 nm (vertical). Locations of localized CSP and SPP excitation wavelengths are shown. Insets show SEM images of the samples. (From Orbons, S.M. et al., *Optics Letters*, 2008. 33(8): 821–823. With permission.)

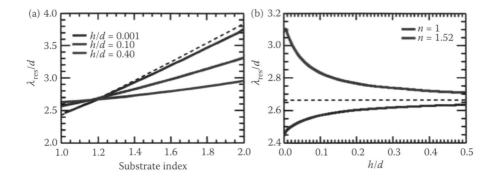

FIGURE 6.13 The location of the transmission maxima associated with the longest localized aperture resonance wavelength as a function of (a) substrate index for structures with films with different thickness, h, and (b) film thickness for structures on substrates of refractive index $n = 1$ and $n = 1.52$. The periodicity of the array is d, and the outer and inner radii of the rings are $0.45d$ and $0.40d$. The resonant wavelength predicted by the relevant equation is shown as a dotted line in (a). The cut-off wavelength for the TE_{11} coaxial waveguide mode is indicated by the dotted line in (b). (From Roberts, A. and L. Lin, *Optical Materials Express*, 2011. **1**(3): 480–488. With permission.)

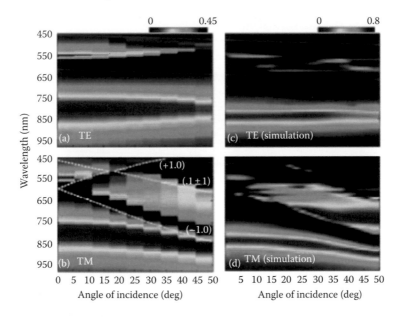

FIGURE 6.16 Angle-dependent light transmission through arrays of cross-shaped apertures in a Ag film formed on a glass substrate ($n = 1.52$). Measured spectra for (a) TE and (b) TM illumination. The white dotted lines shown in (b) represent the calculated dispersion curves of SPPs on Ag/glass interface. Simulated results for (c) TE and (d) TM illumination. The array periodicities are: $p_x = 360$ nm and $p_y = 358$ nm; cross-arm-lengths: $L_x = 240 \pm 6$ nm and $L_y = 241 \pm 6$ nm; width of arm-length L_x:$w_x = 64 \pm 7$ nm, width of arm-length L_y:$w_y = 68 \pm 7$ nm. Subscripts x and y denote different dimensions along the two axes of the array. The total area of the entire array is 85×68 μm².

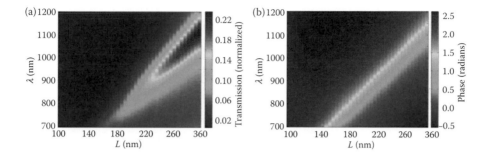

FIGURE 6.17 Simulated results for the (a) normalized transmission and (b) phase variation of light passing through a single cross-shaped aperture formed in a 140-nm-thick silver film as function of wavelength λ and arm-length L. The arm width is fixed at 40 nm.

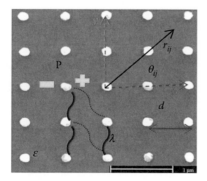

FIGURE 7.1 Light (red waves) scattered from neighboring axial and diagonal subwavelength nanoparticles constructively interferes onto a plasmon dipole (yellow −, +) of one nanoparticle in a square array of lattice constant d with axes oriented in x- and y-directions. Incident light orthogonal to the x–y-plane, polarized in the x-direction has wavelength λ, which approaches the lattice constant. The vector r orients a central particle i relative to a particle j along the diagonal located at angle θ_{ij}.

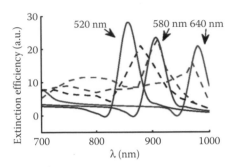

FIGURE 7.3 Extinction spectra from infinite pitched lattices of spherical Au nanoparticles spaced at 520, 580, and 640 nm, respectively. Theoretical predictions (solid) are compared with sketches of experimental spectra (dashed).

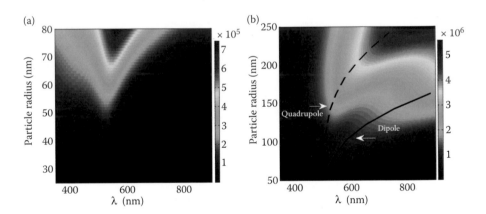

FIGURE 7.4 Magnitude of imaginary dipole and quadrupole components of nanoparticle polarizability [α in Equatuion 7.6] for a single Au nanoparticle with radius ranging from (a) 25 to 80 and (b) 50 to 250 nm and incident radiation between 350 and 900 nm. Loci of maximum values are indicated for dipole (solid line) and quadrupole (dashed line) polarizability.

7 Plasmon Coupling Enhanced in Nanostructured Chem/Bio Sensors

D. Keith Roper, Phillip Blake, Drew DeJarnette, and Braden Harbin

CONTENTS

7.1   INTRODUCTION

Common limitations of existing sensors designed to detect and measure chemical and biological analytes motivate a reevaluation of current paradigms in order to design and fabricate sensor systems with improved capabilities [1]. Existing sensor platforms essentially identify a single analyte. Detection occurs only after the analyte of interest deposits on a sensor surface. A surrogate analyte property such as mass or conductivity is transduced into sensor signal, but at rates and sensitivities limited by equilibrium and diffusion constraints. As a result, sensor costs increase in proportion to utility, but inversely with portability for field use. Recent developments in characterization and fabrication of nanostructured metamaterials offer radical paradigm shifts in sensor design that could provide direct, sensitive distinction of multiple analytes at improved throughput while reducing cost and complexity of sensor platforms.

A range of widely sought, potential field applications of sensors could benefit from improvements in commercially available sensor platforms. Rapid field detection of pathogens with epidemiological significance could allow near real-time monitoring progression of possible pandemics and inform public health decisions to regulate traffic and trade in order to reduce associated morbidity and mortality rates. Identification of infectious or toxic agents in large, undifferentiated supplies of foodstuffs could prevent associated outbreaks of illness and disease and expensive product recall and quarantine. Low-cost, personal, wearable sensing devices could permit individual on-demand or remote electronic monitoring of vital measures of human health to allow rapid intervention in an emergency event, improve disease diagnosis, and support informed treatment of chronic illnesses [2–4]. Automated stand-off detection of hazardous or toxic substances, either in unwelcome environments or bio/chem warfare scenarios [5], could provide rapid warnings for impermissible levels or allow cumulative monitoring over long periods to prevent overexposure [6]. Accurate forensic evaluation of deoxyribonucleic acid from physiological samples taken at scenes of military or criminal significance could speed identification of warfighters in battlefield scenarios or provide supporting evidence to prosecute or defend suspects in judicial investigations. Point-of-care diagnostics to identify genetic disorders or infectious disease could be used to improve specificity and outcomes of therapeutic interventions and to expand general availability of personalized medicine. Potential benefits that could accrue from implementing one or more of these field applications motivate categorical evaluation of performance in existing sensor platforms and processing algorithms as well as investigation of promising alternatives with improved fundamental features.

Existing sensors based on optics [e.g., optical fibers, fluorescence, refractive index (RI)], mechanics (e.g., cantilever), thermometrics, or electromagnetic (EM) fields (e.g., plasmon, metal oxide) generally suffer from essentially three classes of global limitations [1]. First, the platforms for these sensors rarely allow distinction between similar analytes. Sensors frequently detect properties such as mass, resistivity, cross-section, or RI of an analyte or its label. Insufficient contrast relative to competing ligands or low signal-to-noise ratio typically requires physicochemical derivatization and/or reaction at the sensing surface to provide adequate sensitivity or specificity. Real environments and complex sample matrices commonly produce interference and compromise performance of derivatized or reactive surfaces. Secondly, there are issues related to required proximity of the analyte to the sensing surface. Existing sensors usually only detect analytes that are in close proximity to some sensing surface. Thus, the speed and sensitivity of existing sensors are inherently limited by both equilibrium thermodynamics, which drives the analyte from the bulk sample to the sensing surface, and Fickian diffusion, which controls the rate of analyte movement from bulk to surface. Third, due to these constraints, existing sensors have lower economic utility than is often desirable. Derivatization is often costly and lowers sensitivity. High capital costs, maintenance expenses, and regeneration costs generally result from insensitivity, fragility, and nonspecificity, respectively. The degree to which these detrimental characteristics are present in a sensor increases capital and operating costs. Meanwhile, cost and portability are often inversely related. Requirements of current sensors for sample clean-up and concentration in advance of analyte detection also limit sensitivity as well as portability and field usage. Table 7.1 summarizes barriers to performance in existing sensors.

Novel nanostructured metamaterials appear to offer a sensor platform that could overcome the issues of specificity, sensitivity, proximity, and economy that are endemic to existing sensor platforms and designs. Metamaterials are an unusual class of materials that exhibit unique properties that arise from EM interactions in organized material substituents which go beyond physicochemical characteristics that arise from elemental composition, chemical coordination, and physical arrangement of the material components themselves. One particular class of such organized structures is composed of constituent material components the characteristic dimensions of which are <100 nm. This class of structures containing components with size dimensions one-fifth or less than the wavelength of visible light lie just below

TABLE 7.1
Barriers to Performance in Existing Sensors

	Existing Sensors	Barrier
Sensitivity	$\geq 10^{-12}$ M	Near-field (2–100 nm) analyte sensing
	$S_\lambda \leq 1 \times 10^6$ RU RIU^{-1} cm^{-2} A^{-1}	Photo-bleaching
Specificity	Single analyte	Mass, conductivity, or optical label
	Analyte-specific label or receptor	
Throughput	$\leq 5 \times 10^{-5}$ cm s^{-1}	Diffusive transport
		Batch
Stability	<100°C	Labile electrostatic bonds
Portability	>0.1 W power	Low EM coupling
	≥ 10 μL sample	Low sensitivity = high surface area

the diffractive resolution limit and therefore cannot be resolved by conventional light microscopy. Such structures are often referred to as nanostructures. Nanostructures have recently become optically distinguishable using new superresolution techniques. Ordered nanostructures have been demonstrated to provide metamaterial activity in the ultraviolet, visible, and infrared regions of the EM spectrum. Activity of nanostructured metamaterials offers unique possibilities for sensing chemical and biological analytes, the spectroscopic signatures of which generally fall in this range. EM interactions between organized nanoscale subunits of metamaterial nanostructures offer a sensor platform with detection capabilities that transcend conventional optical, photonic, plasmonic, or spectroscopic sensor methods. New computational and micro/spectroscopic tools for characterizing nanostructured metamaterials indicate that significant improvements in sensitivity, specificity, proximity, throughput, and overall economy relative to existing sensor devices may be possible.

Nanostructured materials have typically been fabricated one at a time using state-of-the-art electron beam or atomic force instruments capable of "top-down" assembly and organization of material components with nanometer precision. These expensive instruments require operation by highly skilled personnel in dedicated facilities under extreme environmental conditions such as ultrahigh vacuum and high voltages. Consequently, use of such instruments and techniques is characterized by low fabrication throughput and long developmental cycle times, degradation of labile samples such as soft matter from biological or polymer sources, and prohibitive development and fabrication costs. Newer "bottom-up" fabrication strategies induce continuous self-assembly of material components into templated patterns by balancing local thermodynamic, transport, and surface forces. These strategies require careful and accurate evaluation of local physicochemical composition and EM interactions to ascertain and control the source and magnitude of corresponding competing forces that operate at nanoscopic dimensions. Such strategies expand the availability of nanostructured metamaterials for academic and industrial development and often suggest development of concomitant high-throughput approaches to economic large-scale manufacture of prototypes for broad implementation in chem/bio sensing scenarios.

This chapter provides an overview of rational steps in modeling, design, fabrication and characterization, and analysis of nanostructured metamaterials for sensing applications. Steps in this process include modeling physicochemical properties of ideal sensor platforms to examine their parametric sensitivity; development of novel methods to fabricate sensor prototypes based on nanostructured materials; fabrication of beta-test devices that incorporate platforms containing nanostructured metamaterials; identification of figure(s) of merit to evaluate performance of existing and ideal sensor platforms; and application of microscopic, spectroscopic, statistical, and mathematical evaluation of fabricated prototypes to quantitate their physicochemical properties and also reveal the underlying fundamental sensing characteristics. Selection criteria for methods, materials, instruments, and processes used in each of these steps are presented and discussed. Representative chemical and biological analytes evaluated by early prototypes are considered. Illustrative examples include RI-based sensing of liquids on the basis of localized surface plasmons (LSPs) and surface plasmon polariton (SPP) activity on derivatized gold film in Kretschmann-type configuration. These early evaluations suggest that nanostructured metamaterials offer a future class of sensor platform that utilizes interactions between local and far-field EM potentials to improve sensor potential for direct analyte-specific, multiana-lyte detection in small, portable, low-cost devices that allow stand-off, multispectral detection of analyte-specific features.

7.2 PLASMON SENSORS

Design, characterization, and fabrication of nanostructured metamaterials, to provide a platform with significantly improved capabilities for chem/bio sensing, rely on accurate description and evaluation of their fundamental EM interactions. Nanostructured metamaterials are a new addition to a growing class of chemical and biological sensor platforms that utilize resonant local electric fields induced by incident EM waves on metallic films or nanoparticle (NP) assemblies called surface plasmons [2,3,7,8]. Sensor platforms are based on surface plasmons that occur on planar films and on single or random NP assemblies. SPP sensors employ an oscillating waveform that travels parallel to a dielectric interface with a thin metal film. Localized surface plasmon resonance (LSPR) sensors employ electron oscillations that are geometrically confined to a dielectric interface with a subwavelength (nano) particle, called LSPs [4]. SPP and LSPR sensors measure the sensitivity of the plasmon resonant frequency to small changes in RI adjacent to the metal–dielectric interface. A distinctly different spectral feature—a coupled dipole excitation peak that arises from constructive interference between diffracted radiation and plasmon resonance in ordered NP arrays—was recently proposed as plasmon-based alternative to SPP or LSPR sensors [1].

7.2.1 PLASMON SENSOR PLATFORMS

SPP sensors. SPP sensors (e.g., BIACore) are widely used to analyze pharmaceutical products, perform medical diagnostics, and monitor food and environmental safety [6,9,10]. SPP sensors are usually built in a Kretschmann configuration which matches momentum between the surface plasmon and a laser beam in-coupled to an adjacent quartz prism to create a traveling polariton. Reflected intensity measured as a function of incident angle changes as analyte binds to the surface and increases the effective local RI. Exponential attenuation of plasmon modes in the normal direction away from a gold-plated surface derivatized with analyte-specific capture agents eliminates sensitivity after 200–300 nm. The need for the laser energy to match the SPP wavelength limits detection configurations. However, SPP sensors benefit from a high sensitivity to RI changes of ~2×10^6 response units per RI units (RIUs) for a defined planar gold film area and surface coverage by a particular analyte [1,11–14].

LSPR sensors. LSPR sensors exhibit a tunable resonant frequency that does not require momentum matching. The resonant frequency in LSPR sensors is dependent on the NP size [15–18], shape [19], elemental composition [20–23], surrounding dielectric environment [24–26], arrangement [27], and inter-particle separation [28–30]. NPs for LSPR sensors are synthesized by colloidal growth followed by deposition [31], colloidal lithography [32], sputtering or evaporation [33], and electroless

(EL) plating [34–37]. Silver (Ag) NPs provide more sensitivity than gold (Au) but are readily oxidized [23]. Oxidation shifts the Ag LSPR wavelength like an interacting analyte which complicates analysis. Nobler Au NPs resist chemical oxidation and improve biocompatibility and are easily derivatized using thiol chemistries. LSPR sensors geometrically confine EM energy absorbed from large optical cross-sections to significantly enhance local fields within 5–15 nm of the NP surface. Although reported sensitivities for LSPR sensors are ~2×10^2 nm · RIU^{-1} [13,16,22,23] due largely to smaller plasmon-active adsorptive surfaces of NPs relative to planar films, detection levels are comparable to SPP sensors. This illustrates the need for a more universal metric of sensitivity.

Nanostructured arrays. Nanostructured arrays with a defined long-range order that includes particle-to-particle spacing that approaches the LSPR frequency exhibit an extraordinary spectral feature due to constructive interference between in-phase light scattered from neighboring NPs and incident EM waves that induce the LSP [38–40]. This constructive interference by diffractive scattering amplifies the apparent incident field on an NP in the array compared to an individual NP without coupling. Diffractive coupling between scattered waves and plasmon modes narrows extinction peaks and enhances EM fields up to 10^7 absorption units for spherical Ag NP arrays compared to an isolated NP [41]. Far-field coupling with scattered light is negligible in nonstructured arrays due to out-of-phase destructive interference. This constructive interference will be referred to as diffractive coupled polarization (DCP) or multipole/scattering coupling.

Figure 7.1 illustrates constructive interference from axial and diagonal scattered light (red waves) incident on a plasmon mode (yellow −, +) of one subwavelength NP in a square array with lattice constant d where axes are oriented in x- and y- directions. Incident light orthogonal to the x–y-plane, polarized in the x-direction, has wavelength λ, which approaches the lattice constant. The vector r orients a central particle i relative to a particle j along the diagonal located at angle θ_{ij}. A subset of DCP that occurs in subwavelength particles on which resonant excitation exclusively

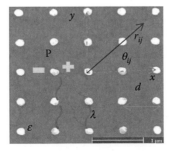

FIGURE 7.1   **(See color insert.)** Light (red waves) scattered from neighboring axial and diagonal subwavelength nanoparticles constructively interferes onto a plasmon dipole (yellow −, +) of one nanoparticle in a square array of lattice constant d with axes oriented in x- and y-directions. Incident light orthogonal to the x–y-plane, polarized in the x-direction has wavelength λ, which approaches the lattice constant. The vector r orients a central particle i relative to a particle j along the diagonal located at angle θ_{ij}.

induces dipoles is called coupled dipole excitation (CoDEx). Changes in local RI around individual NPs influence scattering and affect DCP behavior.

Regular NP arrays. Regular arrays of NPs that lack a long-range order required for DCP may produce near-field coupling that results in reproducible surface-enhanced Raman spectroscopy (SERS) [42,43]. Order-of-magnitude enhancements observed with SERS are a result of near-field coupling between adjacent NPs which occurs when interparticle spacing is less than a few hundred nanometers [44]. Regular arrays of Au NPs also exhibit strong, long-range interactions between particles that cannot be obtained in disordered NP assemblies [45]. Similar to DCP extinction spectra, SERS signal may be amplified by varying particle shape, size, and inter-particle spacing on the regular arrays of Au NPs [1,46]. SERS using plasmon-active nanostructures offers the possibility of direct, analyte-specific sensing in a variety of sampling contexts.

DCP sensors. Sensors based on DCP are anticipated to benefit from (1) the narrow, intense spectral peak that arises due to far-field coupling (see Figure 7.3) and (2) modulation of peak position, intensity, and shape by tuning physicochemical and geometric properties of the array. Development of DCP sensors requires multivariate specification of NP characteristics (size, shape, lattice spacing, and dielectric), RI of the environment, as well as incident EM angle, polarization, and wavelength [47,48]. To date, this specification has relied exclusively on modeling using finite difference time domain (FDTD), discrete or coupled dipole approximations (DDAs or CDAs) to Maxwell's equations, or EM computational methods. Extinction spectra from ordered NP arrays may lack constructive dipolar coupling when one or more of these parameters are internally inconsistent. Previous sensors based on NP arrays measured sensitivity based on LSPR from single particles or aggregate LSPR signal summed from individual elements rather than the singular, more intense coupled spectral feature [49–52].

7.2.2 Describing Nanostructured Plasmon Sensors

Electrodynamic models. Maxwell's equations provide a complete description of the interaction between electric and magnetic fields with matter. Specialization of Maxwell's equations to conditions representative of nanostructured metamaterials results in a range of solutions, from complete computational descriptions such as T-matrix to semi-analytical descriptions such as those based on the DDAs or CDAs. Computational descriptions such as T-matrix fully represent the complexity of a nanostructured metamaterial with high fidelity, but at a large computational cost that limits the extent of the system that can be analyzed and/or the speed at which parameters can be evaluated. Semi-analytical descriptions such as those based on the CDA leverage a sequence of reasonable approximations to obtain useful descriptions of nanostructured metamaterials to retain primary features of fundamental behaviors at conditions for which the underlying approximations remain valid. In exchange for unabridged fidelity of the resulting description, results from these models are obtained at much lower computational cost, which allows for rapid parametric identification of unique and/or optimal configurations of nanostructured metamaterials that could enhance key sensor characteristics such as sensitivity, specificity, or throughput. Semi-analytical solutions may also provide tractable identification of a particular contribution from a distinguishable class of substituents to an overall system property. This important feature can allow characterization of key features or aspects of metamaterials that may be difficult to recognize or distinguish within complete computational descriptions.

Figures of merit. Computational and semi-analytical solutions are used to estimate select figures of merit with which sensor performance, for example, sensitivity, specificity, and/or throughput, can be quantitatively evaluated. Figures of merit are useful in comparing different sensor designs based on nanostructured metamaterials with other sensor platforms and assess relative merits or weaknesses. Conventional figures of merit may require adjustment to account for novel features of new sensor platforms. For example, metamaterials respond to analyte changes in active volumes adjacent to component substituents that are a fraction of the active volume of an SPR sensor with a comparable area, which could ultimately provide significantly higher sensitivity across a much larger sampling region—an important feature in nonproximal or remote stand-off detection. Modification of the conventional figure of merit for sensor sensitivity is therefore necessary to account for this fundamental difference when comparing metamaterials and SPR sensors.

7.2.3 FABRICATING NANOSTRUCTURED PLASMON SENSORS

Identification of a particular design of a nanostructured metamaterial that exhibits enhanced capabilities in a sensing role based on a universal figure of merit estimated by a reliable mathematically based description motivates fabrication of a corresponding prototype to characterize and calibrate its functionality and performance in representative conditions. As an example, increased surface-to-volume ratios in nanostructured metamaterials enhance their responsiveness to slight changes in physicochemical dimensions of their substituent components. This sensitivity requires developing and refining bottom-up fabrication protocols that permit one to fine-tune the geometry and composition of the substituents with nanoscale reproducibility and fidelity. Electron-based scanning and transmission microscopy of bottom-up built prototypes is vital to characterize the nanoscale variability of fabricated features in relation to far-field evaluation of their figures of merit in comparison with predicted behavior. Insights obtained by comparing predicted and measured behavior with micro/spectroscopic characterization of prototype features provide validation of the approaches used and guide modification to improve each respective approach and optimize a preliminary or beta version of a nanostructured metamaterial sensor platform.

7.3 MODELING NANOSTRUCTURED PLASMON SENSORS

7.3.1 COUPLING IN NANOSTRUCTURE ARRAYS

Plasmons. Collective oscillation of unbound electrons concentrated at an interface between a subwavelength metal particle and an adjacent dielectric driven by a resonant, incident EM field and opposed by a Coulombic restoring force is widely

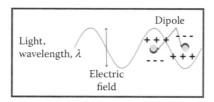

FIGURE 7.2 Light at resonant frequency (λ) induces plasmons, shown by a negative sign (−) and corresponding local enhanced electromagnetic field (dark shading), on two subwavelength nanoparticles spaced at $\lambda/2$.

recognized as an LSP and often referred to *in toto* as a "plasmon." Figure 7.2 illustrates induction of plasmons, shown by a negative sign (−), and corresponding local enhanced EM field (dark shading) on two subwavelength NPs spaced at half the incident light wavelength. Incident photon energy induces the plasmon at a frequency known as the LSPR, which changes theoretically (and consistent with observation) according to size [53–57], shape [58], composition, and arrangement of the NP and their neighbors. Measurement and transduction of this frequency into an electronic signal constitutes LSPR sensing. Electric field distribution on excited subwavelength particles is essentially dipolar with negatively charged oscillating electrons and positively charged stationary nuclei. Larger particles experience perturbations of the electric field distribution which requires description using higher-order electric poles (e.g., quadrupoles).

Localized surface plasmon resonance. In 1908, the first exact solution of Maxwell's equations for scattering of incident EM plane waves onto an individual dielectric sphere was published by Mie [59]. It is currently available in a graphical user interface program based on the Fortran code developed by Bohren and Huffman [60] (MicPlot, v. 4.2.03, philiplaven.com). LSPR has subsequently been modeled effectively as a single dipole on an NP in the quasistatic limit. Plasmon resonance of noble metal particles occurs at visible frequencies corresponding to wavelengths ~400–700 nm and produces a variety of colorimetric phenomena. Gold and silver dust of ~70 nm was diluted by ancient Roman artisans into glass to produce the dichroic Lycurgus Cup whose walls absorb and reflect green wavelengths so that they appear green in light reflected back from in front of it but red in light transmitted from behind it. Dichroic glass in medieval church windows has similar optical properties. Current applications of plasmons rely on the fact that local plasmon fields in subwavelength elements are up to a million fold larger than incident EM energy and significantly enhance Raman scattering, thermal dissipation, and photonic localization, but are limited to within tens of nanometers from the interface and decay with a micron.

Photon diffraction. Photon diffraction from wavelength-scale features, in contrast, persists for meters under regulated geometries and refractive indices to produce sharp bandgaps, edges, and nonlinearities. These features are observable as naturally occurring angle-dependent iridescence in butterflies or soap bubbles and light trapping in algal frustules as well as in lattice-containing photonic crystals such as holey optical fibers. A regular pattern diffracts light with a maximum intensity at angle θ_m due to constructive interference at any wavelength λ given by

$$\lambda = \frac{d \cdot (\sin\theta_m + \sin\theta_i) \cdot \eta}{n},\qquad (7.1)$$

where d is the lattice constant, n is an integer, and θ_i is the angle at which light is incident.

Organizing NP elements into ordered lattices produces photon re-radiation of incident normal ($\theta_i = 0$) light of wavelength

$$\lambda_{i,j} = \frac{d \cdot \sqrt{i^2 + j^2} \cdot \eta}{n},\qquad (7.2)$$

where d is the lattice constant, n is an integer, η is the medium RI, and i and j are the integer multiples corresponding to steps in the x- and y-directions, respectively.

Plasmon–photon coupling. In certain nanostructured materials, plasmonic and photonic effects can hybridize: local electric fields excited on NP surfaces can couple with constructively interfering, long-range diffracted photons to intensify local plasmon fields and produce an extraordinary peak in extinction (i.e., absorption plus scattering) at a wavelength near the particle spacing [39,45,61–63]. Figure 7.3 shows theoretical spectra (solid lines) for square arrays of 90-nm Au spheres spaced at 520, 850, and 640 nm, respectively. The model used the rapid semi-analytical coupled dipole approximation with an effective refractive index weighted by contributions from glass substrate coated by 20 nm indium tin oxide. While incident EM radiation scattered from an individual NP in an array to produce this feature can be described exactly by Mie theory, complete description of the coupling interaction *in toto* requires summing phase retardations from each node that contributes EM radiation to a neighbor. This requires solution of Maxwell's equations to completely describe electrodynamic interactions between electric and magnetic fields with matter, not solely for an individual NP, but for the entire nanostructure array.

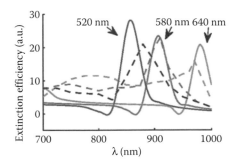

FIGURE 7.3   **(See color insert.)** Extinction spectra from infinite pitched lattices of spherical Au nanoparticles spaced at 520, 580, and 640 nm, respectively. Theoretical predictions (solid) are compared with sketches of experimental spectra (dashed).

7.3.2 MODELING COUPLED DIFFRACTION IN NANOSTRUCTURED ARRAYS

Exact solutions. Exact results from Maxwell's equations may be obtained using FDTD or T-matrix to characterize EM fields that provide sensor activity in nanostructured materials. Both FDTD and T-matrix are well suited to describe structures under illumination by light whose wavelength is on the same order of magnitude as the structure's characteristic dimensions (i.e., the resonance regime). FDTD simulations provide robust general solutions to Maxwell's equations for a wide variety of particle shapes and configurations. The physical structure to be simulated is represented by a computational grid composed of unit cells specifying the structures of the material (e.g., permittivity, permeability, conductivity), and the illumination source is modeled by a Gaussian pulse. Central to the FDTD method is the Yee algorithm [64] which computes the electric and magnetic fields for each unit cell using the curl form of Maxwell's equations. The Yee algorithm does this efficiently by employing a staggered or "leapfrog" stepping scheme to calculate the future field values from the past values (i.e., the FDTD method emerges as a fully explicit method), and thus avoids the computationally intensive task of solving a large system of equations. Since FDTD computes the electric and magnetic fields at every point on the computation grid, the grid must be truncated to allow storage in computer memory. Absorbing boundary conditions (ABCs) can be used for simple simulations to terminate the computational grid. However, ABCs fail when a dielectric is introduced as they depend on the velocity of wave propagation being unity. Perfectly matched layers (PMLs) are the current de facto method for grid truncation. PMLs are essentially a lossy material employed at the boundaries of the simulation domain to absorb outgoing radiation and prevent reflection back onto the scatterer. Efficient simulation of infinite periodic scattering geometries can be achieved using periodic boundary conditions [65].

Despite its versatility, FDTD has several drawbacks which make it unsuitable for certain problem geometries. For example, the rectangular grid used in traditional formulations cannot conform to curved scatterers resulting in "staircasing" errors if the spatial grid is relatively course. Another prohibitive situation arises if the total simulation domain required is much larger than the characteristic dimension of smallest structure of interest. Therefore, calculation of fields in the far-zone of the scatterer is best accomplished using near-to-far-field transforms.

Alternatively, T-matrix solution provides an exact multipole solution [54]. Construction of the T-matrix solution as it pertains to EM scattering can be achieved by the null-field method (NFM), also referred to as the extended boundary condition method. This is in essence an advanced generalization of Mie principles to arbitrary particle geometries. T-matrix solutions have been developed to treat multilayer particles and clusters in both the near- and far-field regime [66–68]. Though construction of the T-matrix solution by NFM is elegant and efficient in theory, its numerical implementation incurs large computation costs due to the number of integrations of spherical Bessel and spherical harmonic functions over the surface of the particle. In addition, round-off errors increase significantly with particle size leading to an ill-conditioned problem. In some cases, quadruple precision arithmetic is required (at an added computational cost) for an accurate calculation of the T-matrix for large or high-aspect-ratio particles [69]. Both FDTD and T-matrix incur large computational costs and additional complexities that limit the number and geometry of elements that is possible to describe in an NP array. This large computational cost slows parametric evaluation of particle and array properties on spectral extinction features.

Approximate solutions. The DDA to Maxwell's equations can approximate arbitrary shapes of arrayed NPs by replacing each particle by multiple dipoles. The CDA provides a solution where nanospheroids have been approximated by single dipoles in the array. Both DDA and CDA provide good agreement between experiment and data [58,63]. To obtain the CDA for subwavelength particles, Green's function solutions of Maxwell's equation may be reduced considering charge continuity and time variance to a set of inhomogeneous wave equations. These equations describe electric fields produced in ordered arrays of nanoinclusions at which moments are excited at resonant frequencies in a homogeneous matrix. In the CDA, an electrodynamic dipole polarization, P_i, is induced at each substituent nanostructure i by incident, electrostatic, and radiative electromagnetism in proportion to the structure's polarizability, α.

Polarizability. Polarizability has been described in CDA using a quasistatic approximation [70], its modified long-wavelength approximation, and a dynamic polarizability introduced by Doyle [71]. With the dynamic polarizability, phase variations that arise from structural elements (e.g., lattice constant, d), the characteristic length of which is in the order of the incident wavelength, can be characterized with an expression obtained from Mie theory. Depolarization and radiative dipole effects are modeled in this polarizability, α, using *dynamic* terms [1,71]

$$\alpha = 4\pi \frac{3iR^3}{2x^3} a_1, \tag{7.3}$$

where R is the particle radius and $x = \eta_0 kR$ is the size parameter with η_0 as the medium RI and k as the vacuum wavenumber. The 4π coefficient is included for conversion to SI units. The term a_1 is the first Mie coefficient corresponding to the electric dipole.

Extinction cross-section. The polarizability is used to calculate the extinction cross-section, C_{ext}, for an array of scattering nodes in which retardation effects are described using the retarded dipole sum, S. The extinction coefficient may be determined by forward scattering in accordance with the optical theorem. For an infinite array this gives

$$C_{ext} = 4\pi \, \text{Im}\left(\frac{\alpha}{1 - \alpha S}\right), \tag{7.4}$$

where the retarded dipole sum is defined by

$$S = \frac{e^{ikr_{ij}}}{4\pi\varepsilon_0} \sum_{i \neq j} \frac{(1 - ikr_{ij})(3\cos^2\theta_{ij} - 1)}{r_{ij}^3} + \frac{k^2 \sin^2\theta_{ij}}{r_{ij}} \tag{7.5}$$

with r_{ij} being the distance from the particle i to the particle j and θ being the angle between a ray connecting these particles and the x-axis of polarization [1]. The retarded dipole sum incorporates EM contributions scattered from neighboring NPs to account for retardation due to optical phase variation across the metamaterial structure. Equation 7.4 reveals that the far-field extinction cross-section for the square array of subwavelength NPs is inversely related to the imaginary component of a difference between material and purely geometric properties. The material properties are the dielectrics of included particles and their surroundings, the particle size, and the wavevector. The geometric lattice properties are particle-to-particle spacing and angle relative to incident polarization. The coefficient of extinction increases sharply as the difference between the retarded dipole sum and the inverse particle polarizability becomes small.

New approximate solution. Neither the DDA nor the CDA distinguish the effects of individual particles or near-, induction, and far-field particle–particle interactions on particular spectral features. It was recently shown that symmetry arguments can be used to transform S into a closed integral form for the axial and diagonal constituents in which components from near-field and induction-zone interactions as well as far-field radiation are explicitly distinguishable. In this form, effects from off-axial/diagonal (OAD) elements may be computed using a truncated sum. The respective contributions from axial, diagonal, and off-axial/off-diagonal substituents are also distinguishable. Transformation of the S based on array symmetry substantially reduces function calls required during matrix evaluation of axial, diagonal, and OAD elements. This results in a rapid semi-analytical CDA (rsa-CDA) description for two-dimensional NP arrays in which dipoles are induced. This description provides an accurate, computationally inexpensive, tractable description of nanostructured metamaterials useful in identifying their EM and far-field optical behavior [1,72–75].

Dynamic polarizability. Recently, a term "dynamic polarizability" was introduced into Equation 7.3 to describe higher-order multipole modes and to rapidly examine the corresponding optical properties [76]. A form of the dynamic polarizability that describes both dipole and quadrupole modes is

$$\alpha = 4\pi \left(\frac{3iR^3}{2x^3} a_1 + \frac{10iR^3}{3x^3} a_2 \right). \tag{7.6}$$

The rsa-CDA containing the dynamic polarizability in Equation 7.6 was benchmarked against exact solutions to Maxwell's equations for single particles (Mie theory) and NP arrays (FDTD) within limits prescribed by the specifications made during reduction of Maxwell's equations to the CDA as well as by computational constraints of FDTD. These specifications include an infinite, square lattice in a homogeneous substrate that is excited by an orthogonal wavevector. Introducing the dynamic polarizability into the rsa-CDA was shown to improve overall agreement with both Mie theory and FDTD for larger particles affected by higher-order multipoles.

Nanostructure polarizability. The imaginary component of particle polarizability in Equation 7.6 is responsible for scattering and attenuation of the radiation with

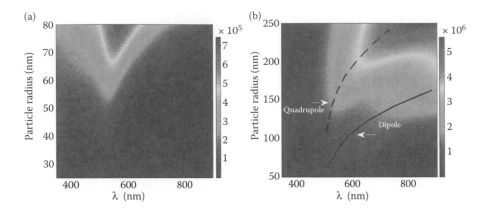

FIGURE 7.4 **(See color insert.)** Magnitude of imaginary dipole and quadrupole components of nanoparticle polarizability [α in Equatuion 7.6] for a single Au nanoparticle with radius ranging from (a) 25 to 80 and (b) 50 to 250 nm and incident radiation between 350 and 900 nm. Loci of maximum values are indicated for dipole (solid line) and quadrupole (dashed line) polarizability.

respect to the incident EM wave. Figure 7.4 is a plot of magnitudes of the respective imaginary dipole and quadrupole components of particle polarizability [α in Equation 7.6] for a single Au NP with radius ranging from (a) 25–80 and (b) 50–250 nm and incident radiation between 350 and 900 nm. As a guide to the eye, Figure 7.4b indicates the loci of maximum values of dipole (solid line) and quadrupole (dashed line) polarizability. Each of these loci red-shifts with a decreasing slope as particle radius is increased. Larger polarizability modes increase scattering and absorption which in turn generate greater extinction magnitudes. Polarizability, as illustrated in Figure 7.4, serves to be a more comprehensive predictor of plasmonic behavior than material dielectric, which is identified as causing plasmonic behavior of noble metals. Plasmon resonance occurs when polarizability attains a maximum, that is, when Equation 7.3 diverges. In the familiar dipole case illustrated in Figure 7.2, this divergence corresponds to the situation when the wavelength-dependent real component of the complex metal dielectric becomes negative at a relative magnitude compared with the external dielectric which satisfies the resonance condition. The sign and relative magnitude of the metal dielectric permits resonance, but at wavelengths for different orders that are determined by the denominator of the a_1 term in Equation 7.3. Thus, plasmon resonance features of both single and arrayed multipoles are more completely described by considering the polarizability rather than the dielectric values alone.

7.3.3 CHARACTERIZING DIFFRACTION COUPLED POLARIZATION IN ARRAYS

Polarizability governs DCP extinction feature. Importantly, rapid implementation of rsa-CDA with dynamic polarizability allowed characterization of important effects on the diffraction coupled peak important for sensing [76]. For example, dynamic polarizability governs the wavelength at which multipole/scattering coupling occurs, especially for particles with $R > 30$ nm. Red-shifts in this wavelength are due primarily to increases in particle polarizability as radius increases and incident wavelength shifts. Larger polarizabilities red-shift the coupled peak further from the lattice constant. Polarizability modulates the intensity of the extraordinary DCP peak in the spectrum of NP arrays.

Figure 7.5 shows how DCP intensity varies as particle radius, lattice constant, and refractive index change for Au nanospheres in square lattices. Array geometries that allow constructive coupling to occur because of phase overlap of scattering elements in the array. Each meshpoint in the figures is a single array geometry with the maximum extinction over the entire incident wavelength range from 400 to 1000 nm in 0.1 nm increments. These simulations were done for both and RI of 1 and 1.33. Increases to RI showed a decrease in the limiting particle radius and lattice constant that support DCP resonances. Interestingly, the loci of maximum intensities of diffraction coupled multipole extinction correspond to a finite number of isometric (not maximal) values of polarizability reduced dramatically by destructive interference from coherent radiation accumulated between OAD nodes in the array. Patterns in the periodicity of maximum values of the coupled peak amplitude changed with particle size and lattice constant showing unique behaviors for each metal dielectric. These results illustrate the effectiveness of the refined rsa-CDA in rapid, facile determination of effects of multipole polarizability on wavelength, intensity, and other spectral characteristics of DCP due to changes in physicochemical particle properties as well as array geometry.

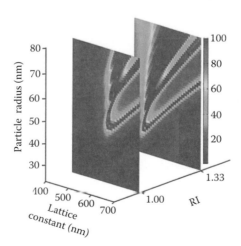

FIGURE 7.5 Maximum extinction values for Au nanospheres across the entire incident range of incident wavelengths for the three sets of parametric changes: radii ranging from 10 to 80 nm, lattice constant ranging from 400 to 700 nm, and refractive index (RI) at either 1.00 or 1.33. Higher extinction (light areas) gives geometries that allow constructive coupling.

7.3.3.1 Features of Diffraction Coupled Polarization Spectral Extinction Peak

Evaluation of DCP spectra using the rsa-CDA illustrated key features of EM inter-actions between incident and retarded scattered fields in square nanolattices on locally polarized nanostructures. The inherent polarizability of NP substituents in metamaterials that is responsible for the appearance of LSPs enhances synchro-nous diffraction between ordered nanospheres in the square lattice to produce these intriguing effects.

Plasmon polarization enhances DCP peak. In Figure 7.6, a minimal extinction peak appears in transmission spectra of wavelength-scale lattices of silica nano-spheres near the 650 nm particle-to-particle lattice constant due to photonic diffrac-tion (inset). Any feature corresponding to LSP is absent from this spectra due to lack of measurable wavelength-specific polarizability arising from real or imaginary parts of silica dielectric at the indicated range of wavelengths. In contrast, extinc-tion efficiency in the transmission UV–vis (T-UV) spectra at wavelengths near the 650 nm particle-to-particle lattice constant in an ordered array containing 70-nm radius Au NP is 500 times larger. This enhancement arises from measurable polar-izability of Au NPs at this frequency. At a resonant frequency of ~530 nm, a cor-responding plasmon resonance peak in the T-UV spectrum of the Au NP lattice is visible. Note the wavelength of the dominant extinction peak in Figure 7.6 corre-sponds to that for axial scattering parallel to the incident x-polarized EM field given by Equation 7.1, where integer values of i and j are zero and one, respectively, d is 650 nm and n is unity.

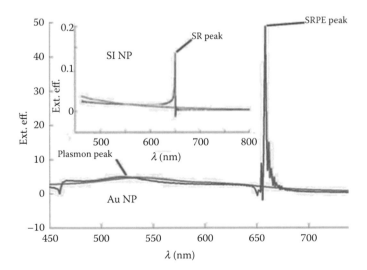

FIGURE 7.6 Extinction spectra for square nanoparticle arrays in vacuum with 140 nm diameter nanoparticles and 650 nm lattice constant. Spectra show a plasmon peak at 530 nm and DCP peak at ~670 nm for Au NP, but only a diffraction peak for silica (Si) nanoparticles (inset).

Scattering enhances DCP peak. Increased scattering from more polarized particles enhances the intensity of the coupled peak. Figure 7.4a shows that polarizability of particles at wavelengths for which plasmon resonance occurs—approximately 525 nm for Au NPs—increases strongly as particle size increases. Increased polarizability increases diffractive light scattered from the particles. Consequently, Figure 7.7 shows that the corresponding magnitude of the T-UV extinction increases with size for subwavelength particles. Square lattices containing particles smaller than ~30 nm continue to exhibit an extinction feature at approximately the lattice constant. Earlier results reported for complete CDA simulations showed that the extinction feature had disappeared at particle radii less than ~33 nm due to postprocessing of calculated spectral points in computed extinction profiles using box-car smoothing.

Tunability of DCP peak. Dependence of the extraordinary extinction feature on coupling between constructive interference in diffractive scattering and the plasmon resonance allows the feature to be red-shifted in the EM spectrum from the plasmon resonant frequency toward the infrared by increasing either particle size or lattice constant. This introduces the possibility of broadband excitation for nanostructured metamaterials. The wavelength of the diffraction coupled excitation can be independently tuned using the size, shape, and dielectric of the nanostructures, the RI of the surrounding medium, and the grating constant.

Low computational cost of the rsa-CDA allows for rapid parametric investigation of nanostructured metamaterials in determining optimal conditions that correspond to desired outcomes for the extraordinary coupled extinction such as magnitude, peak width, and sensitivity to change in parameter values (e.g., NP size, shape, or local dielectric for sensing applications) as illustrated in Figures 7.5 through 7.7. Ideal CDA features for particular device applications can be identified by numerical optimization of particle size, dielectric, grating constant, incident wavelength, and other features.

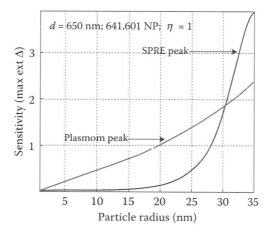

FIGURE 7.7 Peak intensities of DCP (black) and plasmon peaks increase sigmoidally and nearly monotonically, respectively, as particle radius (and polarizability) increases for small subwavelength particles.

7.4 FABRICATING ORDERED NANOSTRUCTURES

Novel features identified in theoretical models of nanostructured materials are realized in technological applications by fabricating prototypes to criteria that are predefined by the models. Existing fabrication approaches often correspond to size-reduced analogies of methods developed to pattern and manufacture integrated circuits. Available routes to fabricate nanostructured metamaterials consist of two general steps: patterning and deposition. In the patterning step, a template is created or applied in order to allow periodic differentiation of each individual nanostructure from its surrounding environment on an underlying substrate. In the deposition step, a material comprising the nanostructure, typically Au or Ag metal, is preferentially introduced at regularly spaced intervals on the underlying substrate by virtue of the patterned template. Top-down methods for patterning and deposition rely on precise, directional, electronic control of EM waves, electrons, ions, or atoms in a macroscale environment to effect a uniform, irreversible, one-dimensional change to a target template or substrate. In contrast, bottom-up methods manipulate local surface forces due to phase, electromagnetism, and chemical composition and equilibrium to direct self-assembly of micro- to nanoscale subcomponents from molecules to particles into patterned or resist or nanostructures.

Figure 7.8 compares steps in top-down patterning and plating by electron beam lithography (EBL) and evaporation, respectively, with bottom-up patterning and plating by nanosphere lithography and EL plating, respectively. Table 7.2 summarizes benefits and limitations of different technologies to fabricate nanostructure arrays. Differences between these methods and their results are significant and are discussed next.

7.4.1 PATTERNING METHODS

Top-down patterning. Available top-down methods for templating ordered nanostructures use precise manipulation of a finely discretized single beam or probe to create high-fidelity nanoscale templates by predominantly electronic or mechanical forces. Electronic methods include lithography using focused beams of electrons (i.e., EBL [18,28,42,77–80]), ions (i.e., focused ion beam, FIB [81,82], reactive ion etching [83]), or interfering light (i.e., laser interference lithography, LIL [84–89]). Mechanical approaches to top-down templating include nanoscale manipulation of a nanoscale element (i.e., atomic force microscopy, AFM [90]) and dip-pen nanolithography (DPN) [91,92] or impressing soft matter with a harder material "negative" that has been templated lithographically (nanoimprint lithography, NIL [93,94]).These methods have been used to pattern regular arrays of Au chains [1,90], elliptical NPs [77], kidney bean-like particles [79], and nanodisks [80,83].

Limitations. Limitations inherent in use of each method to template ordered nanostructures include cost, serial throughput, ease of use, ability to pattern desired shape, commercial viability, and availability of equipment and facilities. EBL allows for precise control of NP shape, size, and interparticle separation but requires a conductive substrate and is slow and costly. Many copies of an optimized pattern can be made but EBL alone cannot form all geometric parameters that must be specified in

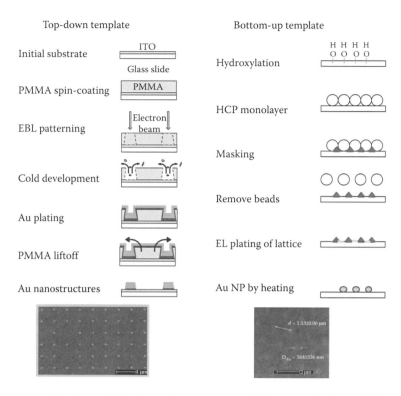

FIGURE 7.8 Top-down fabrication of square arrays of Au nanocylinders by lithographic electron beam degradation of a PMMA resist followed by Au evaporation. Bottom-up fabrication of hexagonal arrays of Au nanospheres by templating nanospheres in a hexagonally close packed (HCP) monolayer followed by electroless (EL) plating over a hydrophobic mask and thermal annealing.

nanostructured platforms. The need for ~36 h to fabricate a $100 \times 100\ \mu m^2$ regular arrays of Au NPs using a \$1.2 million EBL system has been reported [35,36]. DPN uses an AFM style tip to deposit the NP material directly, but the types of material that can be deposited are limited. While work is ongoing to parallelize DPN writing modes [95], DPN, AFM, FIB, and EBL are inherently serial approaches with throughputs limited by substrate resistance to the probe/beam and by translational speed of the probe/beam or stage underlying the sample. Limits of optical diffraction curtail the degree to which size of elements distinguished by LIL can be reduced. In templating nanostructure prototypes, precise control over a wide range of array parameters allowed by EBL favors this approach.

Hybrid patterning. NIL [93,94] methods are promising, scalable approaches to fabricate ordered nanostructures. In NIL, a negative polymeric image of a master pattern originally fabricated by a top-down method is formed to create multiple copies of the original master in an imprint resist. The copies and the negative image are formed by consolidation of a free-flowing polymer around their respective master template. These methods are studied to fabricate microscale optoelectronic elements

TABLE 7.2

Comparison of Different Technologies to Create Nanostructured Arrays

Technology	Description	Benefits	Limitations
Direct			
EBL	Patterning of nanorings with electron beam directly in resist. Used for mask generation	• No mask required • Complex patterns • Well characterized	• Equipment cost • Low throughput • Proximity effects
FIB	Ions etch material away to produce nanoring pattern	• Can directly write metal rings • Less long-range scattering than electrons	• Ion mixing • Implanted dopants • Altered optical properties
DPN	AFM tip deposits "ink"	• Complex patterns	• Substrate-dependent "ink" • Low throughput
NSL	Microspheres are used as a mask for evaporation followed by ion etching	• Inexpensive • Can produce large area patterns	• Limited lattice configurations (hex, square) • Line and pin defects
Indirect (Require a Mask or Pattern from Another Lithography Method)			
EUV	Extreme UV light ($\lambda \sim 10$ nm) used similar to photolithography	• Widely used • High aspect ratios possible	• Diffraction limited • Exotic optic materials
X-ray	X-rays ($\lambda \sim 1$ nm) used to expose resist	• Linewidth independent of substrate	• Expensive source • Complex mask • Mask gap changes resolution
NIL	Master pattern is duplicated in resist via contact	• Rapid • One master produces multiple samples	• Stamp deformation • Highly sensitive to surface irregularities
Angle-resolved EBL	Angle-resolved evaporation on EBL-patterned holes	• Produces rings without directly patterning	• Complex equipment • Height/width are not independent
EL-EBL	Electroless plating on EBL patterned holes	• Produces rings without directly patterning	• Liftoff is difficult

such as lenses, filters, gratings, detectors, lasers, light-emitting diodes, electronic storage, transistors, and integrated circuits. In thermal nanoimprint lithography (TNIL), heating a monomer or polymer above its melting point allows it to flow into a relatively rigid resist to form the template [96,97]. Substrate conformal imprint lithography uses a resist with higher flexibility to reduce the size of defects due to surface variations of the template or inclusions [98]. Step and flash imprint lithography incorporates irradiation to improve local conformation patterned polymer to the negative resist [99]. Microcontact printing (μCP) is a related approach in which a fluidic ink adsorbed or absorbed by a resist is printed onto the surface to be template [100]. Forces due to surface interactions and chemical equilibria and transport phenomena such as thermal heating and mass diffusion of gas, solvents, and molecules determines the time required, achievable line thickness, and aspect ratio of patterns formed in a thermal or optical resist fluid that is redistributed into void spaces in an overlaid stamp in various NIL and μCP approaches.

Bottom-up patterning. Evolving bottom-up methods for templating ordered nanostructures provide simpler, more robust, and cost-effective means to create ordered two- [101] or three-dimensional [102] colloidal crystals which can be used as templates to fabricate metal nanostructures. These methods manipulate local surface forces due to phase, electromagnetism, and chemical composition and equilibrium to orderly self-assemble subcomponents from molecules to particles into nanostructures. In nanosphere lithography (NSL) [33,103], monodisperse micro- to nanoscale silica (Si) or polymer beads suspended in solvent coalesce into ordered colloidal crystals from millimeters to centimeters in scale [104–106] as solvent evaporates. The crystals form a template for a hexagonal NP lattice on an underlying surface whose size and interparticle spacing can be easily tuned by using different sizes of beads. Creation of arbitrary template shapes remains challenging, and defects in resulting arrays are difficult to eliminate. The persistence length of ordered lattices produced by bottom-up methods is determined by complex interactions between chemical- and geometry-induced spatiotemporal variations in solvent evaporation rates, regularity or inconsistency in element dimensions, and surface features of the underlying substrate. Metal nanotriangles and spheres fabricated by NSL have been characterized for use in LSPR spectroscopy and SERS sensing [4,107].

7.4.2 Metal Deposition Methods

Top-down deposition. Common top-down methods for depositing metal onto templated surfaces to form nanostructures are evaporation, sputtering, and electroplating. Vacuum evaporators deposit metal at 10^{-8} Torr by condensation of a vapor obtained using either a heating element or an electron beam. Thickness of the deposited film is monitored for control using a quartz crystal microbalance. Drawbacks include metal spitting—deposition of ~100 nm metal particles on evaporated film surfaces. The directional nature of thermal evaporation and sputtering limits the range of NP shapes that can be deposited. For example, complex tilt and rotation stages are needed to make ring structures in holes patterned by EBL. Consequently, disks, triangles, cylindrical particles, or other shapes can be formed using existing approaches, but not spheres. High vaporization temperatures ostensibly prevent resulting films from forming spheres after subsequent thermal annealing. Overcoming lattice mismatch to obtain good surface adhesion usually requires an adhesive layer of chromium between silica surfaces and gold, which changes the local dielectric and affects optical response characteristics [24].

Bottom-up deposition. A bottom-up alternative to evaporation, sputtering, or electroplating is metal deposition by EL Au plating. EL plating of gold uses sequential introduction of ionic tin (Sn^{2+}), silver (Ag^+), and gold (Au^+), in coordination with pH control and a chemical reducing agent to give strong adhesion of Au island films to the substrate. EL plating results in more cohesive particles with smaller particle size distributions than conventional top-down metal deposition such as evaporation or sputtering. EL plating, however, is not specific to the substrate and when applied to masked surfaces templated by EBL, Au is deposited in an interconnected sheet both on the surface of the electron resist as well as the substrate. Removal of the electron resist breaks off some nanocylinders with the gold film, leaving defects such as partial cylinders or holes in the patterned elements. In EL plating, redox transition of solubilized Au ions is between three common ionization states (0, 1, and 3) and is effected by controlling redox potential of the solvent [108].

Refinements. Uniformity, continuity, and precision of deposited metal films can be improved using a continuous introduction of ions to the surface via bulk flow to allow steady mass transfer of plating solutions to the substrate surface [34]. Steady flow deposition achieves a uniform plating rate since the concentration driving force from the flowing bulk does not vary with time. Models describing metal ion deposition by mass transfer in capillary flow systems for EL plating with gold sulfite give results consistent with optical features measured by T-UV. Films deposited by this method exhibit improved uniformity with an average roughness of 0.64 nm. Random particles derived from the films exhibited improved peak intensity and full-width at half-maxima compared with those created using conventional batch EL deposition. Thermal annealing of EL-plated Au thin films at temperatures from 250 to 800°C transforms the films to thermally stable, spherical NP assemblies whose radii and particle-to-particle spacing are tunable by varying film thickness and annealing temperature. Thermally transformed thin metal films formed spherical Au NPs in random configurations on planar silica substrates [109] and on the inner walls of rectangular borosilicate glass capillaries [30].

7.4.3 HYBRID METHODS

Top-down patterning and bottom-up deposition. Metal deposition by sputtering or evaporation onto patterned electron resist lithographed by EBL can produce shapes such as cylinders that have sharp edges where deposited metal intersects both resist and substrate or superstrate (usually atmospheric gas). But top-down patterning/top-down deposition cannot fabricate metal spheres due to softness in directional deposition of metals such as Au. Recently, rectangular lattices of gold (Au) nanospheres were fabricated on indium tin oxide (ITO)-coated glass substrates by thermally annealing Au nanopillars and nanocylinders deposited via EL plating onto ITO glass substrates patterned via EBL in a novel method [35]. Particle size and shape were controlled by adjusting thickness of the poly(methylmethacrylate) (PMMA) electron resist, e-beam power, and EL plating parameters. Nanostructures thus produced coalesced more strongly than sputtered Au films and attached more stably to ITO substrates, resisting thermal, solvent, and EM exposures. Recently, this hybrid EBL/EL fabrication method was modified by pre-sensitizing the substrate to make EL plating specific to the silica substrate underlying the PMMA resist [36]. This method enhanced size and shape uniformity of individual Au pillars and arrays, resulting in more uniform regular arrays of spherical Au NP patterned with EBL.

Fine fabrication of regular metal NP array platforms for improved optical and plasmonic features in sensing appears possible by this method.

Bottom-up patterning and deposition. Recently, NSL of Si nanobeads was used in conjunction with selective EL plating to deposit Au thin films and hexagonal arrays of Au NP spheres with diameters from 522.5 ± 29.7 to 1116.9 ± 52.6 nm on dimethyldichlorosilane (DMDCS)-lithographed SiO_2 substrates [37]. Controlling availability of surface-associated water in key steps of a proposed silanization reaction allowed tunability of the shape and dimension of resulting nanostructures on the DMDCS-Si bead templated surface. Biological templates such as molecules [110], DNA [111,112], protein [113], yeast cells [114], or bacterial surface layers [115] are used in similar emerging bottom-up lithographic techniques. Biological templates function well in ambient environments but are otherwise labile and their ensuing regular patterns of nanostructures are limited in size to a few nano- to micrometers.

7.5   CHARACTERIZING FABRICATED NANOSTRUCTURE ARRAYS

7.5.1   MICROSPECTROSCOPY

Characterization of spectral features of nanostructured arrays requires optical isolation of a well-defined field of view (FOV) in a fabricated array that is constrained to microscopic dimensions by both limitations of EBL patterning as well as desired minimization of sampled volume. T-UV spectroscopy was selected to corroborate changes in far-field spectral extinction arising from DCP due to tuning of physicochemical and geometric parameters as well as introduction of samples with increasing RI. Microspectroscopy of nanolattice metamaterials was performed in T-UV mode using a customized system. Figure 7.9 shows a schematic representation of the microspectrometer setup. It consisted of a 6-V tungsten microscope light source, a series of lenses for collimating and focusing, a polarization crystal (GT5, ThorLabs, Newton, NJ), a 3-axis micropositioner for flow cell alignment, a 100× microscope objective (NA = 0.70), additional focusing lenses, and a fiber-optic collimator to collect the light with a fiber-optic spectrometer (AvaSpec 2048, Avantes Inc., Colorado). Effects of RI changes on the spectral characteristics of the sample were determined by changing the fluid in the flow cell while recoding spectral measurements. Spectra were analyzed using MATLAB® subroutines including *savfilt*, a generalized Savitzky-Golay smoothing algorithm, and *peakdet*, a maxima detection algorithm [both available from MATLAB Central (Mathworks Inc, Natick, MA)].

7.5.2   ELECTRON MICROSCOPY

Scanning electron microscopy (SEM) is useful to characterize nanoscale features of individual nanostructures as well as long-range order in arrays. Images taken during the steps of EBL patterning, EL metal deposition, and thermal processing of a square array of nanospheres are shown in Figure 7.10. The SEM images were obtained using a Philips XL 30 environmental SEM (FEI, Hillsboro, OR). Equipping this SEM with a Nanometer Pattern Generation System (JC Nabity Lithography Systems, Bozeman, MT) enabled patterning NP arrays with defined features like NP size and spacing.

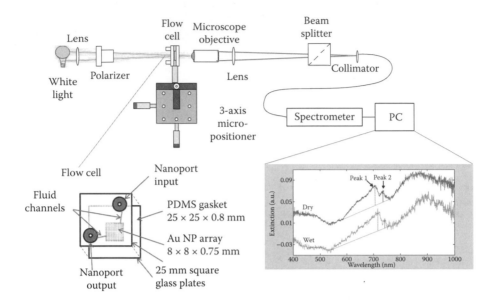

FIGURE 7.9 Microspectroscopy system for T-UV analysis of nanostructured metamaterial arrays. Enlargement of sample containing flow cell and representative spectra are shown. (From P. Blake et al., Enhanced nanoparticle response from coupled dipole excitation for plasmon sensors. IEEE Sensors, 11(12), 3332–3340 © (2011) IEEE.)

Analysis of SEM images was performed using MATLAB to accurately measure size and deviations from circularity of nanostructures in arrays. Standard measures used in particle size and shape analysis were employed to calculate average and individual NP circle equivalent diameter, interparticle spacing, circularity, and elongation from SEM images [116]. Particle diameter was determined by the *regionprops* function built into MATLAB by using the formula

$$D = \sqrt{\frac{4A}{\pi}}, \tag{7.7}$$

where D is the circle-equivalent particle diameter and A is the NP area. Circularity was defined using the ratio

$$\frac{4\pi A}{P^2}, \tag{7.8}$$

where P is the NP perimeter. Circularity approaches 1 for circular objects. Elongation is calculated by

$$1 - \text{aspect ratio} \rightarrow 1 - \frac{\text{minor axis length}}{\text{major axis length}}. \tag{7.9}$$

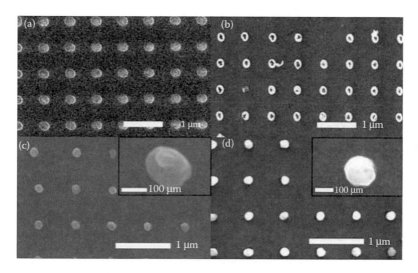

FIGURE 7.10 SEM images of (a) square hole arrays on 280-nm-thick PMMA after e-beam writing ($r = 107.22 \pm 7.04$ nm, $d = 665.56 \pm 5.91$ nm), (b) square arrays of Au nanocylinders after EL Au plating, (c) regular Au NP arrays after one thermal treatment at 800°C for 1 min (3 times), and (d) after additional heating at 800°C for 1 min. (W. Ahn et al. *Journal of Vacuum Science & Technology B*, 28(3), 638–642, 2010.)

Elongation approaches 0 for circular objects. The value of elongation can be transformed to approach 1 for circular objects by subtracting elongation measured in Equation 7.2 from unity to obtain "1 − elongation."

7.5.3 Figures of Merit

Characterization of performance of sensors based on DCP in arrayed nanostructures requires definition of figures of merit for accurate comparison with related sensors on an equivalent basis.

Aggregate sensitivity. A widely used figure of merit to characterize a change in a plasmon-effected spectral feature like a shift in wavelength of the resonant peak maximum to a concomitant change in surrounding dielectric medium is

$$S_A = \Delta RU \Delta \eta_m^{-1}, \tag{7.10}$$

where ΔRU is the change in spectral response and $\Delta \eta_m$ is the change in RI of the surrounding dielectric medium. This form of sensitivity is an *aggregate* response of all NPs in a spectroscopic FOV to a change in RI. It is termed the *aggregate sensitivity*, S_A, hereafter.

Nanoparticle sensitivity. A modification to the aggregate sensitivity was recently introduced to account for the intrinsic instrument sensitivity, active plasmon area exposed to analyte, and spectroscopic signal [1]. This figure of merit provides a consistent basis to compare sensitivity in SPP, LSP, and diffraction coupled dipole sensors. Its value is the magnitude of spectroscopically measured EM wave–NP interaction for each NP. For sensors with equal intrinsic sensitivities and FOV values, this measure reduces to

$$S_{NP} = \Delta\lambda_{max}\Delta\eta_m^{-1}A_{xc}^{-1}N_{NP}^{-1}A, \qquad (7.11)$$

where $\Delta\lambda_{max}$ is the change in peak maximum location, A is the area excited by the incident wave, A_{xc} is the cross-sectional area of an individual NP, and N_{NP} is the total number of NPs in the area A. This NP sensitivity, S_{NP}, has the same units as S_A, but is normalized to that fraction of the surface covered by Au NPs.

7.6   ANALYZING NANOSTRUCTURED ARRAYS

7.6.1   EXTINCTION SPECTRA

Far-field spectral extinction features such as the plasmon resonance and diffraction coupled dipole features obtained computationally using FDTD, CDA, and rsa-CDA appear consistent with similar features in T-UV measurements of ordered Au NPs on condensed substrates [61,62] with characteristic dependence on parameters that affect LSPR as well as those unique to the array (such as special configuration and disorder) [39,45,63]. Variations are attributable to differences between ideal homogeneous constant theoretical and nonideal heterogeneous varying experimental values of particle size, shape, dielectric, and spacing. Importantly, the frequency and intensity of diffraction coupled dipole features are comparable despite the finite size of experimental arrays versus infinite dimensions assumed for computed arrays.

7.6.2   SENSOR SENSITIVITY

An initial comparison of conventional SPR sensors with diffraction coupled dipole sensors used CDA, DDA, and FDTD to compare theoretical with experimental sensitivities to changes in medium RI on a prototype nanostructured metamaterial platform containing Au nanospheres of radius 104 nm spaced at 649 nm [75]. These parameter values were far from values estimated to optimize sensitivity in nanostructured metamaterials (as illustrated in Figure 7.4) relative to conventional plasmon sensors, but they illustrate important features of the different sensor platforms. Sensitivities for EM–NP interactions of LSPR sensors consisting of one single or random Au nanosphere ($r = 104$ nm) immersed from air into water were calculated using Mie simulation to be $S_A = S_{NP} = 140.5$ and 550.1 nm $\cdot$ RIU^{-1}, for quadrupole and dipole peak shifts, respectively. Equal values of aggregate and NP sensitivity result because simulations used a single NP. Similar results for a square array of such nanospheres spaced at 649 nm were calculated using CDA, DDA, and FDTD and are summarized in Table 7.3.

Calculated sensitivities. Calculated sensitivities of diffraction coupled dipoles in square arrays on a per NP basis were about 10 times larger in each case than aggregate sensitivities. Results from CDA and DDA models, which assumed RI changed only in the plasmon-active region within ~100 nm from the surface of the NP were comparable to Mie results for single NP sensitivity. However, diffraction coupled dipole features in NP arrays were dramatically more intense and narrower than dipole or quadrupole peaks for single NP. In practice, this reduces the minimum shift required for detection in practice where signal-to-noise ratio more readily obscures shifts of broader peaks. In contrast, FDTD results that included velocity dampening of light in higher-RI water between NP in the array yielded sensitivities about 7 times larger than those for single or arrayed NP with only surface coverage.

Experimental sensitivities. Experimental sensitivities for diffraction coupled dipoles on these square Au NP arrays were determined (with RI values shown in parentheses) using methanol (1.328), water (1.333), ethanol (1.361), and isopropanol

TABLE 7.3
DCP Sensor Platform Compared with Other LSPR-Based Sensors and Simulation

Reference	Description	S_A (nm · RIU^{-1})	S_{NP} (nm · RIU^{-1})
	Simulation		
CDA	Au sphere square	40	493
DDA	2.37×10^8 NP cm^{-2}	41	503
FDTD	lattice	276	3419
	Sensors		
CoDEx	Au cylinder 2.37×10^8 NP cm^{-2} square lattice	31 ± 2	389
[31]	Au sphere 1.46×10^{11} NP cm^{-2} random	76	371
[119]	Au sphere 1.6×10^{11} NP cm^{-2} random	48	85
[16]	Au sphere	44 ± 3	
	Cube	83 ± 2	
	Rod	$195 \pm 7 - 288 \pm 8$	
	Aspect ratio 2.4–2.6 in solution		
[122]	Ag sphere		160–235
	Triangle		197
	Triangle in array	191	
	Rod		235
[3]	Ag disk	22	22
	–		
	–		
[123]	Au islands Annealed	66–153	66–153
[118]	Au Film SPP	3100	3100

(1.3772) [117]. Experimental values of aggregate and NP sensitivity were determined to be $S_A = 31 \pm 2$ nm $\cdot$ RIU^{-1} and $S_{NP} = 389$ nm $\cdot$ RIU^{-1}, respectively, as shown in Table 7.3. Agreement with CDA and DDA was better than with FDTD, suggesting that quadrupolar and higher-order effects included in the latter could increase sensitivity ~6-fold in Au NP arrays fabricated uniformly at given values of lattice spacing and particle size.

7.6.3 SENSITIVITY VERSUS OTHER SENSOR PLATFORMS

Table 7.3 also compares sensitivities determined for the diffractive coupled dipole platform with other plasmon-based sensor platforms. S_{NP} values for each sensor were calculated using reported values of NP surface density and size in each case. Sensitivity is shape-dependent: nanorods have higher sensitivities than nanospheres [118].

Aggregate sensitivity. Aggregate sensitivity of this unoptimized nanostructured metamaterial platform was 40–75% of values reported for random Au spheres [16,31,119], and 50% higher than that reported for a single Ag disk [120]. Particle distribution and measurement precision could account for this difference. Aggregate SPP sensitivity in Table 7.3 is larger than LSPR or DCM aggregate sensitivity since signal is obtained from a much larger active plasmon-sensing area. Normalization to an equivalent S_{NP} basis (in column 2) brings sensitivity values much closer, with remaining differences due to the significantly larger mass of analytes required to produce signal in SPP sensors than in single-particle sensors. Directly comparing SPP and LSPR sensors shows that large differences in reported sensitivity do not affect the respective ability to detect similar analyte levels [121]. Aggregate and NP sensitivities for SPP and island film sensors are identical since the scaling factor is unity: the sensing area is the same as the plasmon-active area of the sensor surface covered by analytes.

Nanoparticle sensitivity. Normalization based on active plasmon-sensing area in Table 7.3 shows that NP sensitivity of the DCM platform is 10% greater than the highest value reported for random Au spheres and 17 times higher than that reported for a single Ag disk. This initial result appears promising. Ultimate theoretical sensitivities attainable using DCM are higher than individual or random NP due to relatively higher extinction efficiencies compared to LSPR peaks (8 to 33× higher for silver NP) as well as narrower full-width at half-maximum (>50 nm for LSPR vs. <5 nm for DCM) [1]. DCM sensor platforms also reduce the number and size of NPs required for sensing, by lowering the mass and density of analytes required to obtain a detectable signal. This promises a lower limit of detection (LOD) which is important for analytes such as cancer biomarkers present at extremely dilute levels.

7.6.4 MULTISPECTRAL ANALYSIS

Experimental and theoretical data for square NP arrays exhibit multiple spectral features whose wavelength and amplitude shift in response to changes in RI as well as rotation of the incident EM polarization relative to the NP lattice. This could enable multispectral detection of analytes, allowing, for example, detection of an analyte in a matrix that absorbs or scatters strongly in a region of the spectra and interferes with one peak, by using another of the available peaks.

7.6.5 NEAR VERSUS FAR-FIELD SENSING

Unlike SPP and LSPR sensors in which detection is limited to analyte molecules within a few hundred nanometers of the plasmon-active surface, DCM platforms exhibit additional sensitivity to effects of changes in local LSPR properties on far-field scattered light and thus on the maximum wavelength of the diffraction coupled multipole extinction. Furthermore, DCM platforms also exhibit sensitivity to dielectric changes in the far-field region, as seen by comparing FDTD and CDA/DDA results above. DCM platforms could allow distributed sensing among a collection of interactive nodes in a structured metamaterial. Diffractive excitation from incident as well as coherently scattered sources produce an effective polarizability which reflects the difference between the sum of phase-dependent and reciprocal static excitation sources. That the LSPR peak is indistinguishable from noise when the diffraction coupled dipole peak is clearly visible underscores both the robustness and sensitivity of cooperative interactions between individual dipoles.

7.6.6 OVERALL PERFORMANCE

In short, CDA and DDA computations exhibited diffraction coupled multiple extinctions consistent with full FDTD simulation of fabricated nanolattice metamaterials, which was absent in Mie theory. In fabricated nanolattices, the DCM peak was clearly measurable, even when LSPR signal was indistinguishable from noise, revealing more than 10-fold greater LOD inherent with this approach. Both the peak wavelength and amplitude of the CoDEx were sensitive to changes in local RI. Measured experimental sensitivities of fabricated gold nanolattices to changes in local dielectric were linear, consistent with theoretical values, and on par with reported LSPR sensitivities, but lower than silver particles or gold films. On a per NP basis, though, the sensitivities of these first-generation samples were within a factor of 10 of commercially available BIA Core type sensors, before accounting for significantly larger analyte mass required by the latter. A more universal figure of merit for sensitivity in LSPR, SPP, and CoDEx puts various optical and geometrical effects on an equivalent basis. Using the analytic CDA description, we find that nanolattice metamaterials with ideal physical and geometric parameters have sensitivities equivalent to or greater than the best reported plasmon-based sensors.

7.7 SUMMARY

Metal nanostructures organized in periodic arrays constitute metamaterials exhibiting extraordinary spectral features that could provide a novel sensor platform offering analyte-specific, sensitive, stand-off detection; true multispectral analysis with improved LOD and increased SERS; and lower fabrication and analysis cost compared with existing electronic, optical, mechanical, or EM field platforms. A primary extinction feature appears in far-field T-UV spectroscopy at the lattice constant due to EM coupling between the induced multipole components of LSPs and constructive interference from diffracted far-field radiation even when local surface plasmon signal from individual particles is indistinguishable from noise. A rapid, approximate, semi-analytical description of this coupling using a dynamic polarizability accurately characterizes plasmon resonances with dipole and quadrupole origins. Effects of polarizability and diffraction from individual nanostructures on the extraordinary extinction feature are distinguishable using this description. New modeling and fabrication approaches streamline identification and development of prototype diffraction coupled multipole sensor platforms with optimal physicochemical and geometric properties. A recently introduced particle-based sensitivity allows for direct comparison of coupled dipole, LSPR, and SPP sensors on the basis of equal gold-surface analyte coverage. RI-based sensing with an unoptimized prototype square array of Au NPs exhibited aggregate and single-particle sensitivities comparable to reported sensitivities for LSPR sensors and consistent with computational results from CCAs and DDAs. Differences compared to SPP sensors were attributable to different requirements for analyte mass which are not expected to adversely affect the ability to detect dilute analyte solutions. On the contrary, observed and simulated capabilities for sensor platforms of nanostructured metamaterials including inherent selectivity, enhanced sensitivity, multispectral analysis, and far-field sensing offer possible advantages for nanostructured metamaterial sensor platforms. These features could overcome difficulties with specificity, sensitivity, proximity, and economy associated with conventional electronic, optical, mechanical, and EM sensors. For example, SERS using plasmon-active nanostructures offers the possibility of direct, analyte-specific sensing in a variety of sampling contexts. Thus, nanostructured metamaterials provide a promising, competitive platform for acclaimed field applications such as on-site detection of infectious or pandemic pathogens, remote monitoring of wearable sensors, stand-off analysis of hazardous compounds, rapid forensic evaluation, or point-of-care diagnostics.

ACKNOWLEDGMENTS

This work was supported in part by NSF CMMI-0909749, NSF ECCS-1006927, NSF CBET-1134222, the Walton Family Charitable Support Foundation, the University of Arkansas Foundation, and NSF GK-12 program awarded to D. DeJarnette. D. K. Roper directed the work and is the primary author of the text. P. Blake and D. DeJarnette performed simulations and experiments and prepared data and draft text for the manuscript. B. Harbin performed preliminary rsa-CDA and FDTD simulations. The authors would like to thank W. Ahn, G.-G. Jang, A. Russell, and B. Taylor for insights, useful discussions, and contributions to modeling and experiments.

REFERENCES

1. D. K. Roper, W. Ahn, B. Taylor, and A. G. Dall'Asén, Enhanced spectral sensing by electromagnetic coupling with localized surface plasmons on subwavelength structures, *IEEE Sensors Journal*, 10(3), 531–540, Mar. 2010.
2. J. N. Anker, W. P. Hall, O. Lyandres, N. C. Shah, J. Zhao, and R. P. Van Duyne, Biosensing with plasmonic nanosensors, *Nature Materials*, 7(6), 442–453, Jun. 2008.

3. D. A. Stuart, Refractive-index-sensitive, plasmon-resonant-scattering, and surface-enhanced Raman-scattering nanoparticles and arrays as biological sensing platforms, *Proceedings of SPIE*, 5327(847), 60–73, 2004.

4. K. A. Willets and R. P. Van Duyne, Localized surface plasmon resonance spectroscopy and sensing, *Annual Review of Physical Chemistry*, 58, 267–297, Jan. 2007.

5. D. B. Pedersen and E. J. S. Duncan, Surface Plasmon Resonance Spectroscopy of Gold Nanoparticle-Coated Substrates, Technical Report, Defence Research and Development Canda-Suffield, August 2005. (http://www.dtic.mil/cgi-bin/GetTRDoc?AD=ADA440001, accessed July 2013).

6. C. Bertucci, A. Piccoli, and M. Pistolozzi, Optical biosensors as a tool for early determination of absorption and distribution parameters of lead candidates and drugs, *Combinatorial Chemistry & High Throughput Screening*, 10(6), 433–440, Jul. 2007.

7. J. Homola, Surface plasmon resonance sensors for detection of chemical and biological species, *Chemical Reviews*, 108(2), 462–493, Feb. 2008.

8. M. E. Stewart et al., Nanostructured plasmonic sensors, *Chemical Reviews*, 108(2), 494–521, Feb. 2008.

9. A. Subramanian and J. Irudayaraj, Surface plasmon resonance based immunosensing of *E. coli* O157: H7 in apple juice, *Trans. ASABE*, 49(4), 1257–1262, 2006.

10. D. A. Stuart, A. J. Haes, C. R. Yonzon, E. M. Hicks, and R. P. Van Duyne, Biological applications of localised surface plasmonic phenomena, *IEE Proceedings. Nanobiotechnology*, 152(1), 13–32, Feb. 2005.

11. D. G. Myszka, X. He, M. Dembo, T. A. Morton, and B. Goldstein, Extending the range of rate constants available from BIACORE: Interpreting mass transport-influenced binding data, *Biophysical Journal*, 75(2), 583–594, Aug. 1998.

12. T. M. Chinowsky, J. G. Quinn, D. U. Bartholomew, R. Kaiser, and J. L. Elkind, Performance of the Spreeta 2000 integrated surface plasmon resonance affinity sensor, *Sensors and Actuators B: Chemical*, 91(1–3), 266–274, Jun. 2003.

13. A. J. Haes and R. P. V. Duyne, Preliminary studies and potential applications of localized surface plasmon resonance spectroscopy in medical diagnostics, *Expert Review of Molecular Diagnostics*, 4(4), 527–537, Jul. 2004.

14. D. K. Roper, Enhancing lateral mass transport to improve the dynamic range of adsorption rates measured by surface plasmon resonance, *Chemical Engineering Science*, 61(8), 2557–2564, Apr. 2006.

15. A.-S. Grimault, A. Vial, J. Grand, and M. Lamy de la Chapelle, Modelling of the near-field of metallic nanoparticle gratings: Localized surface plasmon resonance and SERS applications, *Journal of Microscopy*, 229(3), 428–432, Mar. 2008.

16. H. Chen, X. Kou, Z. Yang, W. Ni, and J. Wang, Shape- and size-dependent refractive index sensitivity of gold nanoparticles, *Langmuir*, 24(10), 5233–5237, May 2008.

17. L. Gunnarsson et al., Confined plasmons in nanofabricated single silver particle pairs: Experimental observations of strong interparticle interactions, *The Journal of Physical Chemistry B*, 109(3), 1079–1087, Jan. 2005.

18. N. Félidj et al., Multipolar surface plasmon peaks on gold nanotriangles, *The Journal of Chemical Physics*, 128(9), 094702, Mar. 2008.

19. J. Grand et al., Optical extinction spectroscopy of oblate, prolate and ellipsoid shaped gold nanoparticles: Experiments and theory, *Plasmonics*, 1(2–4), 135–140, Sep. 2006.

20. L. E. Kreno, J. T. Hupp, and R. P. Van Duyne, Metal-organic framework thin film for enhanced localized surface plasmon resonance gas sensing, *Analytical Chemistry*, 82(19), 8042–8046, Oct. 2010.

21. Y. Ekinci, A. Christ, M. Agio, O. J. F. Martin, H. H. Solak, and J. F. Löffler, Electric and magnetic resonances in arrays of coupled gold nanoparticle in-tandem pairs, *Optics Express*, 16(17), 13287–13295, Aug. 2008.

22. G. H. Chan, J. Zhao, G. C. Schatz, and R. P. V. Duyne, Localized surface plasmon resonance spectroscopy of triangular aluminum nanoparticles, *Journal of Physical Chemistry C*, 112(36), 13958–13963, Sep. 2008.

23. K.-S. Lee and M. A. El-Sayed, Gold and silver nanoparticles in sensing and imaging: Sensitivity of plasmon response to size, shape, and metal composition, *The Journal of Physical Chemistry B*, 110(39), 19220–19225, Oct. 2006.

24. E. Simsek, On the surface plasmon resonance modes of metal nanoparticle chains and arrays, *Plasmonics*, 4(3), 223–230, Jun. 2009.

25. S. Gao et al., Highly stable Au nanoparticles with tunable spacing and their potential application in surface plasmon resonance biosensors, *Advanced Functional Materials*, 20(1), 78–86, Jan. 2010.

26. T. R. Jensen, M. L. Duval, K. L. Kelly, A. A. Lazarides, G. C. Schatz, and R. P. Van Duyne, Nanosphere lithography: Effect of the external dielectric medium on the surface plasmon resonance spectrum of a periodic array of silver nanoparticles, *The Journal of Physical Chemistry B*, 103(45), 9846–9853, Nov. 1999.

27. F.-L. Hsiao and C. Lee, Computational study of photonic crystals nano-ring resonator for biochemical sensing, *IEEE Sensors Journal*, 10(7), 1185–1191, Jul. 2010.

28. W. Rechberger, A. Hohenau, A. Leitner, J. R. Krenn, B. Lamprecht, and F. R. Aussenegg, Optical properties of two interacting gold nanoparticles, *Optics Communications*, 220(1–3), 137–141, May 2003.

29. K.-H. Su, Q.-H. Wei, X. Zhang, J. J. Mock, D. R. Smith, and S. Schultz, Interparticle coupling effects on plasmon resonances of nanogold particles, *Nano Letters*, 3(8), 1087–1090, Aug. 2003.

30. W. Ahn and D. K. Roper, Transformed gold island film improves light-to-heat transduction of nanoparticles on silica capillaries, *Journal of Physical Chemistry C*, 112(32), 12214–12218, Aug. 2008.

31. N. Nath and A. Chilkoti, A colorimetric gold nanoparticle sensor to interrogate biomolecular interactions in real time on a surface, *Analytical Chemistry*, 74(3), 504–509, Feb. 2002.

32. P. Hanarp, M. Käll, and D. S. Sutherland, Optical properties of short range ordered arrays of nanometer gold disks prepared by colloidal lithography, *The Journal of Physical Chemistry B*, 107(24), 5768–5772, Jun. 2003.

33. B. Sepúlveda, P. C. Angelomé, L. M. Lechuga, and L. M. Liz-Marzán, LSPR-based nanobiosensors, *Nano Today*, 4(3), 244–251, Jun. 2009.

34. G.-G. Jang and D. K. Roper, Continuous flow electroless plating enhances optical features of Au films and nanoparticles, *Journal of Physical Chemistry C*, 113(44), 19228–19236, Nov. 2009.

35. W. Ahn, P. Blake, J. Shultz, M. E. Ware, and D. K. Roper, Fabrication of regular arrays of gold nanospheres by thermal transformation of electroless-plated films, *Journal of Vacuum Science & Technology B: Microelectronics and Nanometer Structures*, 28(3), 638, 2010.

36. P. Blake, W. Ahn, and D. K. Roper, Enhanced uniformity in arrays of electroless plated spherical gold nanoparticles using tin presensitization, *Langmuir*, 26(3), 1533–1538, Feb. 2010.

37. W. Ahn and D. K. Roper, Periodic nanotemplating by selective deposition of electroless gold island films on particle-lithographed dimethyldichlorosilane layers, *ACS Nano*, 4(7), 4181–4189, Jul. 2010.

38. G. Vecchi, V. Giannini, and J. Gómez Rivas, Surface modes in plasmonic crystals induced by diffractive coupling of nanoantennas, *Physical Review B*, 80(20), 1–4, Nov. 2009.

39. A. Evlyukhin, C. Reinhardt, A. Seidel, B. Luk'yanchuk, and B. Chichkov, Optical response features of Si-nanoparticle arrays, *Physical Review B*, 82(4), 1–12, Jul. 2010.

40. B. Auguié and W. Barnes, Collective resonances in gold nanoparticle arrays, *Physical Review Letters*, 101(14), 1–4, Sep. 2008.

41. E. M. Hicks et al., Controlling plasmon line shapes through diffractive coupling in linear arrays of cylindrical nanoparticles fabricated by electron beam lithography, *Nano Letters*, 5(6), 1065–1070, Jun. 2005.

42. E. J. Smythe, M. D. Dickey, J. Bao, G. M. Whitesides, and F. Capasso, Optical antenna arrays on a fiber facet for in situ surface-enhanced Raman scattering detection, *Nano Letters*, 9(3), 1132–1138, Mar. 2009.

43. C. Deeb et al., Quantitative analysis of localized surface plasmons based on molecular probing, *ACS Nano*, 4(8), 4579–4586, Aug. 2010.

44. T. Sannomiya et al., Biosensing by densely packed and optically coupled plasmonic particle arrays, *Small*, 5(16), 1889–1896, Aug. 2009.

45. B. Auguié and W. L. Barnes, Diffractive coupling in gold nanoparticle arrays and the effect of disorder, *Optics Letters*, 34(4), 401–403, Feb. 2009.

46. N. Félidj et al., Gold particle interaction in regular arrays probed by surface enhanced Raman scattering, *The Journal of Chemical Physics*, 120(15), 7141–7146, Apr. 2004.

47. A. Alù and N. Engheta, Effect of small random disorders and imperfections on the performance of arrays of plasmonic nanoparticles, *New Journal of Physics*, 12(1), 013015, Jan. 2010.

48. B. K. Canfield, S. Kujala, K. Jefimovs, T. Vallius, J. Turunen, and M. Kauranen, Polarization effects in the linear and nonlinear optical responses of gold nanoparticle arrays, *Journal of Optics A: Pure and Applied Optics*, 7(2), S110–S117, Feb. 2005.

49. S.-W. Lee, K.-S. Lee, J. Ahn, J.-J. Lee, M.-G. Kim, and Y.-B. Shin, Highly sensitive biosensing using arrays of plasmonic Au nanodisks realized by nanoimprint lithography, *ACS Nano*, 5(2), 897–904, Feb. 2011.

50. H. Li, X. Luo, C. Du, X. Chen, and Y. Fu, Ag dots array fabricated using laser interference technique for biosensing, *Sensors and Actuators B: Chemical*, 134(2), 940–944, Sep. 2008.

51. J. S. Yu et al., Characteristics of localized surface plasmon resonance of nanostructured Au patterns for biosensing, *Journal of Nanoscience and Nanotechnology*, 8(9), 4548–4552, Sep. 2008.

52. Y.-S. Shon, H. Y. Choi, M. S. Guerrero, and C. Kwon, Preparation of nanostructured film arrays for transmission localized surface plasmon sensing, *Plasmonics*, 4(2), 95–105, Mar. 2009.

53. T. Klar, M. Perner, S. Grosse, G. von Plessen, W. Spirkl, and J. Feldmann, Surface-plasmon resonances in single metallic nanoparticles, *Physical Review Letters*, 80(19), 4249–4252, May 1998.

54. L. Zhao, K. L. Kelly, and G. C. Schatz, The extinction spectra of silver nanoparticle arrays: Influence of array structure on plasmon resonance wavelength and width, *The Journal of Physical Chemistry B*, 107(30), 7343–7350, Jul. 2003.

55. P. K. Jain, K. S. Lee, I. H. El-Sayed, and M. A. El-Sayed, Calculated absorption and scattering properties of gold nanoparticles of different size, shape, and composition: Applications in biological imaging and biomedicine, *The Journal of Physical Chemistry B*, 110(14), 7238–7248, Apr. 2006.

56. M. K. Kinnan and G. Chumanov, Plasmon coupling in two-dimensional arrays of silver nanoparticles: II. Effect of the particle size and interparticle distance, *The Journal of Physical Chemistry C*, 114(16), 7496–7501, Apr. 2010.

57. S. Link and M. A. El-Sayed, Size and temperature dependence of the plasmon absorption of colloidal gold nanoparticles, *The Journal of Physical Chemistry B*, 103(21), 4212–4217, May 1999.

58. C. L. Haynes et al., Nanoparticle optics: The importance of radiative dipole coupling in two-dimensional nanoparticle arrays, *The Journal of Physical Chemistry B*, 107(30), 7337–7342, Jul. 2003.

59. G. Mie, Contributions on the optics of turbid media, particularly colloidal metal solutions, *Annalen der physik*, 25, 377–445, 1908.
60. C. F. Bohren and D. R. Huffman, Eds., *Absorption and Scattering of Light by Small Particles*. Weinheim, Germany: Wiley-VCH Verlag GmbH, 1998.
61. Y. Chu, E. Schonbrun, T. Yang, and K. B. Crozier, Experimental observation of narrow surface plasmon resonances in gold nanoparticle arrays, *Applied Physics Letters*, 93(18), 181108, 2008.
62. V. G. Kravets, F. Schedin, and A. N. Grigorenko, Extremely narrow plasmon resonances based on diffraction coupling of localized plasmons in arrays of metallic nanoparticles, *Physical Review Letters*, 101(8), 1–4, Aug. 2008.
63. S. Zou and G. C. Schatz, Narrow plasmonic/photonic extinction and scattering line shapes for one and two dimensional silver nanoparticle arrays, *The Journal of Chemical Physics*, 121(24), 12606–12612, Dec. 2004.
64. K. S. Yee, Numerical solution of initial boundary value problems involving Maxwell's equations in isotropic media, *IEEE Transactions on Antennas and Propagation*, 14(3), 302–307, May 1966.
65. A. Taflove and S. C. Hagness, *Computational Electrodynamics: The Finite-Difference Time-Domain Method*, 3rd ed. Norwood, MA: Artech House, 2005.
66. A. Doicu and T. Wriedt, Near-field computation using the null-field method, *Journal of Auantitative Spectroscopy and Radiative Transfer*, 111(3), 466–473, Feb. 2010.
67. B. Peterson and S. Ström, T matrix for electromagnetic scattering from an arbitrary number of scatterers and representations of $E(3)$, *Physical Review D*, 8(10), 3661–3678, Nov. 1973.
68. B. Peterson and S. Ström, T-matrix formulation of electromagnetic scattering from multilayered scatterers, *Physical Review D*, 10(8), 2670–2684, Oct. 1974.
69. M. I. Mishchenko and L. D. Travis, T-matrix computations of light scattering by large spheroidal particles, *Optics Communications*, 109(1–2), 16–21, Jun. 1994.
70. K. L. Kelly, E. Coronado, L. L. Zhao, and G. C. Schatz, The optical properties of metal nanoparticles: The influence of size, shape, and dielectric environment, *The Journal of Physical Chemistry B*, 107(3), 668–677, Jan. 2003.
71. W. T. Doyle, Optical properties of a suspension of metal spheres, *Physical Review B*, 39(14), 9852–9858, May 1989.
72. D. K. Roper, W. Ahn, B. Taylor, and Y. Dall'Asén, Optoplasmonic gold nanoparticle assembly for sensing, spectroscopy and heat transfer, in *International Symposium on Spectral Sensing Research*, June 2008, Hoboken, NJ, 23–27.
73. D. K. Roper, W. Ahn, P. Blake, B. Harbin, G. G. Jang, and B. Taylor, Subwavelength nanoparticle ordered structures for biological microscopic and spectral analysis, in *238th ACS National Meeting*, August 2009, Washington DC, 16–20.
74. D. K. Roper, W. Ahn, B. Taylor, and P. Blake, Subwavelength nanoparticle ordered structure for enhanced sensing, optoelectronics and DNA analysis, in *237th ACS National Meeting*, March 2009, Salt Lake City, UT, 22–26.
75. P. Blake, J. Obermann, B. Harbin, and D. K. Roper, Enhanced nanoparticle response from coupled dipole excitation for plasmon sensors, *IEEE Sensors Journal*, 11(12), 3332–3340, 2011.
76. D. DeJarnette, D. K. Roper, and B. Harbin, Geometric effects on far-field coupling between multipoles of nanoparticles in square arrays, *Journal of the Optical Society of America B*, 29(1), 88–100, Dec. 2011.
77. R. M. Bakker et al., Enhanced localized fluorescence in plasmonic nanoantennae, *Applied Physics Letters*, 92(4), 043101, 2008.
78. P. P. Pompa et al., Metal-enhanced fluorescence of colloidal nanocrystals with nanoscale control, *Nature Nanotechnology*, 1(2), 126–130, Nov. 2006.

79. R. M. Bakker, V. P. Drachev, H.-K. Yuan, and V. M. Shalaev, Enhanced transmission in near-field imaging of layered plasmonic structures, *Optics Express*, 12(16), 3701–3706, Aug. 2004.

80. M. Salerno et al., Plasmon polaritons in metal nanostructures: The optoelectronic route to nanotechnology, *Opto-Electronics Review*, 10(3), 217–224, 2002.

81. A. Dhawan et al., Fabrication of nanodot plasmonic waveguide structures using FIB milling and electron beam-induced deposition, *Scanning*, 31(4), 139–146, 2009.

82. C. H. Lin, L. Jiang, Y. H. Chai, H. Xiao, S. J. Chen, and H. L. Tsai, A method to fabricate 2D nanoparticle arrays, *Applied Physics A*, 98(4), 855–860, Jan. 2010.

83. Y. B. Zheng, V. K. S. Hsiao, and T. J. Huang, *Active Plasmonic Devices based on Ordered Au Nanodisk Arrays*. Tucson, AZ: IEEE, 2008, 705–708.

84. M. Ellman et al., High-power laser interference lithography process on photoresist: Effect of laser fluence and polarisation, *Applied Surface Science*, 255(10), 5537–5541, Mar. 2009.

85. N. D. Lai, W. P. Liang, J. H. Lin, C. C. Hsu, and C. H. Lin, Fabrication of two- and three-dimensional periodic structures by multi-exposure of two-beam interference technique, *Optics Express*, 13(23), 9605–9511, Nov. 2005.

86. U. Geyer, J. Hetterich, C. Diez, D. Z. Hu, D. M. Schaadt, and U. Lemmer, Nano-structured metallic electrodes for plasmonic optimized light-emitting diodes, in *Proceedings of SPIE*, 2008, 7032, 70320B-1–70320B-10.

87. A. F. Lasagni, D. Yuan, P. Shao, and S. Das, Periodic micropatterning of polyethylene glycol diacrylate hydrogel by laser interference lithography using nano- and femtosecond pulsed lasers, *Advanced Engineering Materials*, 11(3), B20–B24, Mar. 2009.

88. N. D. Lai, Y. D. Huang, J. H. Lin, D. B. Do, and C. C. Hsu, Fabrication of periodic nanovein structures by holography lithography technique, *Optics Express*, 17(5), 3362–3369, Mar. 2009.

89. F. Ma, M. H. Hong, and L. S. Tan, Laser nano-fabrication of large-area plasmonic structures and surface plasmon resonance tuning by thermal effect, *Applied Physics A*, 93(4), 907–910, Jun. 2008.

90. S. A. Maier, M. L. Brongersma, P. G. Kik, S. Meltzer, A. A. G. Requicha, and H. A. Atwater, Plasmonics—a route to nanoscale optical devices, *Advanced Materials*, 13(19), 1501–1505, Sep. 2001.

91. H. Zhang, S.-W. Chung, and C. A. Mirkin, Fabrication of sub-50-nm solid-state nano-structures on the basis of dip-pen nanolithography, *Nano Letters*, 3(1), 43–45, Jan. 2003.

92. H. Zhang, N. A. Amro, S. Disawal, R. Elghanian, R. Shile, and J. Fragala, High-throughput dip-pen-nanolithography-based fabrication of Si nanostructures, *Small*, 3(1), 81–85, Jan. 2007.

93. S. Y. Chou, P. R. Krauss, and P. J. Renstrom, Nanoimprint lithography, *Journal of vacuum science & technology B: Microelectronics and Nanometer Structures*, 14(6), 4129–4133, Nov. 1996.

94. J.-M. Jung, H.-W. Yoo, F. Stellacci, and H.-T. Jung, Two-photon excited fluorescence enhancement for ultrasensitive DNA detection on large-area gold nanopatterns, *Advanced Materials*, 22(23), 2542–2546, Jun. 2010.

95. K. Salaita, Y. Wang, J. Fragala, R. A. Vega, C. Liu, and C. A. Mirkin, Massively parallel dip-pen nanolithography with 55 000-pen two-dimensional arrays, *Angewandte Chemie*, 45(43), 7220–7223, Nov. 2006.

96. K. Perez Toralla, J. De Girolamo, D. Truffier-Boutry, C. Gourgon, and M. Zelsmann, High flowability monomer resists for thermal nanoimprint lithography, *Microelectronic Engineering*, 86(4–6), 779–782, Apr. 2009.

97. M. Zelsmann, K. Perez Toralla, J. De Girolamo, D. Boutry, and C. Gourgon, Comparison of monomer and polymer resists in thermal nanoimprint lithography, *Journal of Vacuum Science & Technology B*, 26(6), 2430–2434, 2008.

98. R. Ji et al., UV enhanced substrate conformal imprint lithography (UV-SCIL) technique for photonic crystals patterning in LED manufacturing, *Microelectronic Engineering*, 87(5–8), 963–967, May 2010.

99. S. Chauhan, F. Palmieri, R. T. Bonnecaze, and C. G. Willson, Feature filling modeling for step and flash imprint lithography, *Journal of Vacuum Science & Technology B*, 27(4), 1926–1932, 2009.

100. I. Bergmair, M. Mühlberger, E. Lausecker, K. Hingerl, and R. Schöftner, Diffusion of thiols during microcontact printing with rigid stamps, *Microelectronic Engineering*, 87(5–8), 848–850, May 2010.

101. N. D. Denkov, O. D. Velev, P. A. Kralchevsky, I. B. Ivanov, H. Yoshimura, and K. Nagayama, Two-dimensional crystallization, *Nature*, 361(6407), 26, Jan. 1993.

102. P. Jiang, J. F. Bertone, K. S. Hwang, and V. L. Colvin, Single-crystal colloidal multilayers of controlled thickness, *Chemistry of Materials*, 11(8), 2132–2140, Aug. 1999.

103. C. R. Yonzon, D. A. Stuart, X. Zhang, A. D. McFarland, C. L. Haynes, and R. P. Van Duyne, Towards advanced chemical and biological nanosensors—an overview, *Talanta*, 67(3), 438–448, Sep. 2005.

104. P. Jiang and M. J. McFarland, Large-scale fabrication of wafer-size colloidal crystals, macroporous polymers and nanocomposites by spin-coating, *Journal of the American Chemical Society*, 126(42), 13778–13786, Oct. 2004.

105. H. Li, J. Low, K. S. Brown, and N. Wu, Large-area well-ordered nanodot array pattern fabricated with self-assembled nanosphere template, *IEEE Sensors*, 8(6), 880–884, Jun. 2008.

106. F. Pan, J. Zhang, C. Cai, and T. Wang, Rapid fabrication of large-area colloidal crystal monolayers by a vortical surface method, *Langmuir*, 22(17), 7101–7104, Aug. 2006.

107. C. L. Haynes and R. P. Van Duyne, Nanosphere lithography: A versatile nanofabrication tool for studies of size-dependent nanoparticle optics, *The Journal of Physical Chemistry B*, 105(24), 5599–5611, Jun. 2001.

108. G.-G. Jang and D. K. Roper, Balancing redox activity allowing spectrophotometric detection of Au(I) using tetramethylbenzidine dihydrochloride, *Analytical Chemistry*, 85(5), 1836–1842, Jan. 2011.

109. W. Ahn, B. Taylor, A. G. Dall'Asén, and D. K. Roper, Electroless gold island thin films: Photoluminescence and thermal transformation to nanoparticle ensembles, *Langmuir*, 24(8), 4174–4184, Apr. 2008.

110. S. Sotiropoulou, Y. Sierra-Sastre, S. S. Mark, and C. A. Batt, Biotemplated nanostructured materials, *Chemistry of Materials*, 20(3), 821–834, Feb. 2008.

111. J. Zhang, Y. Liu, Y. Ke, and H. Yan, Periodic square-like gold nanoparticle arrays templated by self-assembled 2D DNA nanogrids on a surface, *Nano Letters*, 6(2), 248–251, Feb. 2006.

112. J. Zheng, P. E. Constantinou, C. Micheel, A. P. Alivisatos, R. A. Kiehl, and N. C. Seeman, Two-dimensional nanoparticle arrays show the organizational power of robust DNA motifs, *Nano Letters*, 6(7), 1502–1504, Jul. 2006.

113. R. A. McMillan, C. D. Paavola, J. Howard, S. L. Chan, N. J. Zaluzec, and J. D. Trent, Ordered nanoparticle arrays formed on engineered chaperonin protein templates, *Nature Materials*, 1(4), 247–252, Dec. 2002.

114. S. Chia, J. Urano, F. Tamanoi, B. Dunn, and J. I. Zink, Patterned hexagonal arrays of living cells in sol–gel silica films, *Journal of the American Chemical Society*, 122(27), 6488–6489, Jul. 2000.

115. M. Mertig, R. Wahl, M. Lehmann, P. Simon, and W. Pompe, Formation and manipulation of regular metallic nanoparticle arrays on bacterial surface layers: An advanced TEM study, *The European Physical Journal D*, 16(1), 317–320, Oct. 2001.

116. S. Almeida-Prieto, J. Blanco-Méndez, and F. J. Otero-Espinar, Image analysis of the shape of granulated powder grains, *Journal of Pharmaceutical Sciences*, 93(3), 621–634, Mar. 2004.

117. J. A. Dean, *Lange's Handbook of Chemistry*, 15th ed., J. A. Dean, ed., New York, NY: McGraw-Hill, 1999, 1.211, 1.255, 1.302.

118. L. S. Jung, C. T. Campbell, T. M. Chinowsky, M. N. Mar, and S. S. Yee, Quantitative interpretation of the response of surface plasmon resonance sensors to adsorbed films, *Langmuir*, 14(19), 5636–5648, Sep. 1998.

119. O. Kedem, A. B. Tesler, A. Vaskevich, and I. Rubinstein, Sensitivity and optimization of localized surface plasmon resonance transducers, *ACS Nano*, 5(2), 748–760, Feb. 2011.

120. D. A. Stuart, A. J. Haes, A. D. McFarland, and R. P. Van Duyne, Refractive-index-sensitive, plasmon-resonant-scattering, and surface-enhanced Raman-scattering nanoparticles and arrays as biological sensing platforms. *Proceedings of SPIE*, 2004, 60–73.

121. M. Svedendahl, S. Chen, A. Dmitriev, and M. Käll, Refractometric sensing using propagating versus localized surface plasmons: A direct comparison, *Nano Letters*, 9(12), 4428–4433, Dec. 2009.

122. A. D. McFarland and R. P. Van Duyne, Single silver nanoparticles as real-time optical sensors with zeptomole sensitivity, *Nano Letters*, 3(8), 1057–1062, Aug. 2003.

123. T. Karakouz, D. Holder, M. Goomanovsky, A. Vaskevich, and I. Rubinstein, Morphology and refractive index sensitivity of gold island films, *Chemistry of Materials*, 21(24), 5875–5885, Dec. 2009.

8 Advances in Surface Plasmon Resonance Sensing

Brent D. Cameron and Rui Zheng

CONTENTS

8.1 INTRODUCTION TO SURFACE PLASMON RESONANCE SENSORS

8.1.1 BRIEF OVERVIEW

Surface plasmon resonance (SPR) is an optical-based sensing technology that allows for real-time qualitative and quantitative measurement of molecular interactions without the need of labeling procedures. The first report of this phenomenon was in 1902 with the observation that dark troughs occurred in the spectrum of diffracted light when a metal diffraction grating was illuminated with a multispectral light source (Wood 1902). Since this initial observance, many groups have studied SPR and developed methods to theoretically and experimentally characterize this behavior. The use of SPR systems for chemical detection was demonstrated in the 1980s which led to its eventual transition, in the 1990s, for use as a robust bioanalytical tool for probing biomolecular interactions. To date, SPR has become a popular core technology integrated into several research-based biosensing devices due to its ultrasensitivity and potentially inexpensive instrumentation when compared to many other high-performance sensing approaches (e.g., Raman spectroscopy, liquid chromatography, mass spectrometry, etc.).

SPR methods are based on an optical phenomenon that occurs at a metal–dielectric interface. Light at an appropriate wavevector can resonate with free electrons in the metal. When this occurs, electromagnetic waves known as surface plasmons become present along the metallic surface. Under this condition, the light leaving the surface will become attenuated in intensity. Since the resonant condition is dependent on the refractive index in the vicinity of the metal surface, often either the incidence angle or the wavelength of light is used as the measurement parameter (i.e., the component used to determine the conditions for which resonance coupling occurs).

8.1.2 Current Advances in SPR Sensing

To date, numerous approaches and techniques for SPR sensing have been reported on in literature. The most common configuration is based on the Kretschmann design which is a reflectance-type measurement. This approach has proved to be both reliable and robust which is why it has remained the preferred choice for use in most commercial SPR bioanalytical systems. Recent research, however, has been heavily focused on further improving SPR sensitivity and functionality. Through the use of localized surface plasmon resonance (LSPR) techniques, which rely on innovative plasmonic structures and materials including various metallic nanoparticle geometries, significant enhancement in sensitivity has been possible (up to a factor of 7) (Bolduc and Masson 2011). This has led to demonstration of ultralow detection limits extending down to the femtomolar range (Kim et al. 2010). Other configuration designs have also been introduced to provide greater functionality, including an extension of traditional SPR sensing geared toward image-based measurements (commonly referred to as SPRi). SPRi has allowed such measurements to begin transitioning toward high-throughput assessment, demonstrating the capability of monitoring multiple biomolecular interactions on "chip-type" platforms (Hu et al. 2010; Lautner et al. 2010; Ouellet et al. 2010).

Regardless of the configuration, the SPR technique through proper surface modification can provide significant insights into affinity and biomolecular interaction kinetics as well as providing specific and sensitive measurements of chemical concentration in complex samples. A main advantage of SPR is that it can be performed in real time without the need for labeling steps. In the past two decades, different SPR sensors have been developed and are becoming increasingly popular in biological studies, health science research, drug discovery, clinical diagnosis, and environmental and agricultural monitoring. This chapter will provide a fundamental background of SPR physics and configurations, functionalization approaches, and overview of current applications.

8.2 SPR SENSING CONFIGURATIONS

To achieve optimal performance, SPR-sensing configurations should include three key components. The core component is an optical platform that is capable of demonstrating high sensitivity to changes in the refractive index (RI) of the evaluation medium, which is normally some type of fluid. Since SPR, by itself, is not a selective technique, sensor performance is often heavily dependent on the use of secondary molecular recognition components (i.e., an opto-chemical transducer) immobilized on the sensing surface consisting of a deposited thin gold or silver nanometer layer of metal or particle cluster array. Analyte detection with SPR is carried out in real time through analysis of the time course of the sensor response; therefore, the final component provides controlled sample delivery to the sensor surface. This plays an integral part in the SPR measurement and is normally achieved through some type of microfluidic flow structure.

In general terms, Figure 8.1 depicts a typical SPR sensor setup that relates properties of the light signal into respective analyte concentration levels. The input into this system is a fluid or sample to be interrogated. If the sample is complex in nature (i.e., containing numerous analyte species), the target molecule(s) of interest will interact with specific recognition receptors on the SPR-sensing surface that are able to bind the target. The result is an enhanced localized change in the RI at the surface of the optical sensor. This RI change, which is related to the analyte concentration, must then be detected by the optical system. In general, an optical sensor is a device which monitors an optical parameter that is correlated to the quantity of interest. For SPR sensors, a surface plasmon is excited at the interface between the metal surface and a dielectric medium by an incoming light wave. A change in the RI in the dielectric medium will produce a corresponding change in the propagation constant of the surface plasmon. This change is dependent on the coupling condition between the light wave and that of the surface plasmon. Therefore, by monitoring the parameters of the optical wave which affect coupling, the localized change in RI can directly be related back to the analyte concentration or the binding interactions that are occurring at the SPR-sensing surface. Based on the specific light wave characteristic that is monitored, SPR sensors are normally classified as being angular, wavelength, intensity, phase, or polarization modulated.

8.2.1 PROPAGATING SPR

Among all the different SPR modulations, the angular-based setup founded on Kretschmann configuration is the most common. Nowadays, it is an interesting

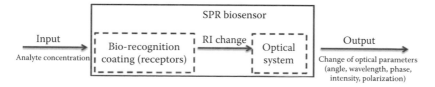

FIGURE 8.1 A typical SPR sensor layout.

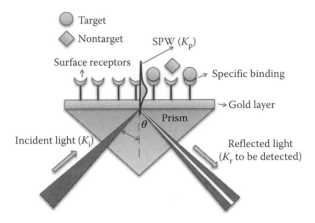

FIGURE 8.2 A typical Kretschmann SPR sensor setup.

observance that, although numerous commercial SPR systems are available, their designs are very similar to that which was originally presented by Liedberg et al. (1983). Figure 8.2 depicts a typical Kretschmann SPR sensor configuration with angular modulation. In this setup, using a prism as an optical coupler, a focused monochromatic incident light beam is totally reflected at the metal–dielectric interface. An electromagnetic field component, commonly termed an "evanescent wave," excites surface plasmons within the metal layer. The strength of coupling between the incident wave (K_i) and the resulting surface plasmon wave (SPW or K_p) is observed at multiple angles of incidence of the incoming light wave. The angle of incidence which yields the strongest coupling (i.e., maximum absorbance) can be identified by monitoring the reflected light using a simple light detector array which is used as the sensor output. This output can then be calibrated to the corresponding RI value at the sensor surface on the sample side. Figure 8.3 shows a typical angle-based SPR sensor response.

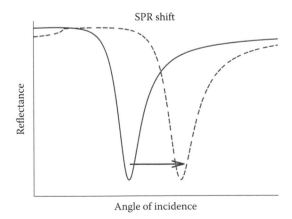

FIGURE 8.3 A typical Kretschmann SPR response profile. A change in the sample RI causes a corresponding shift in the SPR angle (i.e., the dip location of minimum reflected intensity).

For a more detailed analysis, the incident wavevector K_i is given by

$$K_i = \left(\frac{2\pi}{\lambda}\right) n \sin \theta_i, \tag{8.1}$$

where K_i is a component of the incident light wavevector parallel to the prism interface, θ_i is the incident light angle, λ is the wavelength of the incident light, and n is the RI of the prism. The wavevector, K_p, of the surface plasmon is described by

$$K_p = \left(\frac{2\pi}{\lambda}\right) \sqrt{\frac{\varepsilon_1 \varepsilon_2}{\varepsilon_1 + \varepsilon_2}}, \tag{8.2}$$

where ε_1 and ε_2 are the dielectric permittivity constants of the metal film and sample medium, respectively. SPR occurs when

$$K_i = K_p. \tag{8.3}$$

It should also be noted that if the incident angle is fixed and a multispectral light source is reflected from the surface, a reduction in a specific portion of the reflected light spectrum will be observed where resonance occurs. This forms the fundamental basis of wavelength-modulated SPR when compared to angular modulation.

For biosensing measurements, often it is the variation or change in the RI of the sample medium at the gold or metal surface that is of interest. Since the RI of the prism is constant, a change in the resonance condition should be observed when a corresponding change in the RI of the sample medium occurs. Such a condition transpires during binding events which alters the RI in the vicinity of the surface plasmon wave which is a very localized event within just a few nanometers near the metal surface. In this way, it is possible to monitor the binding of the biomolecules that occurs at the sensing surface. Gold is the most common metal used in commercial SPR sensors, although silver is also popular for customized SPR sensors. The preference of gold is commonly tied to the wide availability of possible surface immobilization approaches that can be used to study different binding events. In regards to the thickness of the metal layer, this is governed by the penetration depth of the evanescent field within the metal surface. For planar gold, this is ~200 nm (Aslan and Geddes 2009). Therefore, without the use of other enhancement approaches, the detectable range of a typical gold-layer SPR sensor is constrained by this factor. This range, however, is capable of covering most biological binding effects.

The increase in mass associated with a binding event causes a proportional increase in the RI which is observed as change in the response. For binding studies, response characterization is more commonly used instead of the direct SPR output as shown in Figure 8.3. In this case, the response is usually defined in response units (RU) where $1 \text{ RU} = 10^{-6}$ refractive index units (RIU). Most commercial systems based on the Kretschmann configuration have sensitivities between 10^{-5} and 10^{-6} RIU. A sample response curve is depicted in Figure 8.4. During the sample loading,

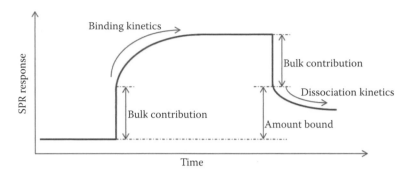

FIGURE 8.4 Schematic sensorgram illustrating the contribution of the bulk RI to the SPR signal in comparison with the most common experimental parameters. Depending on the SPR configuration and corresponding analysis, the SPR response can either be given in RU (response units) m° (angle shift), or nm (wavelength shift).

the analyte is transferred to the sensing surface by convection and diffusion. At this point, its concentration may change due to complex formation(s). This change in concentration can be minimized by guaranteeing a high mass transport coefficient which is normally achieved through the use of a microfluidic flow chamber. The response increase during analyte injection is mainly caused by the secondary formation of analyte–ligand complexes at the surface related to the binding kinetics. This is followed by a decreased response related to dissociation of the complexes. Following the acquisition of a response curve, fitted binding interaction models can be used to determine such factors as kinetic or affinity constants.

8.2.2 SPR IMAGING

SPR imaging (SPRi) allows for simultaneous analysis of multiple biomolecular interactions in real time. The basic physics of an SPRi system is the same as the Kretschmann configuration. A general illustration of an SPRi system is shown in Figure 8.5. In this system, the complete biochip surface is illuminated with a broad beam of monochromatic polarized light. The reflected light from the biochip is then spatially captured using a CCD camera or similar imaging device. Through the use of multiple exposures or video, any changes occurring at the SPR sensor surface can be continuously interrogated. Such an approach has the potential to monitor a multitude of real-time kinetic responses. SPRi has been demonstrated for the multiplexed detection of low-molecular-weight protein biomarkers (Lee et al. 2006), studying interactions of proteins with their associated antibodies (Yuk et al. 2006), and conformational structural changes that occur with protein denaturation (Huang and Chen 2006). Continuous-flow multichannel microfluidic devices have also been integrated to SPRi systems to achieve high-throughput protein array analyses/interactions (Natarajan et al. 2008a,b). Integrated microfluidic arrays have been fabricated with techniques such as soft lithography with more than 200-element addressable sample chambers with picoliter volumes. (Luo et al. 2008; Ouellet et al. 2010; Piliarik et al. 2010). In such configurations, it has also been shown that it is possible to perform

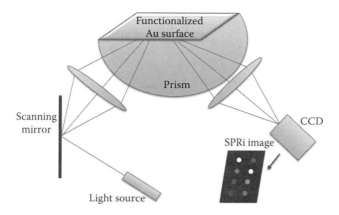

FIGURE 8.5 A generalized SPRi configuration.

serial dilutions through a series of integrated microvalves. New and innovative fluid-delivery strategies have also been demonstrated, such as the use of electro-osmotic flow strategies which is a pumpless technique directly integrated into the sensing platform (Krishnamoorthy et al. 2010). Such configurations with their multiplexed capability and ultralow sample volumes have demonstrated the potential to exploit both the sensitivity and specificity of SPR as a potential candidate for high-through-put biomarker screening, absolute protein expression profiling, and drug-discovery applications (Dong et al. 2008). This demonstration of SPRi illustrates its potential as an alternative to label-based detection techniques while offering nearly compa-rable sensitivity at the picomolar level.

8.2.3 LOCALIZED SPR

LSPR is not dependent on the planar surfaces used in propagating surface plasmon resonance (PSPR) systems. LSPR approaches exploit state-of-the-art nanofabrica-tion techniques to allow for high-performance prism-less SPR systems to be devel-oped. To date, there have been numerous approaches demonstrated for developing nanoparticle arrays, such as top-down lithography using electron beam (Sosa et al. 2003; Grand et al. 2006) and focused ion beam methods (Ohno et al. 2007), colloidal lithography formation of nanoholes (Rindzevicius et al. 2005), nanodisks (Hanarp et al. 2003), and nanorings (Aizpurua et al. 2003), and soft lithography for stamp imprinting periodic nanostructures (Delamarche et al. 1997; Schmid and Michel 2000; Odom et al. 2002a,b; Choi and Rogers 2003; Hua et al. 2004; Kwak et al. 2005; Rogers and Nuzzo 2005; Henzie et al. 2006). One unique and pioneering approach to develop consistent LSPR biosensors was reported by Van Duyne's group to cre-ate triangular silver nanoparticle arrays. The approach utilized a technique called nanosphere lithography (NSL; Hulteen et al. 1999; Jensen et al. 1999a,b, 2000). In this approach, a single layer of polystyrene nanospheres is coated onto a flat surface and then a thin layer of silver is physically vapor deposited on top of the nanospheres. The triangular geometrical gap between the nanospheres is then filled with silver during the deposition process. At the end, by removing the layer of nanospheres, a

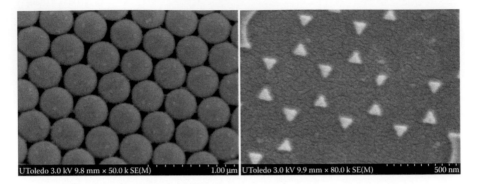

FIGURE 8.6   SEM images demonstrating NSL. Left: a single layer of polystyrene nanospheres. Right: the resulting metal nanoparticle array after removal of the polystyrene particles following the physical vapor deposition procedure.

relatively consistent triangular silver nanoarray is left on the substrate. An example of this process is provided in Figure 8.6. Such an array can function as an extremely sensitive nanoscale affinity biosensor (Haes and Van Duyne 2002) and can be uniformly reproduced with controllable parameters based on the template size.

As has been reported, noble metal nanoparticles exhibit a strong UV–vis absorption band that is absent in the spectrum of the bulk metals. This absorption band is present when the incident photon frequency is resonant with the collective oscillation of the conduction electrons; this is known as LSPR. A typical gold nanoparticle-based LSPR biosensor and its response curve are shown in Figure 8.7. The LSPR excitation manifests itself as wavelength-selective absorption with extremely large molar extinction coefficients. The enhanced local electromagnetic fields near the surface of the nanoparticles are responsible for the intense signals observed in all surface-enhanced spectroscopies, such as surface-enhanced Raman scattering (Haes and Van Duyne 2004) as well as PSPR. Although PSPR spectroscopy is presently the preferred method for commercial SPR instrumentation, LSPR spectroscopy offers

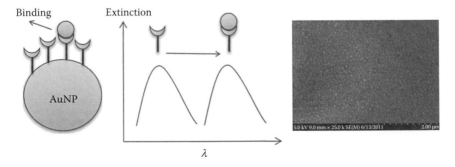

FIGURE 8.7   Schematic representation of an AuNP-based LSPR biosensor. Left: biosensing scheme based on an AuNP, onto which a chemical layer is deposited to selectively capture a molecule in the presence of interfering agents. Middle: the extinction spectrum red shifts with binding of molecules to the AuNP; Right: SEM imaging of a uniformly distributed AuNP layer that is capable of LSPR sensing.

many of the same advantages for sensing plus some additional benefits (Yonzon et al. 2004). A few of the similarities are that both approaches can provide real-time kinetic and thermodynamic data for binding processes and they have comparable if not better performance when measuring near field changes in the RI resulting from a molecular adsorption layer (Haes et al. 2004; Whitney et al. 2005). It should be noted, however, that PSPR methods do provide improved sensitivity to changes in the bulk RI (Haes and Van Duyne 2004). An obvious advantage of LSPR is seen through the use of single nanoparticles, where sensitivities have been reported down to the zeptomole range (McFarland and Van Duyne 2003). Currently, the trend in the SPR field appears to be moving toward LSPR sensing due to the added advantage of potentially using smaller sample volumes, more cost-effective instrumentation, and wavelength optimization.

8.2.4 SPR SENSING WITH OTHER CONFIGURATIONS

Although the conventional Kretschmann SPR configuration is simple, sensitive, and robust, its physical extensions are reaching a limit and the design is not amenable for further miniaturization and integration. Moving toward other designs over the past-decade optical fibers (Sharma et al. 2007, 2009; Sharma and Mohr 2008) and waveguide structures (Hamamoto et al. 2006) have been investigated as alternatives to the prism designs as a method to transmit the surface plasmon excitation and reflected light. However, making their way to a commercialized product will take time. Early fiber-based SPR biosensors commonly use single-mode fibers with a polished and metal-coated tip (Potyrailo et al. 1998; Cooper 2002). Variations in the SPR are measured by either monitoring the back-reflected light from the fiber or the diffracted light taken from the polished end. To improve the light coupling efficiency to the surface plasmon mode, Kurihara et al. (2004) had shown that the fiber tip can be enhanced through the use of micro-fabrication techniques. Using this approach, they etched a microcone into the polished end of the fiber to create a microprism. Performance with this method was limited; however, it was demonstrated to a resolution in the order of 10^{-4} RIU. Through further customization of the microprism design, the operational spectral range can be adapted such as for operation in the near-infrared region. In this case, Masson et al. (2006) were able to achieve up to a magnitude of 10 enhancement in sensitivity. Other milestone work for improving performance was through the use of polarization-preserving fibers (Piliarik et al. 2003; Piliarik and Homola 2009). Such designs utilize broadband radiation depolarization and polarization separation to improve sensor performance and have yielded sensitivities near that of commercial SPR systems (i.e., in the order of 10^{-6} RIU). Such fiber-based designs, however, have encountered issues for practical implementation. Some of the design issues raised by others (Hoa et al. 2007) include problems with the difficult alignment needed while inserting the fiber into a planar substrate containing a microfluidic path or chamber as well as the precise alignment needed to achieve optimal optical light coupling. Regardless of the issues, the future outlook of such SPR configurations are promising especially with the recent application of fiber-optic SPR sensors based on metal nanoparticles (Hamamoto et al. 2006) and photonic crystals (Skorobogatiy and Kabashin 2006).

8.3   OPTICAL-CHEMICAL TRANSDUCING MEDIUMS FOR SPR

Although SPR is very sensitive to subtle changes in the RI of the media in the vicinity of the metal-sensing surface(s), SPR by itself is not a selective technique. Therefore, a key component to develop useful biological and chemical SPR sensors is integrating an optical-chemical transducing medium to the sensor. This normally entails immobilization of biomolecular-recognition elements achieved through a chemical modification (e.g., aptamer, antibody, molecular imprinted polymer, etc.) at the metal–media interface (Hoshino et al. 2008; Kim et al. 2010; Ouellet et al. 2010; Zheng and Cameron 2011; Zheng et al. 2011). The surface modification serves multiple purposes. The most obvious of which is that it allows the surface to selectively capture the desired analyte in the region for which the SPR interaction occurs. Capturing or adsorbing the target also enhances the relative RI change at the surface, thereby further improves the sensitivity. This is especially important for applications involving small molecules present at low concentrations. To date, most commercial SPR biosensors are capable of detecting in the order of 1 pg/mm^2 of absorbed analytes of which surface functionalization plays an integral role. The sensitivity for SPR systems is normally indicated in terms of detectable RIU changes, and most commercial systems have RIU sensitivities between 10^{-5} and 10^{-6}. In addition to sensitivity considerations, through optimizing the surface chemistry, it is possible to avoid other nonspecific interactions, such as undesired protein adsorption and/or to repel other molecules that may lend to measurement error or false-positive detections.

8.3.1   SELF-ASSEMBLED MONOLAYERS

Self-assembled monolayers (SAMs) are organized layers of immobilized molecules that when prepared can be highly stable, densely packed, and ordered. Their structure begins with one end of a specific molecule known as the head group. This group is often specific and has a reversible affinity for a certain substrate which is often a noble metal. The secondary linkage, normally an alkane chain, serves as a spacer. The purpose of the spacer is to provide a well-defined thickness as well as acting as a physical barrier. It is also capable of altering the electronic conductivity as well as other local properties. Connected to the spacer is the terminal or tail group (e.g., –OH, –NH$_3$, or –COOH groups). These groups enable specific binding for certain molecules, such as antibodies, aptamers, or other attachments. These will be discussed in the subsequent sections. To date, the most common substrate employed, especially in SPR sensing, is gold as there are many possibilities that allow for different structural variations. As for head groups, thiols (–SH) are commonly used with gold due to their selective binding and they also allow for the incorporation of a wide range of functional groups or spacers.

In 2010, Kumbhat et al. (2010) demonstrated a SAM-based covalent immobilization of a bioreceptor conjugate of a dengue antigen and bovine serum albumin for SPR. Other recent examples of applying SAMs for SPR sensing include a peptide SAM (Bolduc et al. 2010, 2011), a double-layer SAM for supporting DNA in SPRi (Chen et al. 2009b), and a double-layer SAM for attaching antibodies (Lehnert et al. 2011). Even though SAMs can enable selective binding effects, it is well recognized that for complex samples, the selectivity of SAMs is not comparable to other receptors such as antibodies and aptamers. Therefore, in these cases, SAMs are used merely as a linkage layer for the receptors, helping the immobilization of the receptors and improving the performance of the sensor. Furthermore, a SAM can provide additional structure to the immobilization of the receptor, thus helping optimize the SPR response to the target.

Two of the more common SAM methods applied to SPR involve 3-mercaptopropionic acid (MPA) and 11-mercaptoundecanoic acid (MUA) (Zheng et al. 2011). These are linear molecules that have one thiol head and one carboxyl head. By immersing gold into the solution of MPA or MUA, a carboxyl group can be introduced onto the metal surface. Following treatment with 1-ethyl-3-(3-dimethylaminopropyl)carbodiimide (EDC) and N-hydroxysuccinimide (NHS), known as the EDC/NHS coupling method, protein or any other amine-containing molecules can subsequently be immobilized onto the gold to enable label-free detection. Furthermore, gold nanoparticles can be used to enhance the sensitivity of the PSPR sensors. As a recent example, a double-layer SAM containing AuNP for the detection of human IgG was demonstrated by Wang et al. (2011b). In this case, the response for the target molecules was doubled, whereas the nontarget response remained unchanged, thereby demonstrating an improvement in both sensitivity and selectivity.

Lastly, through surface modification, it is possible to reduce nonspecific interaction, such as undesired protein adsorption, and/or to repel other molecules that may result in measurement error. One such modification that has shown promise is through the use of polymer brushes, which exhibit greatly reduced surface fouling (Zhang et al. 2008; Hu et al. 2010). Efforts to functionalize such brushes through the terminal groups of polyethylene glycol for covalent binding has also been demonstrated (Trmcic-Cvitas et al. 2009). Thiol monolayers that contain polar head groups and binary patterned peptide SAMs can also reduce undesired surface adsorption caused by bulk proteins found in biological serum (Bolduc et al. 2010).

8.3.2 Antibodies

For protein kinetic studies, SPR does not require the use of labels or tracers when compared to some biochemical immunoassay methods such as ELISA. This reduces the required number of steps as well as provides real-time analysis of various interactions. In addition, there is a possibility of regenerating the sensor surface, thereby providing further potential to reduce analysis cost. Based on these advantages, SPR has gained considerable recognition as a potential analytical tool for such studies.

SPR analysis combined with antibody approaches offers several attractive sensing possibilities. Antibodies can serve as the recognition element for a wide variety of foreign targets which are also known as antigens. They are able to achieve this with a high degree of specificity and selectivity. To date, antibodies are the most widely used receptors in biosensors. A common approach to use antibodies with SPR sensing is to immobilize the antibody onto the sensing surface through some linkage. The surface is then loaded through a flow system with the target (i.e., antigen). The interaction of antigens with immobilized antibodies on the surface will elicit a corresponding change in the RI. Such a change, as previously described, can be easily detected and measured by SPR in real time. Some examples of using antibodies combined with SPR sensors include various protein detection and antibody–antigen interaction studies, such as C-reactive protein (Meyer et al. 2006), bovine troponin (Liu et al. 2003), myoglobin and cardiac troponin (Masson et al. 2004), tumor antigen (Campagnolo et al. 2004), human hepatitis B virus (Chung et al. 2005), and several others.

It should be noted, however, that the most critical part of the sensor development is to immobilize the antibody onto the sensing surface without losing its recognition functionality. Through the use of amine coupling, antibodies can be attached to the sensing surface with minimal loss of the antigen-binding capacity. In this case, the majority of monoclonal antibodies can be regenerated with minimal problems through the use of low pH glycine–HCl. Another widely used approach, as discussed in Section 8.3.1, is to modify the gold surface with a thiol-based self-assembly monolayer combined with MUA–MPA and EDC-NHS treatment (Rajesh et al. 2012).

8.3.3 APTAMERS

Aptamers are artificial oligonucleotides which are often used to mimic antibodies, since they also have demonstrable high affinity and selectivity toward various target compounds. The targets can range from small molecules, such as drugs and dyes, to significantly complex biological molecules such as enzymes, peptides, and proteins. Custom aptamers can be identified from random oligonucleotide libraries for specific target compounds by an *in vitro* iterative process called Systematic Evolution of Ligands by Exponential Amplification (SELEX) (Ellington and Szostak 1990; Bock et al. 1992). Aptamers can form a 3D structure serving as receptors specific to their target molecules similar to antibodies. They also have a number of advantages over antibodies, such as a tolerance to wide ranges of pH and salt concentrations, heat stability, ease of synthesis, and cost efficiency. The specificity and affinity of aptamers are comparable to, if not higher than, antibodies. Aptamers are also capable of being reversibly denatured for the release of target molecules, which make them perfect receptors for biosensing applications.

As reported by Zheng et al. (2011), aptamers are well suited for protein monitoring, as demonstrated through their development of an amine-terminated thrombin-binding aptamer SPR sensor. The sensor concept is further illustrated in Figure 8.8. Other groups have also successfully developed aptamer-based SPR sensors for the detection of proteins (Ostatná et al. 2008; Polonschii et al. 2010). These sensors use a SAM as a linker and coadsorbent thiol-modified aptamers combined together to form an aptamer–SAM matrix onto a gold surface. These methods have also demonstrated their potential as a reliable and easy approach for protein monitoring. It should be noted, however, that the cost of the thiol-modified aptamers is considerably higher compared to nonmodified or amine-modified aptamers.

The uniformity and density of the SAM modifications are also not guaranteed from sample to sample which can affect repeatable SPR performance. Recently, an MPA SAM-based coupling approach was reported to provide for a more stable and repeatable modification using amine-modified aptamers (Zheng et al. 2011). Furthermore, for aptamer-based SPR sensors, better performance may possibly be

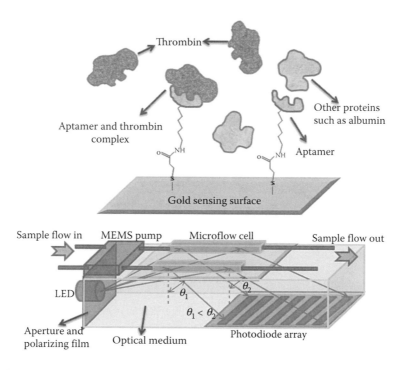

FIGURE 8.8 A portable design of thrombin detection SPR sensor chip.

achieved through the use of a mixed-length spacer layer (Chou et al. 2004), such as MUA combined with MPA. Such layers can be used to improve sensitivity and specificity (Lee et al. 2005). In addition, such configurations could potentially aid in creating and maintaining the 3D shape of the immobilized aptamers, which can be another important factor that can impact performance.

Through the insertion of hydrophilic groups, such as ethylene oxide onto the 5′-end of the aptamer, it is possible to further reduce nonspecific protein binding. However, by using a two-step immobilization process for which the SAM is first immobilized followed by the aptamer, this should result in a more cost-effective and controllable method when compared to adding all of the modifications to the aptamer monomer at once. Therefore, by designing the SAM (i.e., optimizing the hydrocarbon length and introducing coadsorbents) (Love et al. 2005), it is still possible to use simple amine-modified aptamers as a receptor to improve the sensitivity and selectivity.

Since the surface density of immobilized aptamer components can also impact the ability to capture the desired target and the surface coverage is mainly governed by the relative aptamer concentration during immobilization, relatively high aptamer concentrations are required to ensure complete occupation of the sensing surface to avoid nonspecific binding. However, aptamers that are too tightly packed can also hinder the ability of the target molecule to reach its respective binding sites if neighboring sites are occupied, especially for large molecules such as proteins. This may also inhibit the aptamers from folding into secondary structures. Therefore, by optimizing the aptamer immobilization concentration, the SPR sensor performance can also be improved. In this case, a two-step immobilization method can prove useful, such as providing variable aptamer spacing by adjusting the MPA SAM density or by coincubating ethanolamine and the aptamer at various molar ratios.

8.3.4 MOLECULAR IMPRINTING POLYMERS

Molecularly imprinted polymers (MIPs) are occasionally referred to as plastic antibodies. This is because a polymer is formed in the presence of a template molecule, which is later extracted. After the template is removed, complementary cavities are left behind which retain the exact memory of the target analyte in relation to its size, shape, and chemical group orientation. The MIP formation process is illustrated in Figure 8.9. As previously mentioned, the functional mechanism of an MIP is similar to antibodies, in that the affinity is due to the specific 3D structural binding. MIP methods are some of the most adaptable approaches for creating tailor-made sensor films based on artificial receptor-imprinted cavities. Because of their cost effectiveness, relatively simple preparation, and extreme stability, MIPs are attractive for use as a surface modifier in plasmonic-sensing approaches.

For SPR sensors, a thin-film MIP is normally coated onto the metal surface; however, the layer must be extremely thin (i.e., nanometer scale) as the target must be in the vicinity of the surface plasmon response. Upon adsorbing the target molecules, the physical properties, such as the density, solvent adsorbed, and polymer matrix volume, will vary. This will elicit a corresponding change in the SPR response. To date, a multitude of MIP variations has been coated onto SPR surfaces to demonstrate both high sensitivity and selectivity in sensing applications. Some of these include MIP microbeads (Lavine et al. 2007), peptide nanoparticles (Hoshino et al. 2008), thin films (Matsui et al. 2005, 2009; Tokareva et al. 2006), and hydrogels (Wang et al. 2011a).

Nanoparticles, especially those made of gold, have also been embedded into thin polymer films to enhance the sensitivity of the sensor (Matsui et al. 2005). In this case, the MIP swells by incorporating water into the matrix during analyte binding. This will cause significant changes in the dielectric constant near the Au substrate surface. More importantly, this swelling will result in greater separation distance

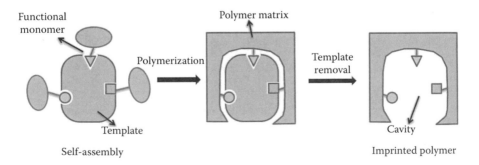

FIGURE 8.9   The typical formation of a molecular imprinted polymer.

between the Au nanoparticle within the polymer gel and the Au layer on the SPR sensor surface. This can result in significant changes in the degree of the SPR angle shift when compared to that without the nanoparticles.

8.3.5 Graphene

Other modifications such as the addition of graphene layers onto an SPR sensing surface can provide further stability due to its high impermeability (Choi et al. 2011). Such modifications are extremely important if SPR measurements are to be made directly on blood, plasma, or derivatives, as commonly there are several undesirable components present such as platelets that can hinder the ability to obtain accurate measurements. Graphene is a newer class of single-atom-thick and 2D carbon nanostructures. This material has recently received considerable attention from the physics, chemistry, and materials science communities. As for SPR sensing, graphene has been successfully integrated to an SPR sensor for applications including blood protein monitoring (Wang et al. 2011c) and for the real-time measurement of hydrogen peroxide (Mao et al. 2011). Graphene serves as a good stabilizer for immobilizing receptors. It is also gaining attention as an alternative to the use of SAMs depending on the application area. Such carbon nanostructures have demonstrated great potential; however, considerably more investigational research will be required as related to SPR, in addition to making its fabrication more cost comparable to existing methods.

8.4 SPR SENSING APPLICATIONS

8.4.1 Measurement of Physical Quantities

One of the oldest applications for SPR is the measurement of physical quantities. As previously discussed in Section 8.2, SPR is extremely sensitive to any change in the environmental RI. Therefore, regardless of the SPR configuration, they are all capable of accurately measuring small RI changes in a variety of different types of media including liquids (Lam et al. 2005; Zeng and Liang 2006), gases (Vukusic et al. 1992; Sharma and Mohr 2008), and solids (Frutos and Corn 1998; Bao et al. 2010) with a high degree of sensitivity. In most applications, a standard RI reference curve is generated to assess the sensitivity of the configuration. If an amount of a substance needs to be measured (i.e., concentration of an analyte) in a simple single-component media, no specific surface modification is required and SPR still exhibits a considerable degree of sensitivity (dependent on the configuration), although surface-modified SPR devices have considerably higher sensitivities. Other studies discussing the use of SPR to make high-resolution angular location (Schaller et al. 1997) and displacement measurements (Margheri et al. 1996) have also been reported. These are related to SPR's sensitivity to the momentum of the incident light (Homola et al. 1999). Accurate light wavelength measurements can also be achieved through monitoring resonance conditions (Chen et al. 2009a) as well as temperature monitoring since the RI is thermally dependent (Grassi and Georgiadis 1999). Furthermore, by coating the SPR surface with a temperature-sensitive material, extremely sensitive thermal measurements can be achieved (Sharma et al. 2009). Similar surface modification approaches can be used to create sensitive, humidity-based SPR sensors (Weiss et al. 1996). Other interesting examples of using SPR for measuring physical quantities include real-time thin-film thickness (Bao et al. 2010; Santillán et al. 2010) and flow condition monitoring (Iwasaki et al. 2006; Loureiro et al. 2007). Due to their high sensitivity and instantaneous response to RI change, SPR sensors normally exhibit better performance compared to most conventional measurement methods.

8.4.2 CHEMICAL SENSING

SPR sensors can be used to determine the concentration of both large and small molecules. As mentioned in the previous section, for pure analytes, SPR can be directly applied to measure the concentration within a certain range without the need for any surface functionalization. An excellent example of this is the measurement of aqueous glucose concentration in simple nonbiological aqueous media (Zhen and Yi 2001; Lam et al. 2005). However, in more realistic conditions, often samples are complex and contain numerous analytes. In this case, surface modification approaches, such as those described in Section 8.3, will need to be utilized or other techniques that either result in the direct adsorption of the chemical species onto the SPR-sensing surface or methods that elicit a secondary chemical reaction affecting the SPR signal.

Many of the applications for chemical sensing are focused on environmental monitoring and involve detections of compounds such as hydrocarbons, heavy metals, dioxins, and various other contaminants found in water supplies (e.g., pesticides, pharmaceutical agents, etc.) (Homola 2008). In 2007, Farre et al. (2007) demonstrated an SPR sensor capable of measuring atrazine in water samples at the level of parts per trillion. This sensor was based on an alkanethiol SAM formed on a gold surface which was capable of being regenerated. Mauriz et al. (2007) using similar immobilization techniques demonstrated a two-channel SPR sensor capable of simultaneous pesticide monitoring of chlorpyrifos, carbaryl, and DDT samples with sensitivities in the 18–50 ng/L range. Some examples of the use of SPR for gas/vapor detection are based on adsorption from various hydrocarbons (Abdelghani et al. 1997; Podgorsek et al. 1997) and tetrachloroethene (Niggemann et al. 1996). Miwa and Arakawa (1996) demonstrated that polyethylene glycol thin-film-modified SPR sensors can provide selective gas detection. Nitrogen dioxide detection was also illustrated by Ashwell and Roberts (1996) through the use of chemisorption of the gas molecules using an active SPR gold layer. Chadwick et al. (1994) described the detection of hydrogen with a surface plasmon approach using a palladium alloy. Most recently, Herminjard et al. (2009) demonstrated an SPR sensor showing enhanced sensitivity for CO_2 detection in the mid-infrared range.

In regards to assessing chemical concentration with SPR, three main approaches are commonly employed. One method is to directly measure the change in the SPR response following a fixed sample injection time over the sensing surface. This type of analysis is used when the analyte is large enough such that it is capable of producing a moderate response even at low molar concentrations. It is also possible to measure the rate of analyte binding at the beginning of sample injection. If the association is limited by mass transport, then the binding rate is directly proportional to analyte concentration and not related to kinetics. Finally, competitive methods can be used for analytes in solution that have a low direct response. This type of an approach is well suited for the analysis of small molecules.

8.4.3 BIOSENSING

SPR spectroscopy and imaging have become well-known label-free optical detection methods that are capable of monitoring surface-binding events in real time. In a traditional sense, biosensing has focused on any technique that senses interactions in aqueous biological samples. Today, however, this has been extended to include surface-based assays such as those employed in SPR. Such approaches can often provide real-time measurements and require less sample volume compared to traditional assays. With the ability to spatially functionalize surfaces through a myriad of techniques and probes as well as the potential to regenerate surfaces, when combined with SPRi and/or LSPR approaches coupled to advanced microfluidic arrays, it appears that high-throughput SPR biosensing is destined to make a significant impact. Furthermore, such approaches may prove quite competitive to more established methods such as ELISA. Many of the recent applications for which these advanced SPR sensors are being developed include creating robust diagnostics-based devices for biomarker sensing and to aid in the assessment or screening involved in therapeutic drug development.

As for a more detailed discussion on assay formats used in these SPR biosensors, most assays can be categorized as either a direct, sandwiched, competitive, or inhibition format. In the case of the direct assay, the recognition elements (e.g., antibodies, aptamers, etc.) are immobilized onto the SPR-sensing surface either directly or through some type of SAM. Such "direct" methods are suitable for target molecules of moderate size (i.e., >10 kDa). If the target molecule is smaller, then normally other types of assay formats are utilized. In the "sandwich" assay approach, a two-step process occurs. Once the target (e.g., antigen) is captured by the immobilized recognition element, as in the "direct" approach, a secondary recognition element is then utilized. The second element will also bind to the target, thereby increasing its size and creating a sandwich complex. The increase in size can facilitate detection and/or improve the detection limit. In "competitive" assays, the approach is slightly different. It is based on two samples, where one is free with the target concentration unknown and the other is the target conjugated to a larger molecule such as a protein with a known concentration. Both will compete for the same recognition elements on the sensor surface. The reduction in the SPR signal is therefore proportional to the target concentration. Finally, "inhibition" format assays are also based on competition binding. The analyte or a derivative is typically attached to the surface as a ligand. The detecting molecule is usually a larger molecule that binds specifically to the analyte. A known amount of the detection molecule is then added to the sample. The mixture is then incubated and allowed to equilibrate. When it is injected over the sensor surface, the remaining free detecting molecules can be measured. Therefore, the detected signal is inversely related to the concentration of the target in the sample. This type of approach is very well suited for low-molecular-weight targets.

8.5 FUTURE TRENDS OF SPR SENSOR DEVELOPMENT

SPR-sensing techniques have made considerable strides over the past three decades and are on the foothold of moving beyond only being used as a tool for bioanalytical research. The most notable advances have been made in the area of surface functionalization and integration of microfluidic sample-delivery platforms. Of all the surface functionalization methods, DNA/RNA-based aptamer modification appears to hold the most promise. Through the use of *in vitro* selection processes, such as SELEX, it is possible to easily screen target compounds to identify viable aptamer candidates for use as biomolecular recognition elements (Kim et al. 2010; Lautner et al. 2010; Polonschii et al. 2010; Zheng et al. 2011). Based on this approach, a bio-specific SPR sensor can be developed for almost any biomarker. Furthermore, due to their nucleic acid composition, aptamers exhibit excellent chemical stability, allow certain kinetic parameters to be modified, are amenable to chemical modification, and denatured aptamers can be regenerated in minutes. Aptamers are also well suited for implementation in microfluidic detection platforms (Xu et al. 2010). SPR measurements have already been extended from well-behaved buffered solutions to more complex media such as plasma and blood (Bolduc et al. 2010; Jha and Sharma 2010; Piliarik et al. 2010; Springer et al. 2010). There are, however, additional challenges that still need to be overcome. Although certain antifouling methods have been demonstrated, these are often dependent on the target properties, thereby making it difficult to develop a "universal" method to address multiple targets in a single platform. Fouling problems can also cause issues with achieving the required limit of detection in nonprocessed samples. Lastly, blood and plasma can also cause issues for the microfluidic platforms; cell or protein adhesion in the microchannels can affect sample delivery, thereby impacting sensor performance. Regardless, the advances in both SPR sensitivity and specificity along with its potential for real-time measurements, low manufacturing cost, and numerous applications offer a promising outlook for its continued development and commercialization in the area of molecular diagnostics.

REFERENCES

Abdelghani, A., J. M. Chovelon, N. Jaffrezic-Renault, C. Ronot-Trioli, C. Veillas, and H. Gagnaire. 1997. Surface plasmon resonance fibre-optic sensor for gas detection. *Sensors and Actuators B: Chemical* 39 (1–3):407–410.

Aizpurua, J., P. Hanarp, D. S. Sutherland, M. Käll, G. W. Bryant, and F. J. García de Abajo. 2003. Optical properties of gold nanorings. *Physical Review Letters* 90 (5):057401.

Ashwell, G. J., and M. P. S. Roberts. 1996. Highly selective surface plasmon resonance sensor for NO_2. *Electronics Letters* 32 (22):2089–2091.

Aslan, K., and C. D. Geddes. 2009. Directional surface plasmon coupled luminescence for analytical sensing applications: Which metal, what wavelength, what observation angle? *Analytical chemistry* 81 (16):6913–6922.

Bao, M., G. Li, D. Jiang, W. Cheng, and X. Ma. 2010. ZnO sensing film thickness effects on the sensitivity of surface plasmon resonance sensors with angular interrogation. *Materials Science & Engineering B* 171 (1–3):155–158.

Bock, L. C., L. C. Griffin, J. A. Latham, E. H. Vermaas, and J. J. Toole. 1992. Selection of single-stranded DNA molecules that bind and inhibit human thrombin. *Nature* 355 (6360):564–566.

Bolduc, O. R., P. Lambert-Lanteigne, D. Y. Colin et al. 2011. Modified peptide monolayer binding His-tagged biomolecules for small ligand screening with SPR biosensors. *Analyst* 136 (15):3142–3148.

Bolduc, O. R., and J. F. Masson. 2011. Advances in surface plasmon resonance sensing with nanoparticles and thin films: Nanomaterials, surface chemistry, and hybrid plasmonic techniques. *Analytical Chemistry* 83 (21):8057–8062.

Bolduc, O. R., J. N. Pelletier, and J. F. Masson. 2010. SPR biosensing in crude serum using ultralow fouling binary patterned peptide SAM. *Analytical Chemistry* 82 (9):3699–3706.

Campagnolo, C., K. J. Meyers, T. Ryan et al. 2004. Real-time, label-free monitoring of tumor antigen and serum antibody interactions. *Journal of Biochemical and Biophysical Methods* 61 (3):283–298.

Chadwick, B., J. Tann, M. Brungs, and M. Gal. 1994. A hydrogen sensor based on the optical generation of surface plasmons in a palladium alloy. *Sensors and Actuators B: Chemical* 17 (3):215–220.

Chen, K., J. Lin, and J. Chen. 2009a. Measuring physical parameters with surface plasmon resonance heterodyne interferometry. *Optik—International Journal for Light and Electron Optics* 120 (1):29–34.

Chen, Y., A. Nguyen, L. Niu, and R. M. Corn. 2009b. Fabrication of DNA microarrays with poly(L-glutamic acid) monolayers on gold substrates for SPR imaging measurements. *Langmuir* 25 (9):5054–5060.

Choi, S. H., Y. L. Kim, and K. M. Byun. 2011. Graphene-on-silver substrates for sensitive surface plasmon resonance imaging biosensors. *Optics Express* 19 (2):458–466.

Choi, K. M., and J. A. Rogers. 2003. A photocurable poly(dimethylsiloxane) chemistry designed for soft lithographic molding and printing in the nanometer regime. *Journal of the American Chemical Society* 125 (14):4060–4061.

Chou, C., C. Chen, T. Lee, and K. Peck. 2004. Optimization of probe length and the number of probes per gene for optimal microarray analysis of gene expression. *Nucleic Acids Research* 32 (12):e99–e99.

Chung, J. W., S. D. Kim, R. Bernhardt, and J. C. Pyun. 2005. Application of SPR biosensor for medical diagnostics of human hepatitis B virus (hHBV). *Sensors and Actuators B: Chemical* 111–112:416–422.

Cooper, M. A. 2002. Optical biosensors in drug discovery. *Nature Reviews Drug Discovery* 1 (7):515–528.

Delamarche, E., H. Schmid, B. Michel, and H. Biebuyck. 1997. Stability of molded polydimethylsiloxane microstructures. *Advanced Materials* 9 (9):741–746.

Dong, Y., T. Wilkop, D. Xu, Z. Wang, and Q. Cheng. 2008. Microchannel chips for the multiplexed analysis of human immunoglobulin G-antibody interactions by surface plasmon resonance imaging. *Analytical and Bioanalytical Chemistry* 390 (6):1575–1583.

Ellington, A. D., and J. W. Szostak. 1990. *In vitro* selection of RNA molecules that bind specific ligands. *Nature* 346 (6287):818–822.

Farre, M., F. Martinez, J. Ramon et al. 2007. Part per trillion determination of atrazine in natural water samples by a surface plasmon resonance immunosensor. *Analytical and Bioanalytical Chemistry* 388 (1):207–214.

Frutos, A. G., and R. M. Corn. 1998. Peer reviewed: SPR of ultrathin organic films. *Analytical Chemistry* 70 (13):449A–455A.

Grand, J., P. M. Adam, A. S. Grimault et al. 2006. Optical extinction spectroscopy of oblate, prolate and ellipsoid shaped gold nanoparticles: Experiments and theory. *Plasmonics* 1 (2):135–140.

Grassi, J. H., and R. M. Georgiadis. 1999. Temperature-dependent refractive index determination from critical angle measurements: Implications for quantitative SPR sensing. *Analytical Chemistry* 71 (19):4392–4396.

Haes, A. J., and R. P. Van Duyne. 2002. A nanoscale optical biosensor: Sensitivity and selectivity of an approach based on the localized surface plasmon resonance spectroscopy of triangular silver nanoparticles. *Journal of the American Chemical Society* 124 (35):10596–10604.

Haes, A. J., and R. P. Van Duyne. 2004. A unified view of propagating and localized surface plasmon resonance biosensors. *Analytical and Bioanalytical Chemistry* 379 (7–8):920–930.

Haes, A. J., S. Zou, G. C. Schatz, and R. P. Van Duyne. 2004. A nanoscale optical biosensor: The long range distance dependence of the localized surface plasmon resonance of noble metal nanoparticles. *The Journal of Physical Chemistry B* 108 (1):109–116.

Hamamoto, K., R. Micheletto, M. Oyama, A. Ali Umar, S. Kawai, and Y. Kawakami. 2006. An original planar multireflection system for sensing using the local surface plasmon resonance of gold nanospheres. *Journal of Optics A: Pure and Applied Optics* 8 (3):268.

Hanarp, P., M. Käll, and D. S. Sutherland. 2003. Optical properties of short range ordered arrays of nanometer gold disks prepared by colloidal lithography. *The Journal of Physical Chemistry B* 107 (24):5768–5772.

Henzie, J., J. E. Barton, C. L. Stender, and T. W. Odom. 2006. Large-area nanoscale patterning: chemistry meets fabrication. *Accounts of Chemical Research* 39 (4):249–257.

Herminjard, S., L. Sirigu, H. P. Herzig et al. 2009. Surface plasmon resonance sensor showing enhanced sensitivity for CO_2 detection in the mid-infrared range. *Optics Express* 17 (1):293–303.

Hoa, X. D., A. G. Kirk, and M. Tabrizian. 2007. Towards integrated and sensitive surface plasmon resonance biosensors: A review of recent progress. *Biosensors & Bioelectronics* 23 (2):151–160.

Homola, J. 2008. Surface plasmon resonance sensors for detection of chemical and biological species. *Chemical Reviews* 108 (2):462–493.

Homola, J., S. S. Yee, and G. Gauglitz. 1999. Surface plasmon resonance sensors: Review. *Sensors and Actuators B: Chemical* 54 (1–2):3–15.

Hoshino, Y., T. Kodama, Y. Okahata, and K. J. Shea. 2008. Peptide imprinted polymer nanoparticles: A plastic antibody. *Journal of the American Chemical Society* 130 (46):15242–15243.

Hu, W., Y. Liu, Z. Lu, and C. M. Li. 2010. Poly[oligo(ethylene glycol) methacrylate-co-glycidyl methacrylate] brush substrate for sensitive surface plasmon resonance imaging protein arrays. *Advanced Functional Materials* 20 (20):3497–3503.

Hua, F., Y. Sun, A. Gaur et al. 2004. Polymer imprint lithography with molecular-scale resolution. *Nano Letters* 4 (12):2467–2471.

Huang, H., and Y. Chen. 2006. Label-free reading of microarray-based proteins with high throughput surface plasmon resonance imaging. *Biosensors & Bioelectronics* 22 (5):644–648.

Hulteen, J. C., D. A. Treichel, M. T. Smith, M. L. Duval, T. R. Jensen, and R. P. Van Duyne. 1999. Nanosphere lithography: Size-tunable silver nanoparticle and surface cluster arrays. *Journal of Physical Chemistry B* 103 (19):3854–3863.

Iwasaki, Y., T. Tobita, K. Kurihara, T. Horiuchi, K. Suzuki, and O. Niwa. 2006. Imaging of flow pattern in micro flow channel using surface plasmon resonance. *Measurement Science and Technology* 17 (12):3184–3188.

Jensen, T. R., M. L. Duval, K. L. Kelly, A. A. Lazarides, G. C. Schatz, and R. P. Van Duyne. 1999a. Nanosphere lithography: Effect of the external dielectric medium on the surface plasmon resonance spectrum of a periodic array of sliver nanoparticles. *Journal of Physical Chemistry B* 103 (45):9846–9853.

Jensen, T. R., M. D. Malinsky, C. L. Haynes, and R. P. Van Duyne. 2000. Nanosphere lithography: Tunable localized surface plasmon resonance spectra of silver nanoparticles. *Journal of Physical Chemistry B* 104 (45):10549–10556.

Jensen, T. R., G. C. Schatz, and R. P. Van Duyne. 1999b. Nanosphere lithography: Surface plasmon resonance spectrum of a periodic array of silver nanoparticles by ultraviolet–visible extinction spectroscopy and electrodynamic modeling. *Journal of Physical Chemistry B* 103 (13):2394–2401.

Jha, R., and A. K. Sharma. 2010. Design of a silicon-based plasmonic biosensor chip for human blood-group identification. *Sensors and Actuators B: Chemical* 145 (1):200–204.

Kim, S., J. Lee, S. J. Lee, and H. J. Lee. 2010. Ultra-sensitive detection of IgE using biofunctionalized nanoparticle-enhanced SPR. *Talanta* 81 (4–5):1755–1759.

Krishnamoorthy, G., E. T. Carlen, H. L. deBoer, A. van den Berg, and R. B. Schasfoort. 2010. Electrokinetic lab-on-a-biochip for multi-ligand/multi-analyte biosensing. *Analytical Chemistry* 82 (10):4145–4150.

Kumbhat, S., K. Sharma, R. Gehlot, A. Solanki, and V. Joshi. 2010. Surface plasmon resonance based immunosensor for serological diagnosis of dengue virus infection. *Journal of Pharmaceutical and Biomedical Analysis* 52 (2):255–259.

Kurihara, K., H. Ohkawa, Y. Iwasaki, O. Niwa, T. Tobita, and K. Suzuki. 2004. Fiber-optic conical microsensors for surface plasmon resonance using chemically etched single-mode fiber. *Analytica Chimica Acta* 523 (2):165–170.

Kwak, E., J. Henzie, S. Chang, S. K. Gray, G. C. Schatz, and T. W. Odom. 2005. Surface plasmon standing waves in large-area subwavelength hole arrays. *Nano Letters* 5 (10):1963–1967.

Lam, W. W., L. H. Chu, C. L. Wong, and Y. T. Zhang. 2005. A surface plasmon resonance system for the measurement of glucose in aqueous solution. *Sensors and Actuators B: Chemical* 105 (2):138–143.

Lautner, G., Z. Balogh, V. Bardoczy, T. Meszaros, and R. E. Gyurcsanyi. 2010. Aptamer-based biochips for label-free detection of plant virus coat proteins by SPR imaging. *The Analyst* 135 (5):918–926.

Lavine, B. K., D. J. Westover, N. Kaval, N. Mirjankar, L. Oxenford, and G. K. Mwangi. 2007. Swellable molecularly imprinted poly$N$-(N-propyl)acrylamide particles for detection of emerging organic contaminants using surface plasmon resonance spectroscopy. *Talanta* 72 (3):1042–1048.

Lee, H. J., D. Nedelkov, and R. M. Corn. 2006. Surface plasmon resonance imaging measurements of antibody arrays for the multiplexed detection of low molecular weight protein biomarkers. *Analytical Chemistry* 78 (18):6504–6510.

Lee, J. W., S. J. Sim, S. M. Cho, and J. Lee. 2005. Characterization of a self-assembled monolayer of thiol on a gold surface and the fabrication of a biosensor chip based on surface plasmon resonance for detecting anti-GAD antibody. *Biosensors & Bioelectronics* 20 (7):1422–1427.

Lehnert, M., M. Gorbahn, C. Rosin et al. 2011. Adsorption and conformation behavior of biotinylated fibronectin on streptavidin-modified TiO(X) surfaces studied by SPR and AFM. *Langmuir* 27 (12):7743–7751.

Liedberg, B., C. Nylander, and I. Lunström. 1983. Surface plasmon resonance for gas detection and biosensing. *Sensors and Actuators* 4:299–304.

Liu, X., J. Wei, D. Song, Z. Zhang, H. Zhang, and G. Luo. 2003. Determination of affinities and antigenic epitopes of bovine cardiac troponin I (cTnI) with monoclonal antibodies by surface plasmon resonance biosensor. *Analytical Biochemistry* 314 (2):301–309.

Loureiro, C. C. L., A. M. N. Lima, and H. Neff. 2007. Optical monitoring of micro fluidic flow conditions, employing surface plasmon resonance sensing. *Journal of Physics: Conference Series* 85 (1):012023.

Love, J. C., L. A. Estroff, J. K. Kriebel, R. G. Nuzzo, and G. M. Whitesides. 2005. Self-assembled monolayers of thiolates on metals as a form of nanotechnology. *Chemical Reviews* 105 (4):1103–1169.

Luo, Y., F. Yu, and R. N. Zare. 2008. Microfluidic device for immunoassays based on surface plasmon resonance imaging. *Lab on a Chip* 8 (5):694–700.

Mao, Y., Y. Bao, W. Wang, Z. Li, F. Li, and L. Niu. 2011. Layer-by-layer assembled multilayer of graphene/Prussian blue toward simultaneous electrochemical and SPR detection of H_2O_2. *Talanta* 85 (4):2106–2112.

Margheri, G., A. Mannoni, and F. Quercioli. 1996. *New High-resolution Displacement Sensor based on Surface Plasmon Resonance*, Besancon, France.

Masson, J. F., Y. C. Kim, L. A. Obando, W. Peng, and K. S. Booksh. 2006. Fiber-optic surface plasmon resonance sensors in the near-infrared spectral region. *Applied Spectroscopy* 60 (11):1241–1246.

Masson, J. F., L. Obando, S. Beaudoin, and K. Booksh. 2004. Sensitive and real-time fiber-optic-based surface plasmon resonance sensors for myoglobin and cardiac troponin I. *Talanta* 62 (5):865–870.

Matsui, J., K. Akamatsu, N. Hara et al. 2005. SPR sensor chip for detection of small molecules using molecularly imprinted polymer with embedded gold nanoparticles. *Analytical Chemistry* 77 (13):4282–4285.

Matsui, J., M. Takayose, K. Akamatsu, H. Nawafune, K. Tamaki, and N. Sugimoto. 2009. Molecularly imprinted nanocomposites for highly sensitive SPR detection of a non-aqueous atrazine sample. *Analyst* 134 (1):80–86.

Mauriz, E., A. Calle, J. J. Manclus, A. Montoya, and L. M. Lechuga. 2007. Multi-analyte SPR immunoassays for environmental biosensing of pesticides. *Analytical and Bioanalytical Chemistry* 387 (4):1449–1458.

McFarland, A. D., and R. P. Van Duyne. 2003. Single silver nanoparticles as real-time optical sensors with zeptomole sensitivity. *Nano Letters* 3 (8):1057–1062.

Meyer, M., M. Hartmann, and M. Keusgen. 2006. SPR-based immunosensor for the CRP detection—a new method to detect a well known protein. *Biosensors & Bioelectronics* 21 (10):1987–1990.

Miwa, S., and T. Arakawa. 1996. Selective gas detection by means of surface plasmon resonance sensors. *Thin Solid Films* 281–282:466–468.

Natarajan, S., A. Hatch, D. G. Myszka, and B. K. Gale. 2008a. Optimal conditions for protein array deposition using continuous flow. *Analytical Chemistry* 80 (22):8561–8567.

Natarajan, S., P. S. Katsamba, A. Miles et al. 2008b. Continuous-flow microfluidic printing of proteins for array-based applications including surface plasmon resonance imaging. *Analytical Biochemistry* 373 (1):141–146.

Niggemann, M., A. Katerkamp, M. Pellmann, P. Bolsmann, J. Reinbold, and K. Cammann. 1996. Remote sensing of tetrachloroethene with a micro-fibre optical gas sensor based on surface plasmon resonance spectroscopy. *Sensors and Actuators B: Chemical* 34 (1–3):328–333.

Odom, T. W., J. Love, D. B. Wolfe, K. E. Paul, and G. M. Whitesides. 2002a. Improved pattern transfer in soft lithography using composite stamps. *Langmuir* 18 (13):5314–5320.

Odom, T. W., V. R. Thalladi, J. Love, and G. M. Whitesides. 2002b. Generation of 30–50 nm structures using easily fabricated, composite PDMS masks. *Journal of the American Chemical Society* 124 (41):12112–12113.

Ohno, T., J. A. Bain, and T. E. Schlesinger. 2007. Observation of geometrical resonance in optical throughput of very small aperture lasers associated with surface plasmons. *Journal of Applied Physics* 101 (8):083107.

Ostatná, V., H. Vaisocherová, J. Homola, and T. Hianik. 2008. Effect of the immobilisation of DNA aptamers on the detection of thrombin by means of surface plasmon resonance. *Analytical and Bioanalytical Chemistry* 391 (5):1861–1869.

Ouellet, E., C. Lausted, T. Lin, C. W. Yang, L. Hood, and E. T. Lagally. 2010. Parallel microfluidic surface plasmon resonance imaging arrays. *Lab on a chip* 10 (5):581–588.

Piliarik, M., M. Bockova, and J. Homola. 2010. Surface plasmon resonance biosensor for parallelized detection of protein biomarkers in diluted blood plasma. *Biosensors & Bioelectronics* 26 (4):1656–1661.

Piliarik, M., and J. Homola. 2009. Surface plasmon resonance (SPR) sensors: Approaching their limits? *Optics Express* 17 (19):16505–16517.

Piliarik, M., J. Homola, Z. Maníková, and J. Čtyroký. 2003. Surface plasmon resonance sensor based on a single-mode polarization-maintaining optical fiber. *Sensors and Actuators B: Chemical* 90 (1–3):236–242.

Podgorsek, R. P., T. Sterkenburgh, J. Wolters, T. Ehrenreich, S. Nischwitz, and H. Franke. 1997. Optical gas sensing by evaluating ATR leaky mode spectra. *Sensors and Actuators B: Chemical* 39 (1–3):349–352.

Polonschii, C., S. David, S. Tombelli, M. Mascini, and M. Gheorghiu. 2010. A novel low-cost and easy to develop functionalization platform. Case study: Aptamer-based detection of thrombin by surface plasmon resonance. *Talanta* 80 (5):2157–2164.

Potyrailo, R. A., S. E. Hobbs, and G. M. Hieftje. 1998. Optical waveguide sensors in analytical chemistry: today's instrumentation, applications and trends for future development. *Fresenius' Journal of Analytical Chemistry* 362 (4):349–373.

Rajesh, V. S., S. K. Mishra, and A. M. Biradar. 2012. Synthesis and electrochemical characterization of myoglobin–antibody protein immobilized self-assembled gold nanoparticles on ITO-glass plate. *Materials Chemistry and Physics* 132 (1):22–28.

Rindzevicius, T., Y. Alaverdyan, A. Dahlin, F. Höök, D. S. Sutherland, and M. Käll. 2005. Plasmonic sensing characteristics of single nanometric holes. *Nano Letters* 5 (11):2335–2339.

Rogers, J. A., and R. G. Nuzzo. 2005. Recent progress in soft lithography. *Materials Today* 8 (2):50–56.

Santillán, J. M., L. B. Scaffardi, D. C. Schinca, and F. A. Videla. 2010. Determination of nanometric Ag_2O film thickness by surface plasmon resonance and optical waveguide mode coupling techniques. *Journal of Optics* 12 (4):045002.

Schaller, J. K., R. Czepluch, and C. G. Stojanoff. 1997. *Plasmon Spectroscopy for High-resolution Angular Measurements*, Munich, Germany.

Schmid, H., and B. Michel. 2000. Siloxane polymers for high-resolution, high-accuracy soft lithography. *Macromolecules* 33 (8):3042–3049.

Sharma, A. K., R. Jha, and B. D. Gupta. 2007. Fiber-optic sensors based on surface plasmon resonance: a comprehensive review. *Sensors Journal, IEEE* 7 (8):1118–1129.

Sharma, A. K., and G. J. Mohr. 2008. Theoretical understanding of an alternating dielectric multilayer-based fiber optic SPR sensor and its application to gas sensing. *New Journal of Physics* 10 (2):023039.

Sharma, A. K., H. S. Pattanaik, and G. J. Mohr. 2009. On the temperature sensing capability of a fibre optic SPR mechanism based on bimetallic alloy nanoparticles. *Journal of Physics D: Applied Physics* 42 (4):045104.

Skorobogatiy, M., and A. V. Kabashin. 2006. Photon crystal waveguide-based surface plasmon resonance biosensor. *Applied Physics Letters* 89 (14):143518-3.

Sosa, I., C. Noguez, R. G. Barrera. 2003. Optical properties of metal nanoparticles with arbitrary shapes. *The Journal of Physical Chemistry B* 107 (26):6269–6275.

Springer, T., M. Piliarik, and J. Homola. 2010. Real-time monitoring of biomolecular interactions in blood plasma using a surface plasmon resonance biosensor. *Analytical and Bioanalytical Chemistry* 398 (5):1955–1961.

Tokareva, I., I. Tokarev, S. Minko, E. Hutter, and J. H. Fendler. 2006. Ultrathin molecularly imprinted polymer sensors employing enhanced transmission surface plasmon resonance spectroscopy. *Chemical Communications* 31:3343–3345.

Trmcic-Cvitas, J., E. Hasan, M. Ramstedt et al. 2009. Biofunctionalized protein resistant oligo(ethylene glycol)-derived polymer brushes as selective immobilization and sensing platforms. *Biomacromolecules* 10 (10):2885–2894.

Vukusic, P. S., G. P. Bryan-Brown, and J. R. Sambles. 1992. Surface plasmon resonance on gratings as a novel means for gas sensing. *Sensors and Actuators B: Chemical* 8 (2):155–160.

Wang, J., S. Banerji, N. Menegazzo, W. Peng, Q. Zou, and K. S. Booksh. 2011a. Glucose detection with surface plasmon resonance spectroscopy and molecularly imprinted hydrogel coatings. *Talanta* 86:133–141.

Wang, J., D. Song, L. Wang, H. Zhang, H. Zhang, and Y. Sun. 2011b. Design and performances of immunoassay based on SPR biosensor with Au/Ag alloy nanocomposites. *Sensors and Actuators B: Chemical* 157 (2):547–553.

Wang, L., C. Zhu, L. Han, L. Jin, M. Zhou, and S. Dong. 2011c. Label-free, regenerative and sensitive surface plasmon resonance and electrochemical aptasensors based on graphene. *Chemical Communications* 47 (27):7794–7796.

Weiss, M. N., R. Srivastava, and H. Groger. 1996. Experimental investigation of a surface plasmon-based integrated-optic humidity sensor. *Electronics Letters* 32 (9):842–843.

Whitney, A. V., J. W. Elam, S. Zou et al. 2005. Localized surface plasmon resonance nanosensor: a high-resolution distance-dependence study using atomic layer deposition. *The Journal of Physical Chemistry B* 109 (43):20522–20528.

Wood, R. W. 1902. On a remarkable case of uneven distribution of light in a diffraction grating spectrum. *Philosophical Magazine* 4 (21):396–402.

Xu, Y., X. Yang, and E. Wang. 2010. Review: Aptamers in microfluidic chips. *Analytica Chimica Acta* 683 (1):12–20.

Yonzon, C. R., E. Jeoung, S. Zou, G. C. Schatz, M. Mrksich, and R. P. Van Duyne. 2004. A comparative analysis of localized and propagating surface plasmon resonance sensors: The binding of concanavalin A to a monosaccharide functionalized self-assembled monolayer. *Journal of the American Chemical Society* 126 (39):12669–12676.

Yuk, J. S., H. S. Kim, J. W. Jung et al. 2006. Analysis of protein interactions on protein arrays by a novel spectral surface plasmon resonance imaging. *Biosensors & Bioelectronics* 21 (8):1521–1528.

Zeng, J., and D. Liang. 2006. Application of fiber optic surface plasmon resonance sensor for measuring liquid refractive Index. *Journal of Intelligent Material Systems and Structures* 17 (8–9):787–791.

Zhang, Z., M. Zhang, S. Chen, T. A. Horbett, B. D. Ratner, and S. Jiang. 2008. Blood compatibility of surfaces with superlow protein adsorption. *Biomaterials* 29 (32):4285–4291.

Zhen, W., and C. Yi. 2001. Analysis of mono- and oligosaccharides by multiwavelength surface plasmon resonance (SPR) spectroscopy. *Carbohydrate Research* 332 (2):209–213.

Zheng, R., and B. D. Cameron. 2011. Surface plasmon resonance: recent progress toward the development of portable real-time blood diagnostics. *Expert Review of Molecular Diagnostics* 12 (1):5–7.

Zheng, R., B. W. Park, D. S. Kim, and B. D. Cameron. 2011. Development of a highly specific amine-terminated aptamer functionalized surface plasmon resonance biosensor for blood protein detection. *Biomedical Optics Express* 2 (9):2731–2740.

Index

Printed and bound by CPI Group (UK) Ltd, Croydon, CR0 4YY

18/10/2024

01776236-0005